Invertebrate Immunity

ACADEMIC PRESS RAPID MANUSCRIPT REPRODUCTION

Invertebrate Immunity

Mechanisms of Invertebrate Vector–Parasite Relations

EDITED BY

Karl Maramorosch

Waksman Institute of Microbiology
Rutgers University
New Brunswick, New Jersey

Robert E. Shope

Department of Epidemiology and Public Health
School of Medicine
Yale University
New Haven, Connecticut

Academic Press, Inc. New York San Francisco London 1975

A Subsidiary of Harcourt Brace Jovanovich • Publishers

ACADEMIC PRESS, INC.
111 Fifth Avenue, New York, New York 10003

United Kingdom Edition published by
ACADEMIC PRESS, INC. (LONDON) LTD.
24/28 Oval Road, London NW1

Library of Congress Cataloging in Publication Data
Main entry under title:

Invertebrate immunity.
"The chapters of this book were prepared for presentation at a workshop . . . sponsored by the National Institutes of Health and held in Bethesda, Maryland, April 17 and 18, 1974."
Bibliography: p.
Includes index.
1. Immune response–Congresses. 2. Immunity–Congresses. 3. Invertebrates–Physiology–Congresses. I. Maramorosch, Karl. II. Shope, Robert E. III. United States. National Institutes of Health. [DNLM: 1. Invertebrates–Immunology–Congresses. 2. Parasites–Congresses. QL362 I623 1974]
QR186.I58 592'.02'9 75-9772
ISBN 0-12-470265-1

PRINTED IN THE UNITED STATES OF AMERICA

CONTENTS

CONTRIBUTORS

Robert S. Anderson – Sloan - Kettering Institute for Cancer Research, Donald S. Walker Laboratory, 145 Boston Post Road, Rye, New York

Frederik B. Bang – Department of Pathobiology, The Johns Hopkins University, School of Hygiene and Public Health, Baltimore, Maryland

A. Ralph Barr – School of Public Health, University of California, Los Angeles, California

James T. Blankemeyer – Department of Biology, Temple University, Philadelphia, Pennsylvania

June Stephens Chadwick – Department of Microbiology and Immunology, Queens's University, Kingston, Ontario, Canada

Roy W. Chamberlain – Center for Disease Control, Public Health Service, U.S. Department of Health, Education, and Welfare, Atlanta, Georgia

Eli Chernin – Department of Tropical Public Health, Harvard School of Public Health, Boston, Massachusetts

William R. Harvey – Department of Biology, Temple University, Philadelphia, Pennsylvania

Donald Heyneman – The George Williams Hooper Foundation, University of California, San Francisco, California

Jack Colvard Jones – Department of Entomology, University of Maryland, College Park, Maryland

Phyllis Johnson – Department of the Interior, Oxford Laboratory, Oxford, Maryland

H. Lipke – Department of Biology, University of Massachusetts, Boston, Massachusetts

Karl Maramorosch – Waksman Institute of Microbiology, Rutgers University, New Brunswick, New Jersey

Edward H. Michelson – Department of Tropical Public Health, Harvard School of Public Health, Boston, Massachusetts

Frederick A. Murphy – Center for Disease Control, Public Health Service, U.S. Department of Health, Education, and Welfare, Atlanta, Georgia

A. J. Nappi – Department of Biological Sciences, State University of New York, Oswego, New York

Thomas C. Orihel – Tulane University, School of Public Health and Tropical Medicine, New Orleans, Louisiana

J. D. Paschke – Department of Entomology Laboratory – Purdue University, West Lafayette, Indiana

Lynn M. Riddiford – Department of Zoology, University of Washington, Seattle, Washington

Robert E. Shope – Department of Epidemiology and Public Health, School of Medicine, Yale University, New Haven, Connecticut

CONTRIBUTORS

M. A. Stirewalt – Biomedical Research Institute, American Foundation for Biological Research, Rockville, Maryland

W. Daniel Sudia – Center for Disease Control, Public Health Service, U.S. Department of Health, Education, and Welfare, Atlanta, Georgia

Max D. Summers – The Cell Research Institute, Department of Botany, The University of Texas, Austin, Texas

T. W. Tinsley – Natural Environment Research Council, Unit of Invertebrate Virology, Oxford, England

M. R. Tripp – Department of Biological Sciences, University of Delaware, Newark, Delaware

Franklin G. Wallace – Department of Zoology, University of Minnesota, Minneapolis Minnesota

A. Burns Weathersby – Entomology Department, University of Georgia, Athens, Georgia

Sylvia G. Whitfield – Center for Disease Control, Public Health Service, U.S. Department of Health, Education, and Welfare, Atlanta, Georgia

G. R. Wyatt – Department of Biology, Queen's University, Kingston, Ontario, Canada

PREFACE

The chapters of this book were prepared for presentation at a workshop on invertebrate defense mechanisms sponsored by the National Institutes of Health and held in Bethesda, Maryland, April 17 and 18, 1974.

Invertebrates are represented by over a million species. The organizers of the workshop imposed no limits on the authors' breadth of coverage but there was an understandable bias in the choice of models to those higher invertebrate phyla such as mollusca and arthropoda which are of human disease importance. It is clear however that even the single-celled invertebrates, the protozoa, have effective defenses.

It will be apparent–and indeed this is why the workshop was organized–that much is yet to be learned about invertebrate defenses. This volume was not intended to answer all questions but rather was designed a) to survey the status of research in the field of defense factors in invertebrates (especially insects, ticks, and snails), b) to determine areas in which research is lacking or limited, and c) to identify specific investigations needed for advancement of invertebrate immunology. The authors were urged to bring out imaginative and provocative hypotheses and to suggest experimental approaches which might be applied to solutions.

Major questions were asked. What are the defense mechanisms of invertebrates? Why do some parasites, such as arboviruses, infect the invertebrate for life without cytopathology? Why do others, such as nuclear polyhedrosis viruses, kill their hosts? Why are parasites sometimes species-specific or tissue-specific? Are these specificities under genetic control? Are defense mechanisms comparable in all invertebrate phyla? How does the insect gut-barrier work? What are the parallels between vertebrate and invertebrate immune mechanisms?

The first 6 chapters relate to the barrier imposed by the gut to invading parasites. It is not that other known routes of parasite entry (such as the respiratory system or the integument) were forgotten but the gut is the usual route and must represent the first line of defense. Chapter 3 also addresses the questions of interferon and defective interfering virus particles, an attempt to relate known mechanisms in vertebrates that might have been inherited from invertebrates in the evolutionary scheme.

There follow 4 chapters on the role of invertebrate hemocytes and other phagocytic cells, 4 chapters on humoral defense mechanisms, and 3 on encapsulation, melanization, and hormones.

This book provides the first modern, integrated presentation of phenomena and mechanisms pertaining to immunity in invertebrate animals. Special attention was given to groups where more information was available, but authors focused attention at areas of importance where additional research is needed and expected to provide answers to as yet unresolved questions. Since no comparable book is available, this volume will be virtually indispensable for all interested in comparative immunology, as well as those specifically interested in invertebrate immune responses. It will be of special interest to entomologists, protozoologists, physicians, veterinarians, ecologists, pathologists, research workers, science teachers, and graduate students.

The editors would like to pay special tribute to Dr. George W. Luttermoser, Executive Secretary of the Tropical Medicine and Parasitology Study Section, National Institutes of Health, Bethesda, Maryland, who not only stimulated the arrangement of the workshop, but unselfishly helped in all stages of organization of the conference, including the selection of the participants and discussants. His remarkable skill and energy were responsible for obtaining all manuscripts by the opening date of the conference.

The chairmen of the workshop and editors of this book also express their sincere gratitude to the contributors who came from the United Kingdom, Canada, and the United States, for their effort and care with which they prepared their chapters: to the Division of Research Grants of the National Institutes of Health, Bethesda, Maryland, for financial support and encouragement, to Dr. John E. Scanlon, Chairman of the Tropical Medicine and Parasitology Study Section and to all others who helped in the arrangements. Last, but not least, thanks are due to Academic Press for their prompt production of this volume and for their part in editing and producing the book in a record short time.

It is hoped that the primary purpose of the present book will be achieved, namely to provide a stimulating forum for discussion of new ideas and observations concerning the mechanism of invertebrate immunity and its relevance to human health. The world-wide growth of immunological science is unique in the history of biology and the combined efforts of eminent contributors of this volume will benefit all interested in immunology.

June Stephens Chadwick

Thomas C. Orihel — William R. Harvey — T. W. Tinsley — J. D. Paschke

M. A. Stirewalt

Edward H. Michelson

John E. Scanlon

Daniel J. Sullivan S.J.

Lynn M. Riddiford — H. Lipke — A. J. Nappi

A. Burns Weathersby

Don Hall June Stephens Chadwick

Franklin G. Wallace Marion Brooks

M. R. Tripp

Frederick A. Murphy

Leo A. Jachowski, Jr. Eli Chernin

G. R. Wyatt

A. Ralph Barr

Thomas C. Orihel Alfred W. Senft

I. Invertebrate Gut as a Barrier to Invading Parasites

EPITHELIAL STRUCTURE AND FUNCTION

William R. Harvey and James T. Blankemeyer
Department of Biology, Temple University
Philadelphia, Pennsylvania

The insect midgut plays a major role in the resistance of insects to infection by viruses and parasites. It consists of an acellular peritrophic membrane (see papers by Orihel and Paschke, this workshop), an epithelium, and discontinuous layers of muscle and trachea. The epithelium has two probable roles in resisting infection; it forms an electrically charged barrier to pathogen entry, and it regulates the solute composition of the midgut contents, the intracellular fluids of the midgut epithelium, and the hemolymph. It is necessary for pathogens to pass through, exist in, and divide in these fluid compartments to accomplish invasion of the insect. This paper will examine the regulatory function of the midgut epithelium.

The general concept of an epithelium is emerging from structural studies on junctional structures and from biophysical studies on the mechanism and route of ion transport through epithelial cells. We will examine the role of epithelia by using the midgut of the silk moth, *Hyalaphora cecropia* (L), as an operational model with which other epithelia may easily be compared. Four generalizations apply to epithelia in general. The first is that epithelial cells are mechanically joined together by "adhering" junctional structures. The second is that epithelia impede solute and solvent flow by a mosaic barrier composed of "occluding" belts around the cell and membranes impermeable to solutes at the apical or, less frequently, the basal and lateral regions. The presence of this barrier requires that a specialized mechanism of facilitated diffusion or active transport exist at the barrier structures. The third generalization is that epithelial cells are electrically and metabolically coupled (Kanno and Loewenstein, 1966) which requires that a coupling junction exist between epithelial cells. Finally, the fourth generalization is that the concentrations of three key cations must be regulated accurately within epithelial cells: cellular K must be maintained at approximately .1 M if protein synthesis is to take place; cellular Na must be maintained at a low level if key enzymes are to remain active (Lubin, 1964); cellular Ca must be maintained at about 10^{-6}M if intercellular electrical communication is to exist (Loewenstein, 1966). The structural basis for these generalizations as they pertain to vertebrate epithelia is now quite well known (reviews by McNutt and Weinstein, 1973; Staehelin, 1974). For this reason we will first review junctional structures found within vertebrate epithelia as a model for our subsequent comments on junctional structures of insects and other invertebrates which are not as well understood (Satir and Gilula, 1973). Finally, we will review the evidence demonstrating that electrical coupling between epithelial cells combined with a cellular pathway of active potassium transport across the midgut leads directly to the conclusion that the cation concentration within the midgut cells and in the hemolymph is controlled by the effects of Ca^{++} on the cationic permeability of the basal and lateral membranes and by the kinetic parameters of the K-pump in the apical membranes, rather than by a cellular Na-K exchange pump as is generally accepted (e.g. Keynes, 1969).

EPITHELIAL CELL STRUCTURE

The midgut of the silkworm will serve as an example of the structure of insect midgut epithelia. The epithelium consists of columnar cells and occasional regenerative cells together with goblet cells, which are characteristic only of lepidopterous and tricopterous larvae, arranged in a one-cell-thick layer resting on a thick basement membrane.

A low power electron micrograph of the midgut of the closely related Saturniid, *Anthereae pernyi* is shown in Figure 1, and a diagram of the ultrastructure of the midgut is shown in Figure 2.

When viewed with light optics, the tall (65 μ) columnar cells exhibit an apical brush border which extends over the shorter (35 μ) goblet cells, a large centrally located nucleus, and basal striations. Transmission electron microscopy reveals that the brush border is composed of a regular array of microvilli into which project filaments which are 60-80 Å in diameter. Microtubules and rough endoplasmic reticulum both lying parallel to the long axis of the cell are prominent. The basal plasma membrane possesses deep infoldings which are lined by elongate mitochondria.

Goblet cells are invariably surrounded by columnar cells which implies that there are at least four times as many columnar cells as there are goblet cells present. The apical surface of the goblet cell is invaginated to form a large goblet cavity with large villus-like projections forming a value-like structure apically and with microvillus-like projections extending into the goblet cavity each surrounding an elongate mitochrondrion. The cytoplasmic leaflet of the plasma membrane which lines the goblet cavity, in these projections, contains particles resembling the elementary particles of mitochondria (Fig. 3). These particles (Fig. 4) were first described by Gupta and Berridge (1966) and since then have been shown to be characteristic of insect epithelia which are known to transport potassium (See Berridge and Oschman, 1972). In the midgut the particles are found only on the portion of the plasma membrane forming the goblet projections and nowhere else on the goblet cell or anywhere at all on the columnar cell (Anderson and Harvey, 1966). Junctional structures of the midgut epithelium will be discussed in the next section.

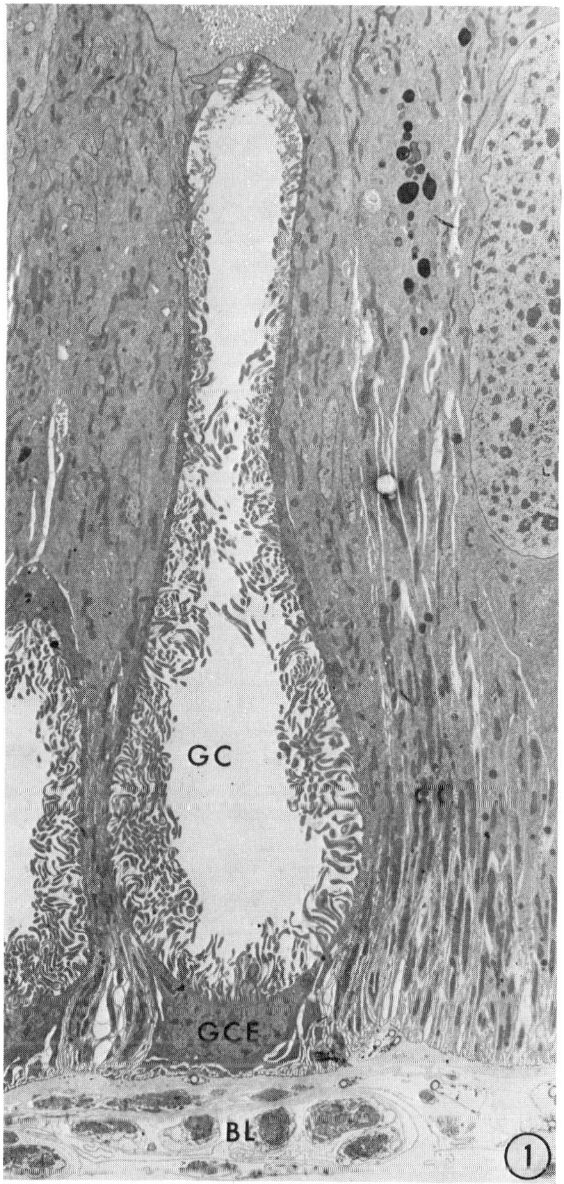

Figure 1. A section through the midgut of A. Pernyi showing a goblet cell (GCE), columnar cell (CC), goblet cavity (GC), and basement lamina (BL). X4000. Courtesy of B. Gupta, Cambridge University, with permission from Academic Press.

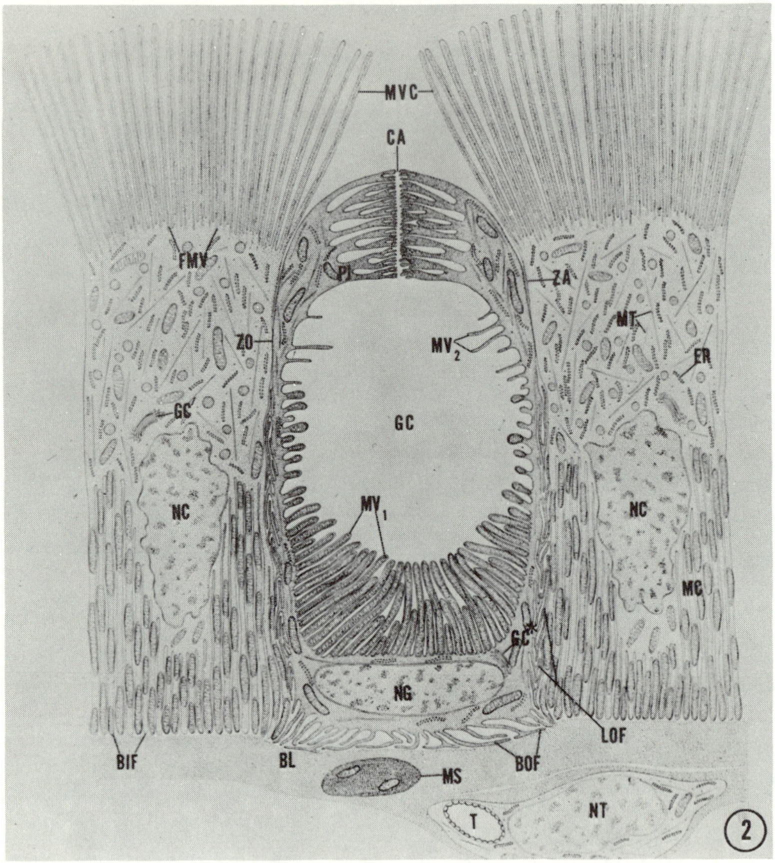

Figure 2. A schematic representation of the cell types comprising the epithelium of the midgut of a mature fifth instar larvae of Hyalaphora cecropia. Microvilli of the columnar cells (MVC); canal (CA) formed by the villus-like units derived from the larger protoplasmic projections (PJ) of the apical portion of the goblet cell; fine filaments within the microvilli of the columnar cell (FMV); zonula adherens (ZA); zonula occludens (ZO; apparent from thin section); microvilli (MV_2); microtubules (MT); endoplasmic reticulum (ER); cavity of goblet cell (GC); Golgi complex of goblet cell (GC'); nucleus of columnar cell (NC); mitochondria-filled cytoplasmic projections that line the major portion of the cavity of the goblet cell (MV_1); mitochondria of columnar cell (MC); Golgi complex of goblet cell (GC); nucleus of goblet cell (NG); basal infoldings of columnar cell (BIF): basement lamina (BL); muscle fiber (MS): basal podocytelike extensions of the goblet cell (BOF); lateral evaginations of the goblet cell (LOF); nucleus of tracheolar cell (NT); tracheole (T). From Anderson and Harvey (1966), with permission from the Journal of Cell Biology.*

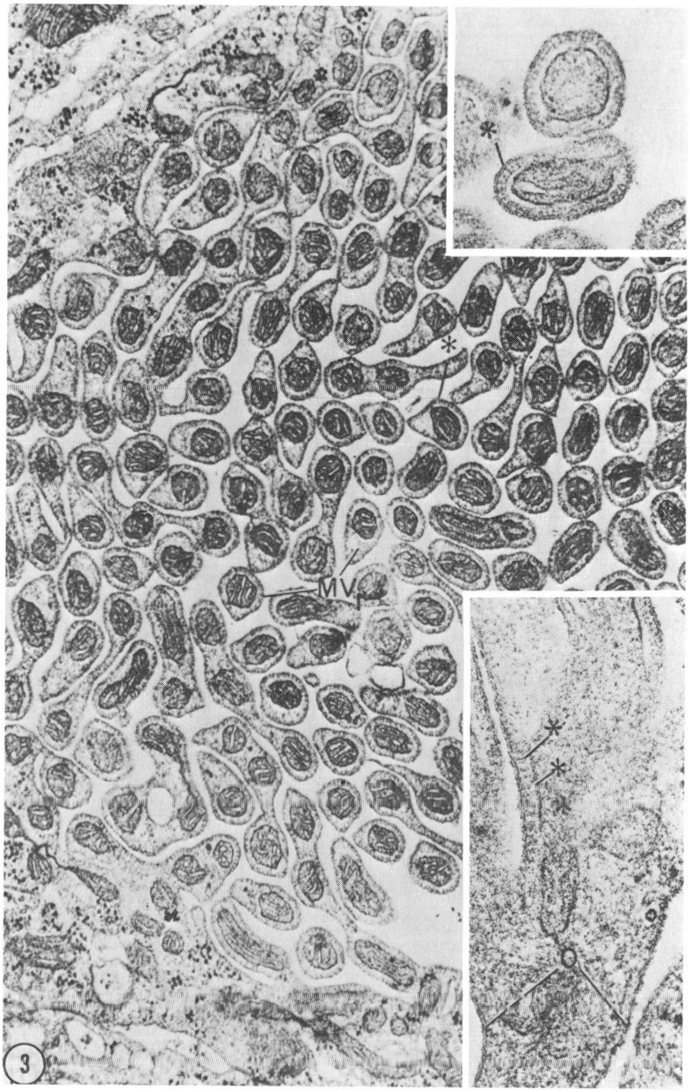

Figure 3. A section through cytoplasmic projections (MV_1) that line the cavity of the goblet cell. Note the spikelike units on the inner portion of the plasma membrane that form these projections (*, Figure 4 and insects). The inset at the lower right of the picture also shows portions of the plasma membrane whose inner membrane is devoid of spikes (0). Figure 4 X 30,000; upper right inset, X 60,000; lower right inset, X 70,000. From Anderson and Harvey (1966), with permission from the Journal of Cell Biology.

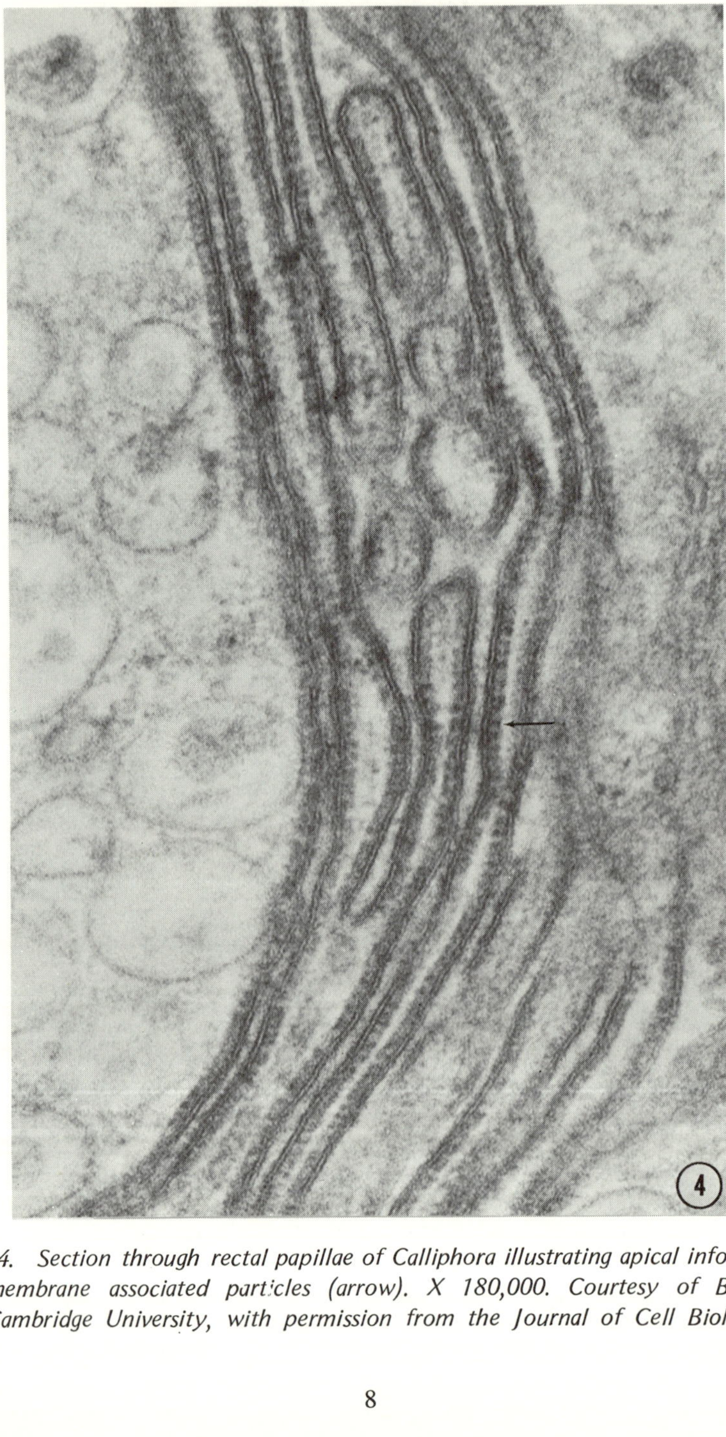

Figure 4. Section through rectal papillae of Calliphora illustrating apical infolding and membrane associated particles (arrow). X 180,000. Courtesy of B. Gupta, Cambridge University, with permission from the Journal of Cell Biology.

JUNCTIONAL STRUCTURES

Functionally, there must be three types of junctional elements corresponding to the three principal functions of epithelia; sealing, adhering, and coupling junctions. The problem is to determine which of the junctions classified by electron microscopy correspond to these functional types. Two technical advances make this difficult task possible: exposure of the epithelium to lanthanum hydroxide from one side prior to fixation, and analysis of freeze-cleaved replicas. These replicas are prepared by freezing tissues very rapidly. The resulting entropy change weakens hydrophobic bonds within the lipid bilayer and results in the exposure of new faces from within the membrane (Branton, 1966, McNutt and Weinstein, 1973). Staehelin's interpretation of the cleavage process is illustrated by Figure 5 in which the new face revealed by the cytoplasmic portion is designated the A face whereas the external portion is designated the B face. A binomial system of junctional nomenclature was introduced by Farquhar and Palade (1963) in which the first term describes the extent of opposition of cell surfaces: *zonula* (belt); *fascia* (sheet); *macula* (spot). The second term loosely describes the closeness of contact between the apposed cell surfaces; *occludens* for junctions with no extracellular space between the surfaces; *adherens* (which is actually a functional term) for junctions with extracellular space between the apposed membranes. Revel and Karnovsky (1967) upset this scheme completely when they found that lanthanum hydroxide would penetrate junctions previously thought to be of the "occludens" type. This "gap" junction which exhibits an intermembrane distance of 20 Å is made of small subunits visible in freeze-cleaved replicas. Clearly, a new workable system of

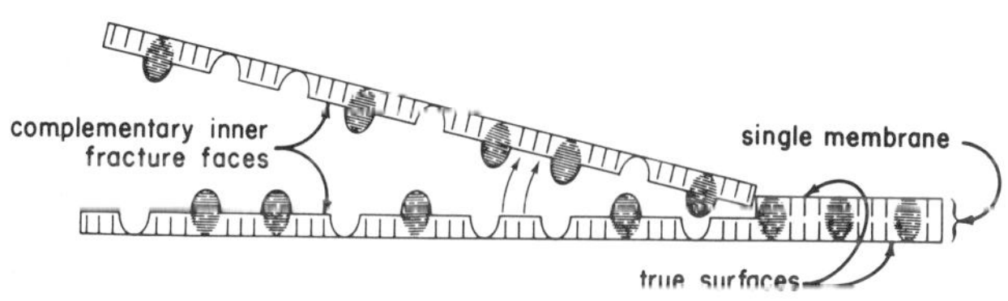

Figure 5. An interpretation of the fracture process of a single non-junctional membrane. The two inner faces produced by a freeze-cleaving exhibit complementary features, either particles or depressions out of which the particles have been fractured. Courtesy of L. A. Staehelin; Staehelin, 1974, with permission from Academic Press.

nomenclature encompassing the discoveries of the last ten years is needed.

The sealing junction of vertebrate epithelia is the *zonula occludens* or "tight junction" (Fig. 6, 7). The trilaminar membranes come into such close contact that the thickness of the junction is somewhat less than the sum of the thicknesses of the two membranes (140Å vs. 150Å). The term membrane fusion has been applied to this junction but McNutt and Weinstein caution that this usage is premature. The *zonula occludens* (see Fig. 8) is impermeable to penetration by lanthanum hydroxide and appears in freeze-cleaved replicas as a system of (sometimes) complementary ridges and grooves on the respective fracture faces (Fig. 7, 9).

Zonula adherens and *macula adherens* are both implicated in the mechanical adhesion of cells. They are characterized by 250 to 350 Å interspaces between apposed membranes and dense plaques and tonofibrils (100 Å) which are associated with these junctions.

Gap junctions (Fig. 10) often show a seven-layered structure in thin sections corresponding to two tri-laminar membranes separated by a 20 Å gap. The junctional thickness is 150 to 190 Å and they are penetrable by lanthanum hydroxide. Fig. 11 is a freeze-cleaved replica of a rat intestinal gap junction showing an A face with many particles (of two types) and a B face with few particles. In higher magnifications a central dot is seen in these particles which is thought to be the site of the coupling channel (see Fig. 42 of Staehelin, 1974). Staehelin's conception of a gap junction is shown in Figure 12.

Although invertebrate epithelia have not been analyzed as thoroughly as those of vertebrates several important differences seem to be emerging. The most striking difference is that a lanthanum impermeable junction which exhibits ridges and grooves in freeze-cleave replicas (tight junction) has not yet been observed in invertebrate tissues. Nevertheless, epithelia such as the silkworm midgut have a very low passive permeability to such small ions as potassium so that an efficient sealing junction must exist. The most likely candidate is the continuous junction (Fig. 13, 14) which has grooves and ridges like tight junctions in freeze-cleave sections but has faint septa, like septate junctions in thin sections. Loewenstein (Loewenstein and Kanno, 1964) originally suggested that the septate junction (Fig. 15) was the coupling junction but so far these junctions have not been found in the absence of gap junctions in coupled cells. Gap junctions of invertebrates exhibit globules within the junctional membrane as seen in cross sections of the crayfish hepatopancreas (Fig. 14). When this junction is treated with colloidal lanthanum and viewed in tangential section, the lanthanum outlines junctional particles which are larger than those of the vertebrate gap junction. The junction between goblet cell and columnar cell of the midgut of *A. pernyi* has been described by Gupta (cited in Berridge and Oschman, 1972) as a gap junction although the prospect that is a continuous junction has not been ruled out (Fig. 16). Table 1 is a summary of the properties of common junctional structures.

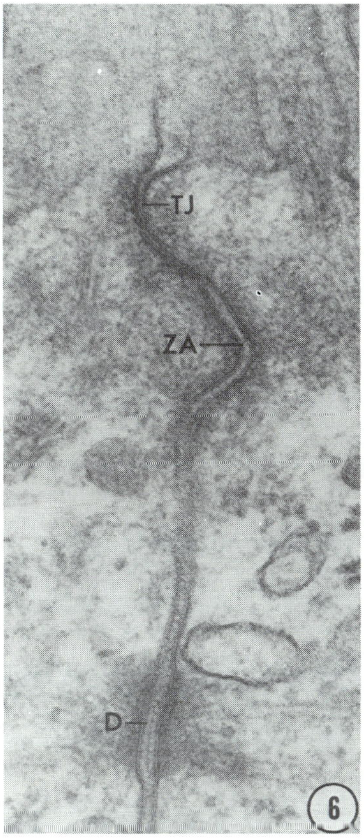

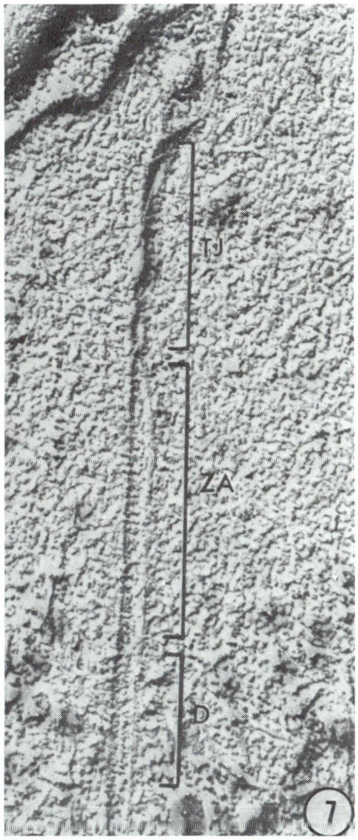

Figure 6. Junctional complex between two epithelial cells of rat intestine. The tight junction (TJ) or zonula occludens is located nearest the lumen and consists of very closely apposed membranes that seem to fuse at certain points (see text). The intermediate junction (zonula adherens, ZA) is found basal to the tight junction. Its membranes are spaced approximately 200 Å apart and are associated with a cytoplasmic thin filament matrix. The spot desmosome (D) or macula adherens is the most basal junctional complex. The inter-membrane distance is about 280 Å and is bisected by a dense stratum. Budles of tonofilaments lie on each side of the desmosome. X 90,000. Courtesy of L. A. Staehelin; Staehelin, 1974, with permission from Academic Press.

Figure 7. Freeze-cleaved junctional complex of the intestinal epithelium of a mouse depicting the same junctional elements as those shown in Figure 6. Tight junction (TJ), zonula adherens (ZA); spot desmosome (D). X 110,000. Courtesy of L. A. Staehelin; Staehelin, 1974, with permission from Academic Press.

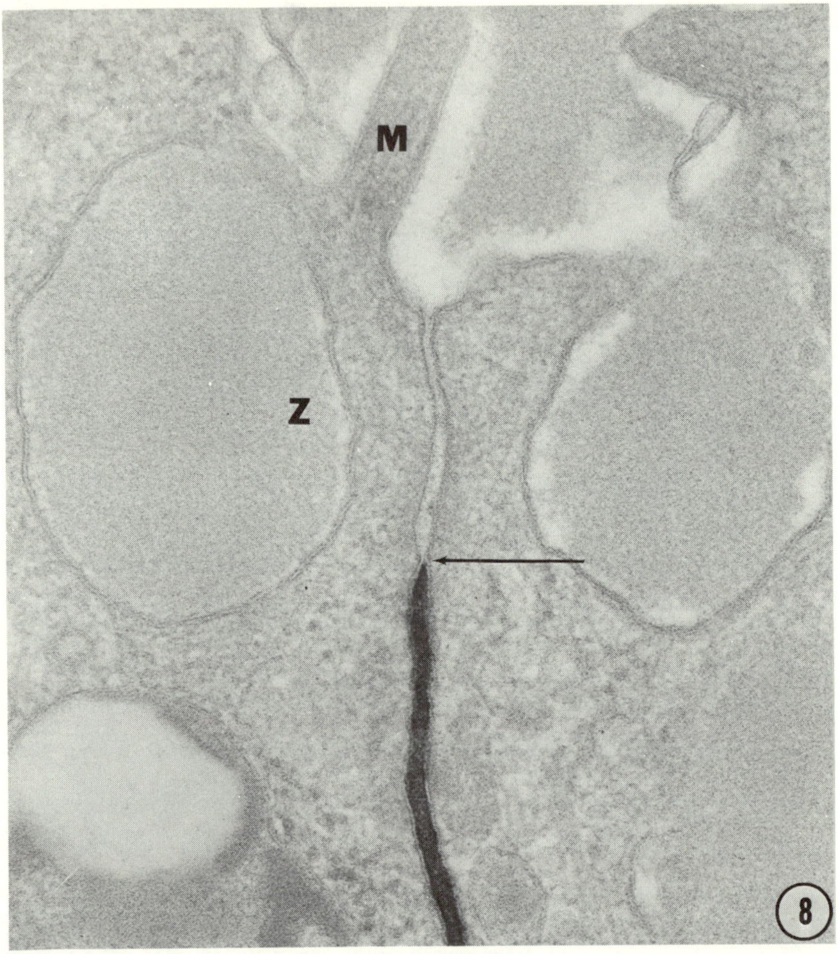

Figure 8. Pancreatic acinar cells from tissue injected with lanthanum hydroxide before fixation. The lanthanum penetrated the inter-cellular space from the vascular pole (bottom) and was stopped in its lumenward flow by tight junction (arrow). Microvillus (M); zymogen granule (Z). Courtesy of D. S. Friend; Friend and Gilula, 1972, with permission from the Journal of Cell Biology.

ELECTRICAL COUPLING IN EPITHELIA

The structural unity of epithelia is paralleled by an electrical unity. The demonstration of an electrical synapse by Furshpan and Potter (1959) was followed by the demonstration of electrical coupling between salivary epithelial cells of *Drosophila* larvae (Loewenstein and Kanno, 1964). Using microelectrodes it was found that the

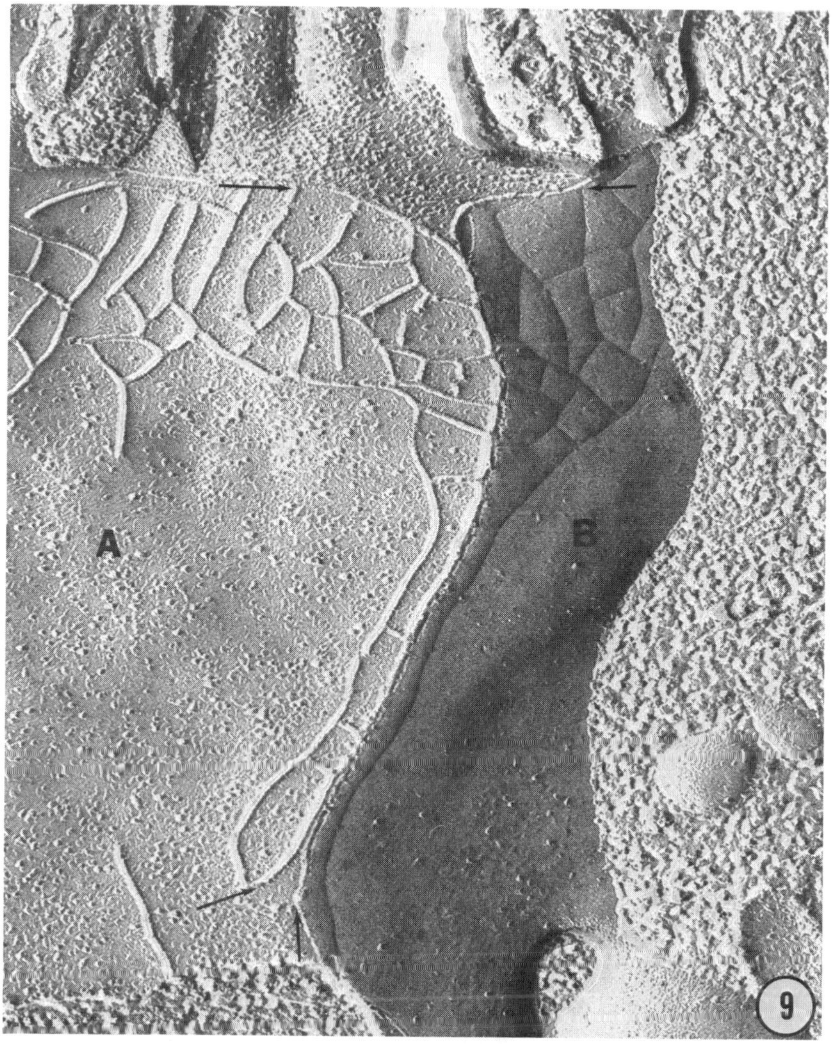

Figure 9. Characteristic extension of the sealing element meshwork of a tight junction. The two most apical seals (top arrows) of the adjoining cells are seen to turn from a horizontal to a vertical orientation in the central region of the triple junction. Due to a tilting of the membrane where these seals come together, they appear to fuse, but towards the basal end of the extension the two sealing elements can be resolved again (bottom arrows). A, A face; B, B face. X 1000,000. Courtesy of L. A. Staehelin; Staehelin, 1974, with permission from Academic Press.

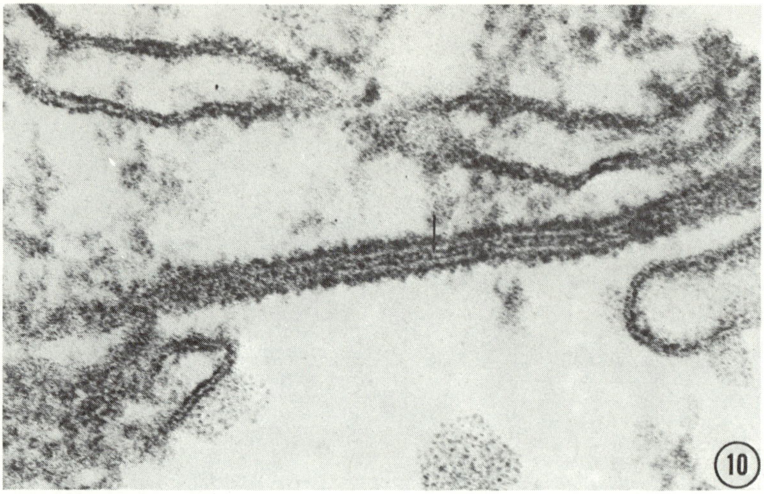

Figure 10. Typical seven-layered gap junction of human embryonic myocardium after fixation with glutaraldehyde and osmium tetroxide and en bloc staining with uranyl acetate. The central gap between the adjacent membranes is marked with an arrow. Courtesy of J. E. Rash, University of Colorado, Boulder, Colorado. X 300,000.

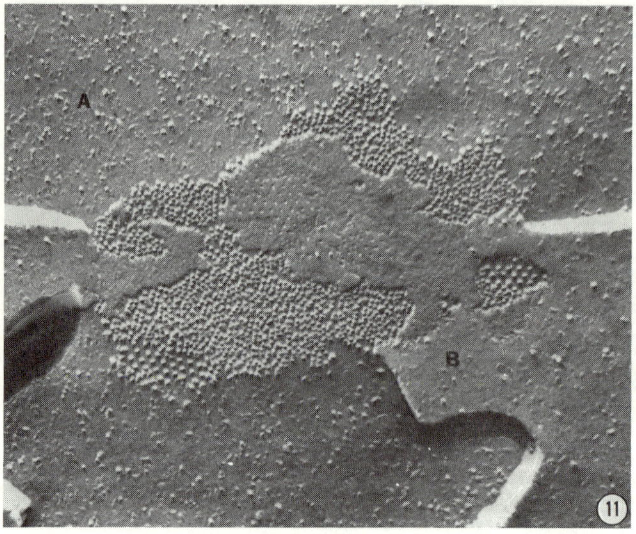

Figure 11. Micrograph of a freeze-cleaved gap junction from rat intestine revealing both A and B face features. The particles are the 80 to 90 Å variety which are only occasionally arrayed hexagonally. X 165,000. Courtesy of L. A. Staehelin; Staehelin, 1974, with permission from Academic Press.

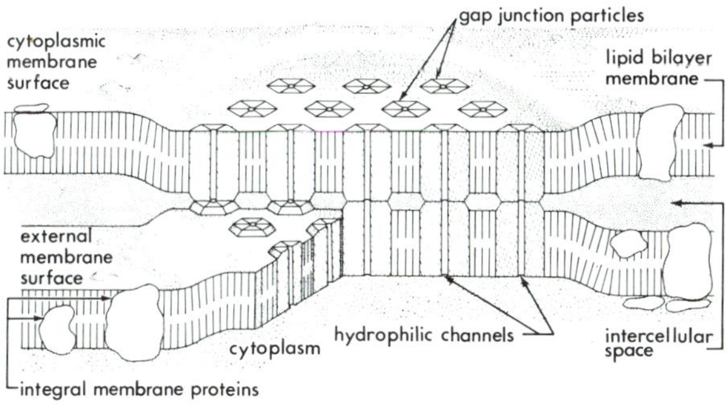

Figure 12. Three dimensional reconstruction of a gap junction. Courtesy of L. A. Staehelin; Staehelin, 1974, with permission from Academic Press.

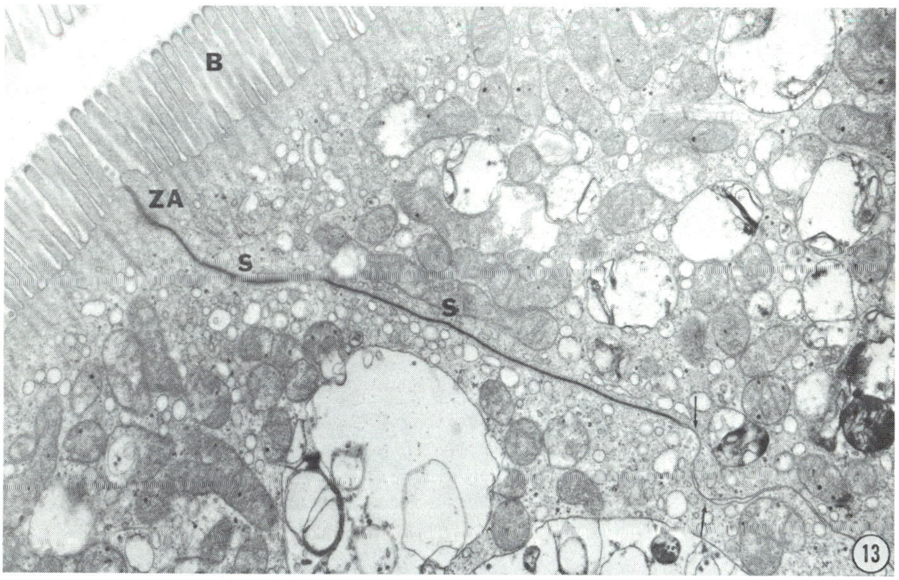

Figure 13. Junctional region between parenchymal cells of the crayfish hepatopancreas (digestive gland). The epithelial cells possess a microvillar brush border (B) and are connected by: (1) an apical zonula adherens (ZA); (2) an extensive zonula continua or modified septate junction which extends for 5-6 μ (S); and a series of many small gap junctions (arrows). X 24,000. Courtesy of N. B. Gilula; Gilula, 1971.

Figure 14. Freeze fracture of the crayfish hepatopancreas. The microvilli of the brush border (B) constitute the apical surface of these cells. Membrane fracture faces are revealed in the zonula adherens (ZA) region, the zonula continua region (ZC), and the gap junction region (G). X 42,000. Courtesy of N. B. Gilula; Gilula, 1971.

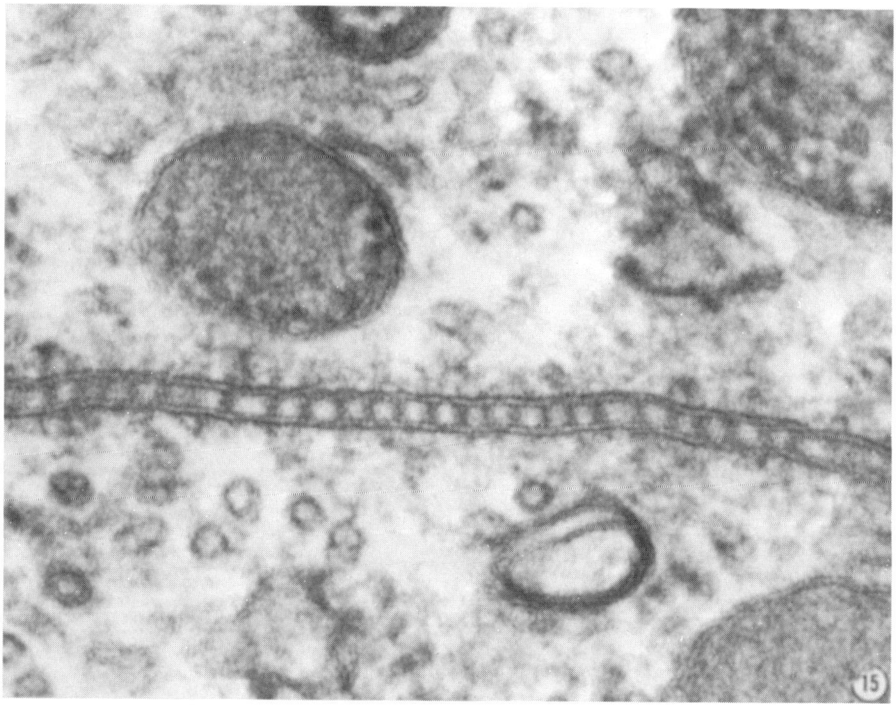

Figure 15. *Micrograph of a septate junction in salivary gland of Calliphora larva. Courtesy of B. Gupta, Cambridge University.*

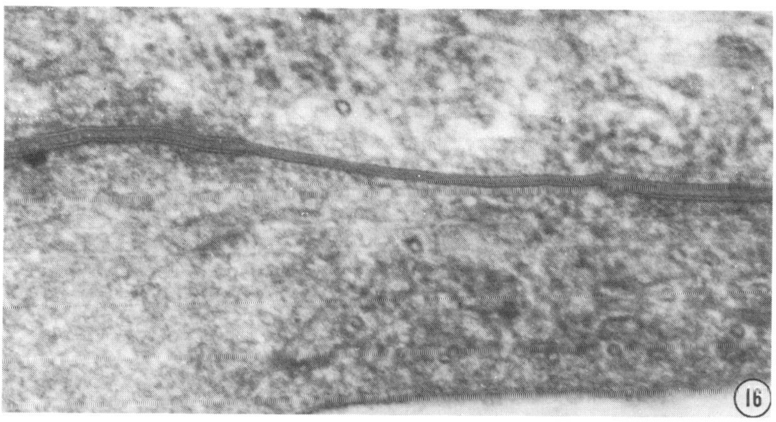

Figure 16. *Micrograph of a junction (see text) between goblet cell and columnar cell of the midgut epithelium of A. pernyi. X 140,000. Courtesy of B. Gupta, Cambridge University.*

TABLE 1

Junctional Complexes

	STRUCTURAL PROPERTIES			FUNCTIONAL PROPERTIES		
LATIN NAME	COMMON NAME	DISTANCE BETWEEN INNER LEAFLETS(Å)	FOUND IN	ADHESION	COMMUNI-CATION	COMMENTS
Zonula occludens	tight	0	vertebrates	yes	no	Ridges seen in freeze-cleaved replicas. No extracellular space with lanthanum penetration. Functions as a sealing element.
Macula and *Fascia occludens*	focal tight junction	0	both	yes	no	Sealing function broken up by discontinuities. Often confused with gap junction. Has a fibrillar structure in freeze-cleaved replicas.
Macula occludens	gap	20-40	both	no	yes	Strong evidence exists to support communication as the primary role of the gap junction (especially small subunit or nexus). Lanthanum penetration is necessary to see extracellular space. Freeze-cleaving shows particles of 80-100 Å in type A.
Zonula, Fascia adherens	belt, sheet desmosome	150-250	both	yes	no	Has 70 Å actin filaments associated with the thread-like complexes.
Macula adherens	spot desmosome	220-350	both	yes	no	Associated with 100 Å tonofilaments that are not actin. Has cytoplasmic plaques at the *Macula adherens* site.

TABLE 1 cont.

	STRUCTURAL PROPERTIES			FUNCTIONAL PROPERTIES		
LATIN NAME	COMMON NAME	DISTANCE BETWEEN INNER LEAFLETS(Å)	FOUND IN	ADHESION	COMMUNI-CATION	COMMENTS
	septate desmosome	150-170	both	yes	no (?)	Has electron dense bands of "septae" that lie between the inner membrane leaflets. Lanthanum partially penetrates septate junctions.
Zonula cortinua		110-150	invertebrates	?	?	Appears to have septae but also has ridges in freeze-cleaved replicas.

voltage applied to one cell appears in neighboring cells with an attenuation no greater than that expected if the cytoplasm of the cells were free of lateral membranes. This electrical transmission between cells is carried by small ions but molecules such as fluorescein (MW = 376) and Procion Yellow (MW = 500) (Loewenstein and Kanno, 1964, Payton et al., 1969) were observed to pass between cells. As mentioned above, Loewenstein and Kanno (1964) originally suggested that the septate junction was the coupling junction in insects. Later work has established that gap and septate junctions coexist in insects which indicates that the gap junction is probably the coupling junction (Oschman and Berridge, 1970; Rose, 1971). This interpretation received elegant support from the genetic studies of Gilula et al. (1972) on mutant cell lines cultured from hampsters. Both electrical and metabolic coupling were found in cells possessing gap junctions but were absent in mutants lacking these junctions.

K-TRANSPORT IN SILKWORM MIDGUT

It has been commonly assumed that the route of cation transport across epithelia is through the cells (Koefoed-Johnsen and Ussing, 1958), but the complex structure of the frogskin model and the small size of its sodium pool have made a clear demonstration difficult leaving an extracellular pathway still possible (Cereijido and Rotanno, 1968). Similarly Zerahn (1973) has argued against a cellular route for K-transport through the insect midgut. However, Wood (1972) presented evidence that the K-pool is too large to be extracellular and Harvey and Wood (1972, 1973) have shown that the pool occupies approximately two-thirds of the tissue K and is located before the K-pump. Recently Wood and Harvey (unpublished results) have shown by direct measurements in double-label experiments that the size of the kinetic pool is indeed large. Previously, Wood et al. (1969) had shown that the pump is located in the apical membranes because the large apical potential step was oxygen dependent. Since the pool is so large that it must be in all of the epithelial cells the pump must be located in the apical membranes of the cells rather than between the cells. The particles in the apical projections of the goblet cell discussed earlier are likely candiates for the actual pumping sites. Wood and Harvey propose that potassium enters the midgut epithelial cells passively from the hemolymph-side, mixes with the K in the cytoplasm, diffuses from columnar to goblet cell across a coupling junction, and is pumped out of the latter by pumps located in the plasma membrane of the apical projections. The consequence of this route is that the transepithelial K-pump regulates not only the hemolymph K-concentration but regulates the intracellular K-concentration as well. These results provide a rationale for attempts to alter the intracellular ionic environment and render it hostile to pathogens without removing the vector insect from the ecosystem.

PROSPECTS

The small size of many vector insects makes biophysical measurements difficult but is an advantage for structural studies. Hecher et al. (1971a, b) have studied the ultrastructure of the midgut of the mosquito *Aedes aegypti* (L) in order to understand the relation of ultrastructure to the digestion of ingested blood; females digest the meal whereas males do not. Recent unpublished work by Brooks has indicated that the potassium concentration in which symbionts from insect cells are transferred must mimic the high cellular K concentrations whereas the culture medium for the insect cells must mimic the lower hemolypmh K concentrations. Her work suggests that regulated cellular and hemolymph ionic environments may be crucial for pathogens. These two lines of research indicate a pathway to follow in attempts to control pathogens transmitted by insects. An understanding of the structure and function of cellular junctions of insects is a necessary step along this pathway.

ACKNOWLEDGMENT

This research was supported in part by a research grant (AI-05903) from the National Institute of Allergy and Infectious Diseases, U. S. Public Health Service. We thank Dr. Andrew Staehelin for access to his manuscript (Staehelin, 1974) and for prints and slides of several of his illustrations, Drs. Friend, Gilula, Gupta, and Rash for prints and slides of micrographs, and Dr. Wood for reading this manuscript.

REFERENCES

Anderson, E. A., and Harvey, W. R. (1966). *J. Cell Biol., 31*:107-134.

Berridge, M. J., and Oschman, J. L. (1973). "Transporting Epithelia", Academic Press, New York.

Branton, D. (1966). *Proc. Nat. Acad. Sci. USA, 55*:1048-1056.

Cereijido, M., and Rotunno, C. A. (1968). *J. Physiol.* (London) *190*:(3)481-497.

Farquhar, M. G., and Palade, G. E. (1963). *J. Cell Biol., 17*:375-412.

Friend, D. S., and Gilula, N. B. (1972). *J. Cell Biol. 53*:758-776.

Furshpan, E. J., and Potter, D. D. (1959). *J. Physiol.* (London) *145*:289-325.

Gilula, N. B. (1971). "Studies on the Septate Junction". Ph.D. thesis, Univ. of California, Berkeley.

Gilula, N. B., Reeves, O. R., and Steinbach, A. (1972). *Nature, 235*:262-265.

Gupta, B. L., and Berridge, M. J. (1966). *J. Cell Biol., 29*:376-382.

Harvey, W. R., and Wood, J. L. (1972). In "Role of Membranes in Secretory Processes" (L. Bolis, R. D. Keynes, and W. Wilbrandt, eds.), pp. 310-331. North-Holland Publishing Company, Amsterdam.

Harvey, W. R., and Wood, J. L. (1973). In "Transport Mechanisms in Epithelia" (N. A. Thorn and H. H. Ussing, eds.). Alfred Benzon Symposium *V.* pp. 342-357. Munskgaard, Copenhagen.

Hecker, H., Freyvogel, T. A., Briegel, H. Steiger, R. (1971a). *Acta. Tropica., 28*:80-104.

Hecker, H. Freyvogel, T. A., Briegel, H. Steiger, R. (1971b). *Acta. Tropica., 28*:275-290.

Kanno, Y., and Loewenstein, W. R. (1966). *Nature, 212*:629-630.

Keynes, R. D. (1969). *Quart. Rev. Biophysics., 2*:177-281.

Koefoed-Johnson, V., and Ussing, H. H. (1958). *Acta. Physiol. Scand., 12*:298-308.

Loewenstein, W. R. (1966). *Ann. N.Y. Acad. Sci., 137*:440-471.

Loewenstein, W. R. (1967). *J. of Colloid and Interface Science, 26*:34-46.

Loewenstein, W. R., and Kanno, Y. (1964). *J. Cell Biol., 22*:565-586.

Lubin, M. (1964). In "The Cellular Functions of Membrane Transport" (J. F. Hoffman, ed.), pp. 193-211. Prentice-Hall, Englewood Cliffs, New Jersey.

McNutt, N. S., and Weinstein, R. S. (1973). In "Progress in Biophysics and Molecular Biology" (J. A. V. Butler and D. Noble, eds.), *26*, pp. 47-101. Pergamon Press, Oxford and New York.

Oschman, J. L., and Berridge, M. J. (1970). *Tissue and Cell 2*:281-310.

Payton, B. W., and Bennett, M. V. L., and Pappas, G. D. (1969). *Science, 166*:1641-1643.

Revel, J. P., and Karnovsky, M. J. (1967). *J. Cell Biol., 33*:C7.

Rose, B. (1971). *J. Memb. Biol., 5*:1-19.

Satir, P., and Gilula, N. B. (1973). *Ann. Rev. Entomol., 18*:143-166.

Staehelin, L. A. (1974). In "International Review of Cytology" (G. H. Bourme and J. F. Danielli, eds.), Academic Press, New York (In press).

Wood, J. L. (1972). Ph.D. Thesis, Cambridge University, London.

Wood, J. L., Farrand, P. S., and Harvey, W. R. (1969). *J. Exp. Biol.*, *50*:169-178.

Zerahn, K. (1973). In "Transport Mechanisms in Epithelia" (N. A. Thorn and H. H. Ussing, eds.), Alfred Benzon Symposium *V*, pp. 360-367. Munksgaard, Copenhagen.

INTERACTIONS OF VECTOR WITH VERTEBRATE PATHOGENIC VIRUSES

Frederick A. Murphy, Sylvia G. Whitfield,
W. Daniel Sudia and Roy W. Chamberlain*
Center for Disease Control
Public Health Service
U.S. Department of Health, Education, and Welfare
Atlanta, Georgia

Among the viral infections of arthropods, a division has been made between those viruses which alternately cycle through plants, arthropods, and vertebrates. This division, which in many ways is arbitrary, is useful here to limit coverage to the "arboviruses," those viruses maintained in nature principally via biological transmission between hematophagous arthropods and *vertebrate* hosts. Coverage is also restricted by the narrowness of our knowledge in this area: at present more than 300 distinct arboviruses are known, which fall into more than 28 serogroups (W.H.O. Scientific Group, 1967) and five major physiocochemically defined groups (with many unclassified viruses left over) (Casals, 1971), but in only a few instances have attempts been made to describe infectious processes within the arthropod vectors in any detail. Infectious processes in arthropods have primarily been studied by sequential assay of transmissibility and sequential titration of arthropod organs and tissues (Chamberlain and Sudia, 1961). These approaches have been complemented by application of immunofluorescence techniques to arthropod tissues (Doi, 1970; Maguire, 1973). We have used thin section electron microscopy of organs of *Aedes triseriatus* mosquitoes infected with eastern equine encephalitis (EEE) virus (Whitfield et al., 1971) and of *Culex pipiens* mosquitoes infected with St. Louis encephalitis (SLE) virus (Whitfield et al., 1973) in an effort to provide a better understanding of the relationship of arboviruses with their natural vectors.

In view of the obviously successful perpetuation of arboviruses in nature, their life cycles seem inordinately tenuous. A life cycle impasse, occurring as a failure of viral spread or proliferation, may appear with many variations; two such impasses have received most attention. First, in vertebrates, failure of infection to yield a viremia sufficient to infect a feeding arthropod seems to be the key characteristic distinguishing an abortive cycle from a progressive or amplifying cycle (Schaeffer and Arnold, 1954). Second, in arthropods, failure of infection has been attributed to the presence of a

*U.S. Department of Health, Education and Welfare, Ft. Collins, Colorado.

"threshold" or "gut barrier" (Chamberlain et al., 1954; 1961). This relative or absolute "barrier," whether physiologic or anatomic, intervening in the course of viral movement from an infected blood meal in the gut lumen or at stages of viral adsorption, penetration, or infection of gut epithelium, is sophisticated in the viral specificity and sensitivity exhibited by particular arthropods. In part, the barrier may gain its effectiveness by acting at that stage in the viral life cycle where the smallest number of infectious units must "carry the ball," and where adverse conditions for viral survival are concentrated. In order to pursue, experimentally, the negative concepts of "threshold" or "gut barrier," it is necessary to explore the alternative tack; that is, it is necessary to study successful, progressive infection in the arthropod gut. Besides defining arthropod defense mechanisms, this tack emphasizes gut infection as the key to parenchymal viral invasion and spread to other target organs including salivary gland epithelium. We cannot predict where other potential impasses in arbovirus life cycles may act, but a weak link at any stage may provide insight into means for disease control.

ARBOVIRUSES IN MOSQUITOES

The Midgut. In mosquitoes a viremic blood meal bypasses the gut diverticula and goes directly to the abdominal midgut. Sequential organ titration, despite its inherent insensitivity, has indicated that multiplication of several Group A, B, and Bunyamwera viruses in abdominal viscera of vector species immediately follows an eclipse phase (LaMotte, 1960; Chamberlain and Sudia, 1961; Altman, 1963; Thomas, 1963; Janzen et al., 1970; Peers, 1972). In the most carefully studied model, that of Japanese encephalitis (JE) virus infection of *C. pipiens* (LaMotte, 1960), abdominal midgut titers reached a maximum in less than 1 week after ingestion of an infected meal and before significant involvement of other organs. Titers declined to very low levels over the following 3 weeks. The added anatomic specificity of the immunofluorescence technique indicated that JE viral antigen localization initially occurred in the posterior part of the midgut (the site receiving most or all of ingested infected blood), and that only later did infection spread to the anterior portion (Doi, 1970). Finally, viral antigen disappeared from the gut entirely. In other virus-mosquito systems an early rise of midgut infectivity titer was followed by a persisting "plateau" level (Peers, 1972). In refractory arthropod species, particular viruses decline in titer at rates consistent with the concept that harsh thermal and chemical environment "sterilizes" the digesting blood meal. Although it is at present impossible to define the host mechanisms within an arthropod species which prevent or suppress viral replication in gut epithelium, it is clear that maximal viral concentration in the epithelium is transient at best and less than considered theoretically possible in view of the number of cells available and the yield capability per cell per unit of time.

Experimental observations: At the ultrastructural level, the midgut of *C. pipiens* and *A. triseriatus* mosquitoes is seen to consist of a single layer of columnar cells with elaborate microvillous luminal margins, tight lateral junctions and a smooth outer margin separated from the hemocoele by a complex multilayered basement lamina (Whitfield et al., 1971, 1973). It was possible to visualize SLE virus particles in the distended lumen of the midgut of *C. pipiens* after a feeding upon viremic mice; the particles (sparsely scattered) were found at 1, 4, and 8 hours, and were located between the cells of the blood meal. In one instance, 8 hours after feeding, virus was found in cytoplasmic vacuoles of a gut epithelial cell. Thereafter, no evidence of viral presence or infection was detected until day 6, and no evidence of a "peritrophic membrane" was found at any time. On day 6, mature virus particles in small numbers occurred primarily within cisternae of the endoplasmic reticulum of gut epithelial cells (Fig. 1) and even fewer particles were located extracellularly between cells and under the basement lamina. The *concentration* (number of particles per cell) of intracisternal SLE virus particles increased between days 6 and 8, then remained constant through day 12, and finally declined through the rest of the 32-day study. There were also changes in the *proportion* of gut cells exhibiting evidence of infection at different time periods. More and more cells were observed to be infected between days 6 and 12; then, with a maximum of one in five cells infected, this proportion remained constant until the end of the experiment. Virus particles produced from this gut infection appeared in extracellular spaces from day 9 onward; most significantly, some of these particles were associated with basement lamina and its juxtaposition with the hemocoele (Fig. 2). Some particles penetrated the basement lamina en route to the hemocoele. Other membranous structures characteristically associated with Group B arbovirus morphogenesis were spatially and temporally related to virus particle formation in early phases of infection, but accumulated in an asynchronous manner later.

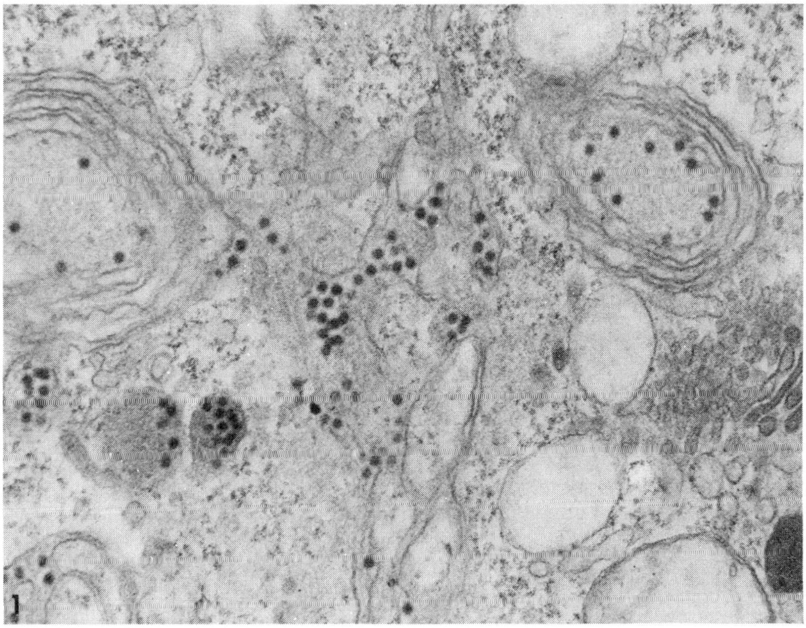

Fig. 1. SLE virus in the endoplasmic reticulum of a midgut epithelial cell of Culex pipiens 8 days after infection. X56,000.

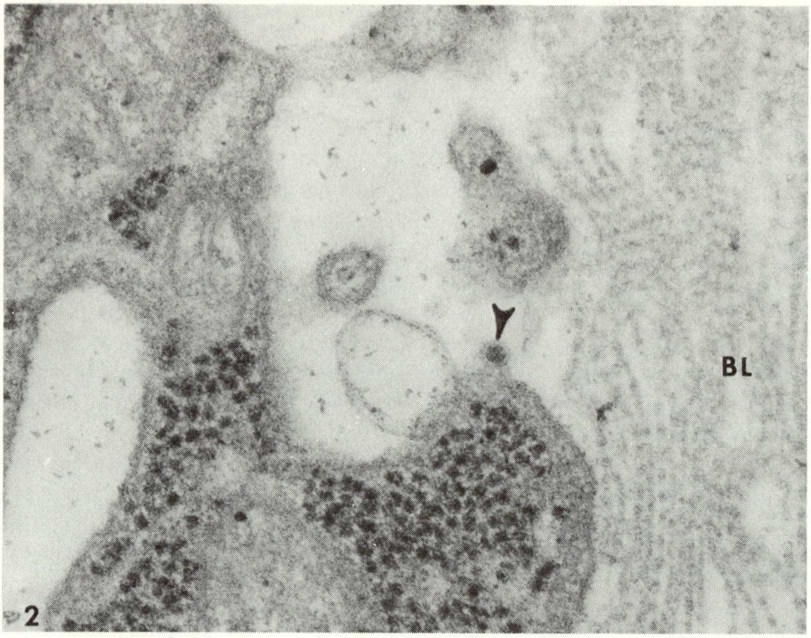

Fig. 2. One SLE virus particle (arrow) at the outer edge of a midgut epithelial cell directly beneath the basement lamina (BL) and hemocoele. X93,000.

EEE virus in *A. triseriatus* midgut epithelium underwent an infection course rather similar to that of SLE virus in *C. pipiens*. The productive phase of EEE infection yielded concentrations of virus particles per infected cell which were greater than those of SLE virus (Fig. 3), and this yield per gut cell remained constant from day 13 to day 31 post infection. Viral nucleocapsid concentration, which was high at day 13, declined through the harvest series (Fig. 4). The proportion of gut cells exhibiting evidence of EEE infection was also higher than in the SLE experiments; we estimated that a rather constant infection ratio of one gut cell in three persisted from day 13 to 31. It is worthy of note that throughout this study period, and in the similar time course of our SLE virus experiments, no gut epithelial cell pathology was discernible by either light or electron microscopy.

The two studies of SLE and EEE viruses in arthropod midgut complemented each other in showing morphologically the progressive involvement of increasing numbers of cells and substantial virus yields. Both studies also revealed the yield of progeny virus to be somewhat directional: more virus particles left the infected cells on their outer (hemocoele) margins than on the inner (luminal) margins (Fig. 5). This enrichment of virus particles in proximity to the gut basal lamina and surrounding hemocoele matches the active "flow" of nutrients from blood meal to parenchyma across gut epithelium. In this gut infection process, however, both of the viruses reached "plateau"

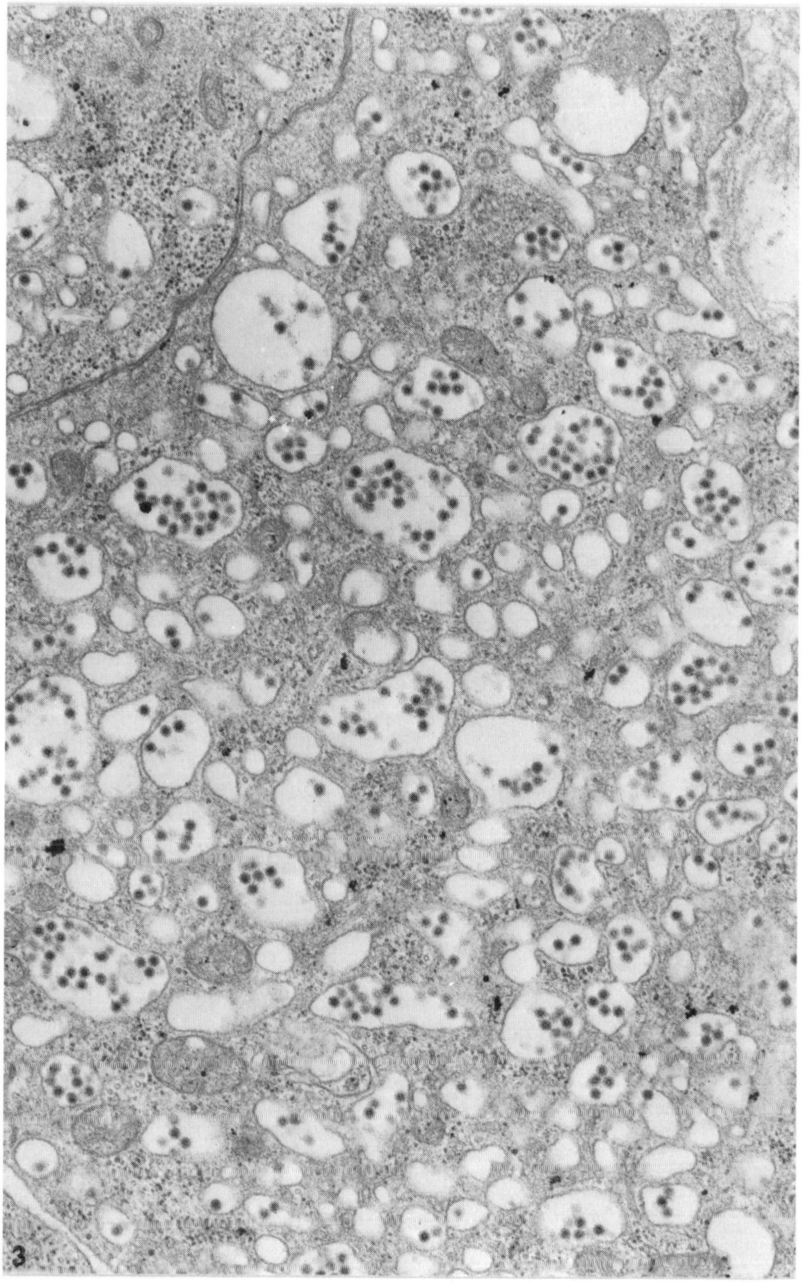

Fig. 3. EEE virus in the reticulum of a midgut epithelial cell of Aedes triseriatus 13 days after infection. X33,000.

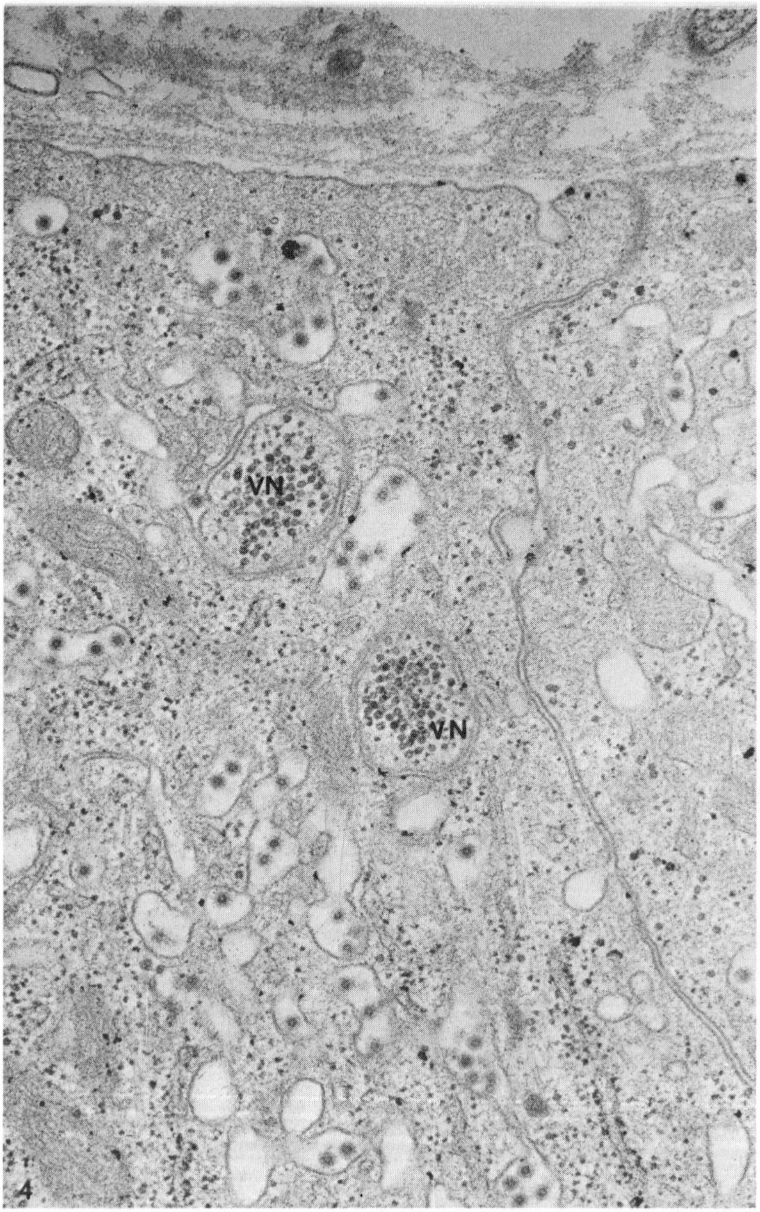

Fig. 4. EEE viral nucleocapsids (VN) accumulating focally in the cytoplasm of a midgut epithelial cell 13 days after infection. Fewer nucleocapsids were seen than in infected vertebrate cells. X46,000.

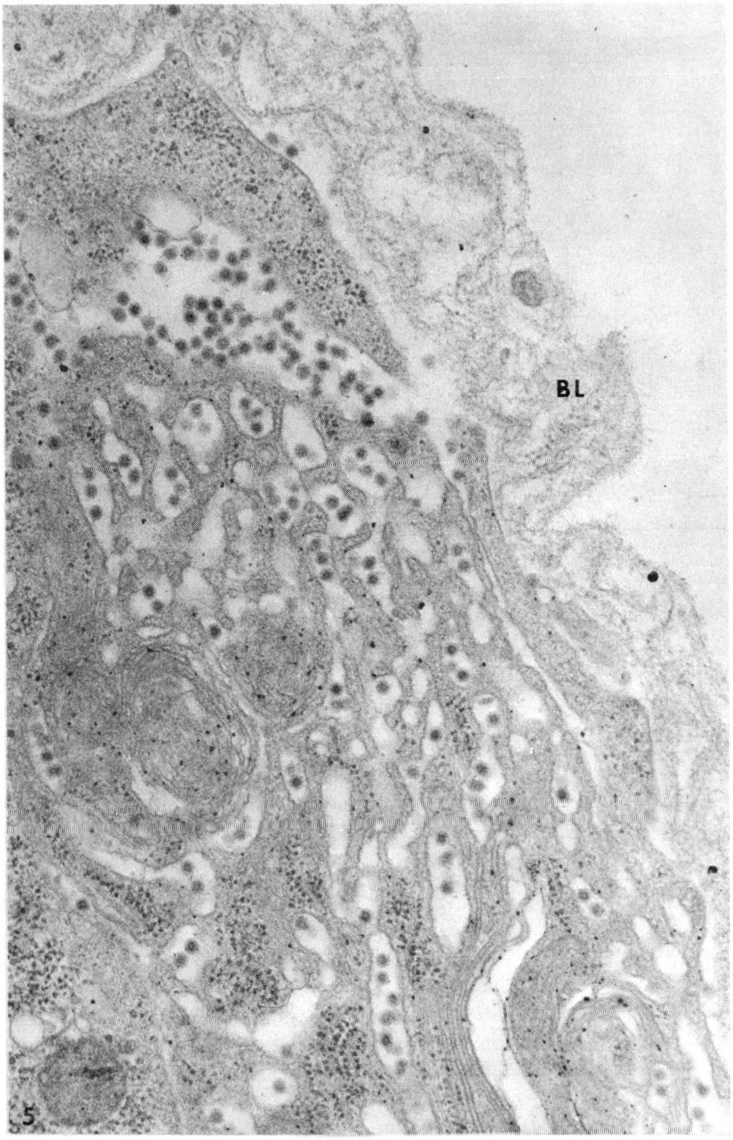

Fig. 5. EEE virus particles concentrated at the outer margin of a midgut cell beneath the basement lamina (BL). There was preferential accumulation at basal margins of gut cells and luminal margins of salivary gland cells, possibly indicating a "directional flow." X44,000.

or declining levels before all of the gut cells became involved, thus precluding production of virus in the extraordinary concentrations which climax EEE and SLE infections in other systems such as vertebrate cell cultures and newborn mouse brain (Murphy and Whitfield, 1970; Murphy et al., 1968). Thus, electron microscopic evidence, like that obtained earlier by arthropod titrations, advances a concept of a "limited" or "homeostatic" infection in the arthropod gut. The possible nature of the infection limitation will be discussed further.

In some arbovirus-arthropod systems, virus apparently can pass through the gut into the hemocoele in much less time than would be expected if replication occurred. As examples of such presumed direct passage, Miles et al. (1973) found Whataroa virus in the hemolymph of *Aedes australis* as early as 4 hours after feeding, and Boorman (1960), working with Uganda S virus in *Aedes aegypti*, showed hemolymph to be infectious from 10 minutes onward. These observations are consistent with some exceptionally short extrinsic incubation periods which have been reported; for example, the transmission of western equine encephalitis (WEE) virus by the bite of *Culex tarsalis* occurred as early as 4 days after an infected blood meal (with this virus-vector combination, 10-14 days is more usual). Such a short extrinsic incubation period is comparable to that expected after direct injection of virus into the hemocoele, which brings about a more immediate salivary gland infection than ingestion of a virus meal (Thomas, 1963). The implication of this passive, rapid movement of virus from the gut lumen to the hemocoele primarily relates to potentially explosive cycling and vertebrate epidemics. In the present context, however, a bypass of the virus growth cycle in gut may be seen also as a circumvention of limiting "host defense" mechanisms active at the level of the gut.

The Salivary Glands. There are at least two pathways of virus from the arthropod midgut to the salivary glands: one is by way of the hemolymph, and the other, by progressive sequential infection of contiguous organs. Chamberlain and Sudia (1961) have reviewed the considerable evidence of the hemolymph pathway. Arthropod organ titrations have repeatedly shown simultaneous involvement of various organs soon after virus reaches the hemolymph, and direct injection of virus into the mosquito hemocoele usually shortens the extrinsic incubation time and increases transmissibility, even in nonvector species. As evidence of a pathway other than via the hemolymph, Maguire (1973) has recently shown by immunofluorescence that Whataroa virus proceeds sequentially through organs of *Aedes australis* mosquitoes. Midgut, abdominal fat, thoracic ganglia, cerebral ganglia, and finally, salivary glands (from day 16 onward) were progressively infected. Viral antigen disappeared from these organs in the same sequence and by day 21 could be detected only in salivary glands. When Maguire infected mosquitoes by injection instead of feeding, immunofluorescent viral antigen localized in the salivary glands 4 to 6 days earlier, and organs did not exhibit the same sequential involvement. Whatever the pathway, salivary gland infection of vector species usually reaches a level greater than that in other organs. This is true whether assay is by immunofluorescence (Doi, 1970; Gaidamovich et al., 1973), organ titration (LaMotte, 1960; Peers, 1972), or electron microscopy (Bergold and Weibel, 1962; Janzen et al.,

1970; Larsen and Ashley, 1971).

Experimental observations. The paired salivary glands of *A. triseriatus* and *C. pipiens*, like those of other mosquitoes, each consist of three lobes connected to a common chitinized duct at the hypopharynx (Wright, 1969; Janzen and Wright, 1971). Each lobe is made up of a single layer of cuboidal epithelial cells around a central lumen containing a cuticular central duct structure. In our electron microscopic study of EEE virus infection in *A. triseriatus* mosquitoes (Whitfield et al., 1971), virus was first found in salivary glands 9 days after an infected feeding. In this and a subsequent 13-day harvest, virus particles were sparse and a very small proportion of the epithelial cells was infected. At all later harvest times, EEE virus infection remained remarkably focal, involving only a few cells in each lobe examined. However, at these sites of infection, increasingly high concentrations of mature virus particles were contained within lumina of endoplasmic reticulum and vacuoles and in extracellular spaces (Fig. 6). From day 21 onward the glandular lumen and its convoluted diverticular extending into the apical end of epithelial cells became so densely packed with virus particles that paracrystalline arrays were formed (Fig. 7). Large parts of these diverticular of the glandular lumen were totally impacted by virus crystals.

In our electron microscopic study of SLE virus in *C. pipiens* (Whitfield et al. 1973), virus particles were initially found in salivary glands 8 days after the infecting blood meal. Between days 8 and 12 increasing numbers of virus particles were found within the cisternae of the endoplasmic reticulum of salivary gland epithelium (Fig. 8); extracellular particles were found primarily within the diverticular of the glandular lumen. The number of extracellular virus particles continued to increase dramatically from day 12 until the end of the experiment at day 32; by day 19 very large numbers of virus particles were dispersed throughout the secretory space (Fig. 9), and by day 25 clustering of virus particles into crystalline arrays had reached a remarkable magnitude. The closely packed virus particles occurred along the apical margins of the epithelial cells, and in fewer instances massive aggregates filled a significant volume of the secretory space (Fig. 10). Such massive accumulation, exceeding that found in any in vivo or in vitro vertebrate host system, remained very focal although virus particles extended through the whole length of the central chitinous duct (Fig. 11). In no instance were more than 1 in 20 epithelial cells infected in a particular site.

In both the EEE and SLE studies, a distortion of cytoplasmic architecture and displacement by products of viral infection were the only variances from normal glands of uninfected mosquitoes; specific viral cytopathic changes, such as generally seen in susceptible vertebrate host cells, were never evident. Both of the virus infections also exhibited directional accumulation of virus in glandular lumen as a result of synthetic and transport activities at the apical end of the plasma membrane of infected cells. Very little virus accumulated in lateral or basal extracellular spaces except in some EEE-infected *A. triseriatus* very late in infection. As was found in the gut, this directional "flow" and accumulation matches the direction of physiologic salivary secretory product movement from the host cells. Moreover, the limitation of both virus infections to a

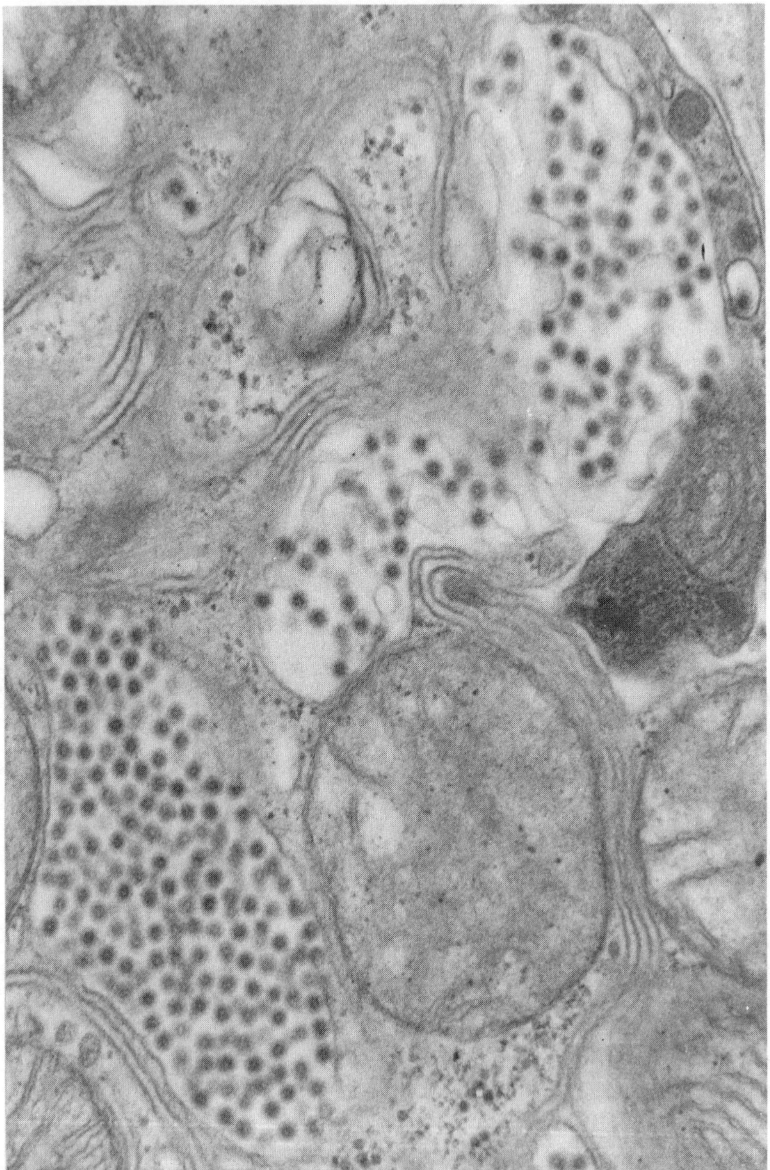

Fig. 6. EEE virus within the lumina of endoplasmic reticulum of a salivary gland epithelial cell at 19 days after infection. Cellular architecture is normal. X69,000.

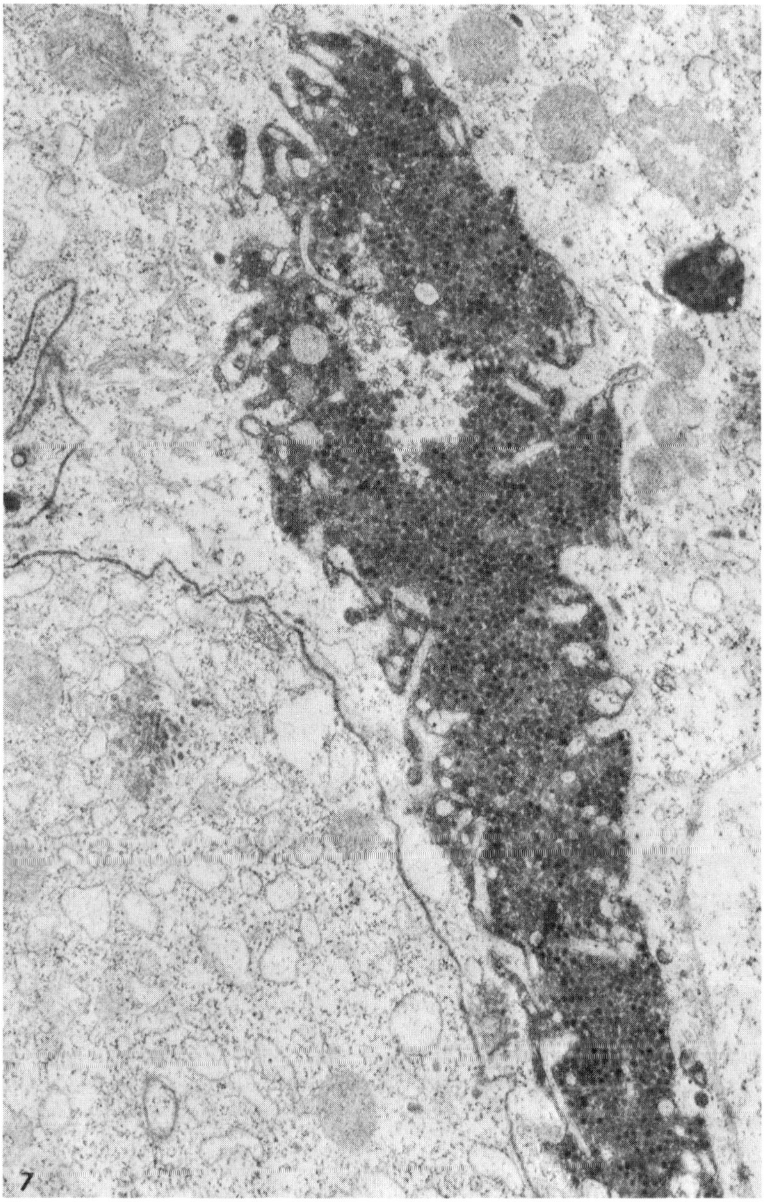

Fig. 7. EEE virus particles crowded into a diverticulum of the salivary gland lumen at 21 days after infection. Virus particles appear permanently impacted in crystalline arrays. X21,000.

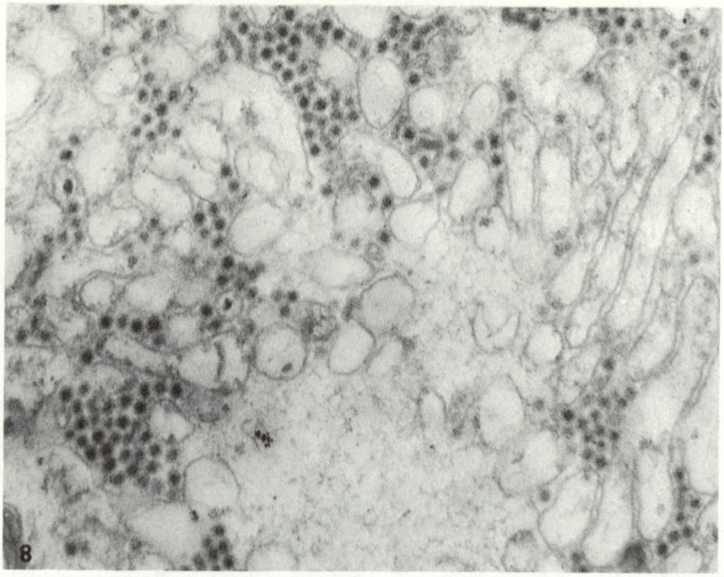

Fig. 8. SLE virus within the endoplasmic reticulum of a salivary gland epithelial cell 8 days after infection. This is the major site of morphogenesis of this virus. X60,000.

relatively small proportion of the salivary gland cells, even in the continued presence of the large numbers of progeny virus particles yielded from the infected cells, emphasizes our lack of understanding of the nature of potential "homeostatic" or "host defense" mechanisms in mosquitoes. It is apparent that some limiting mechanism restricts the involvement of a large proportion of the cells and is a likely explanation of the apparent interference between attenuated and virulent strains of West Nile virus reported by Rozeboom and Kassira (1969). It may also explain the capacity for dual infections by heterologous arboviruses as studied by Lam and Marshall (1968a; 1968b) and Chamberlain and Sudia (1957), since there seems to be ample cells to support more than one virus system.

Other Organs and Tissues. In the mosquito infected with an arbovirus, all organs and tissues are exposed to virus via hemolymph or via intercellular spread. It is apparent that the different organs of a vector species vary in their susceptibility to a particular virus, but in general most of the organs become infected at one time or another and yield virions in only modest amounts. This was difficult to prove by organ titration because there was always the question of contamination of the test sample by virus-infected hemolymph. However, the immunofluorescence technique has produced convincing results by identifying antigen in situ. By this method, Doi (1970) demonstrated JE virus in various organs of *Culex spp*.; Maguire (1973), working with *Aedes australis* infected with Whataroa virus, definitely showed infection of fat bodies,

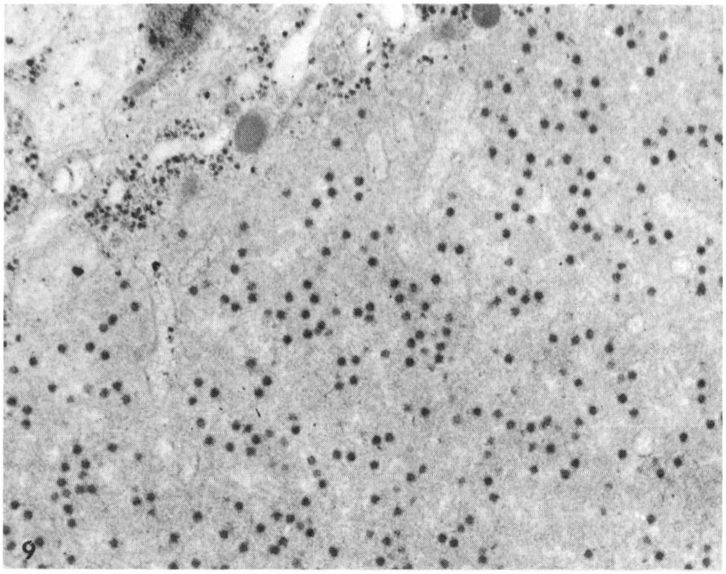

Fig. 9. SLE virus dispersed within the salivary gland lumen at day 19. Increasing concentrations of virus particles occurred in saliva despite the limitation of infection to relatively few epithelial cells. X56,000.

thoracic and cerebral ganglia, compound eyes, Malpighian tubules, and ovarian sheaths. Neural tissues are particularly susceptible, as shown in dengue virus-infected *Aedes* species of the Pacific (Maguire, 1973; Rosen, 1973).

Experimental observations. By electron microscopy, EEE and SLE viruses have been visualized in abdominal and thoracic muscles (Fig. 12), Malpighian tubules, and in the ovarian sheath (Fig. 13), but not in deeper ovarian tissues or eggs of *A. triseriatus* or *C. pipiens*, (Whitfield et al., 1971, 1973). Venezuelan equine encephalitis (VEE) virus has, likewise, been demonstrated by electron microscopy in cerebral ganglia, Malpighian tubules and ovaries, but not in ovarian follicle cells or eggs (Larsen and Ashley, 1971). No immunofluorescent or electron microscopic studies have yet been undertaken in mosquito-virus systems where transovarial viral transmission may occur, for example, LaCrosse virus in *A. triseriatus* (Watts et al., 1973) and Keystone virus in *A. atlanticus* D. & K. (Eldridge et al., 1974).

It is presumptous on the basis of current data to draw conclusions about the relative levels of infection in various organs as a reflection of possible homeostatic activity. Nonetheless, rather severe limitations to viral proliferation in most mosquito organs, including the midgut are apparent, and relatively few cells are involved. Only ganglia and salivary glands yield extraordinary virus concentrations, and our electron microscopic studies of infected salivary glands show a sparing of a high proportion of

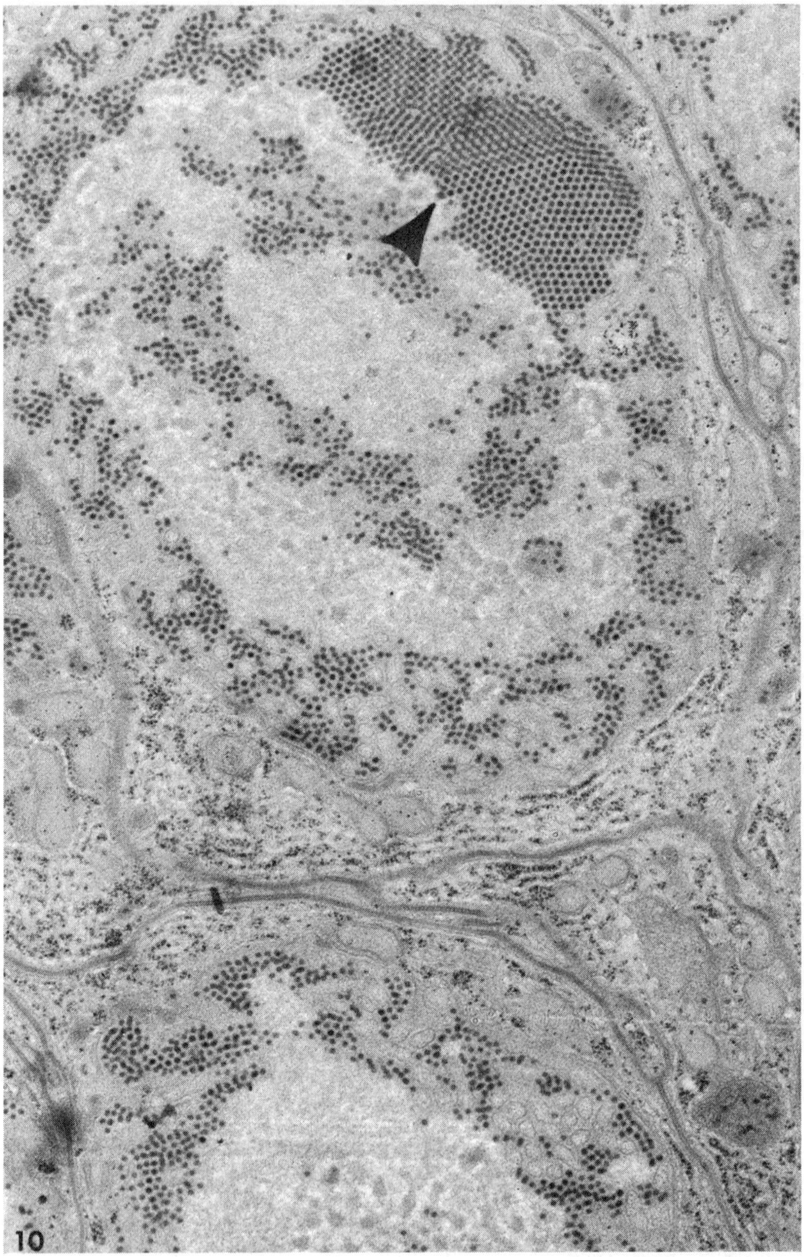

Fig. 10. SLE virus particles in crystalline aggregate (arrow) within a diverticulum of the salivary gland lumen at 25 days after infection. X29,000.

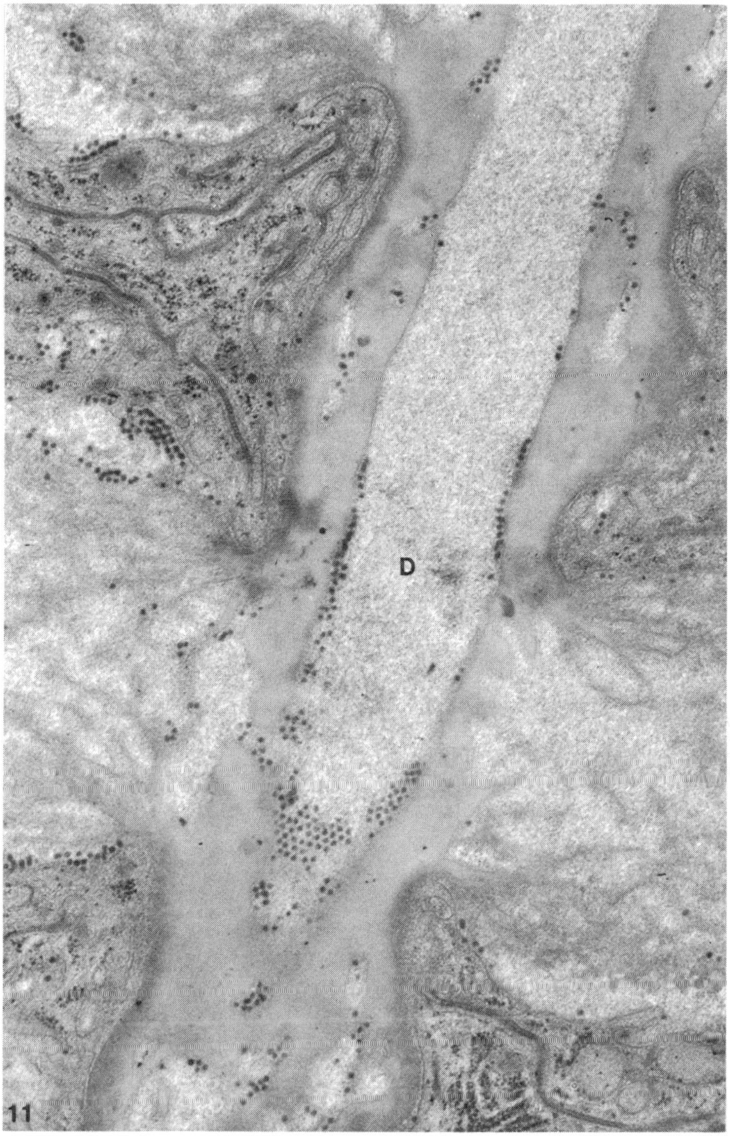

Fig. 11. SLE virus within the chitinous central duct (D) of the salivary gland at 25 days after infection. Virus particles extended through the whole length of the duct, although infected cells were restricted to a few foci. X29,000.

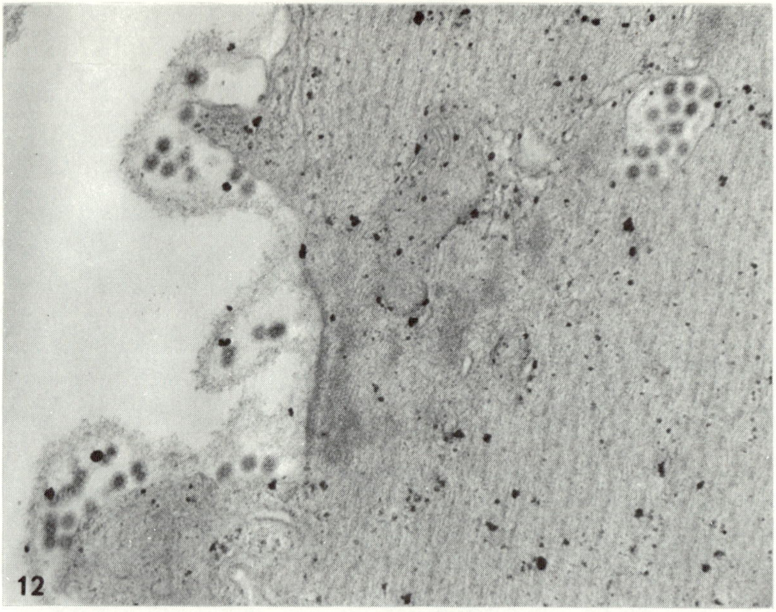

Fig. 12. EEE virus within sarcoplasmic reticulum and under a loose basement lamina of an abdominal muscle in Aedes triseriatus 12 days after infection. X47,000.

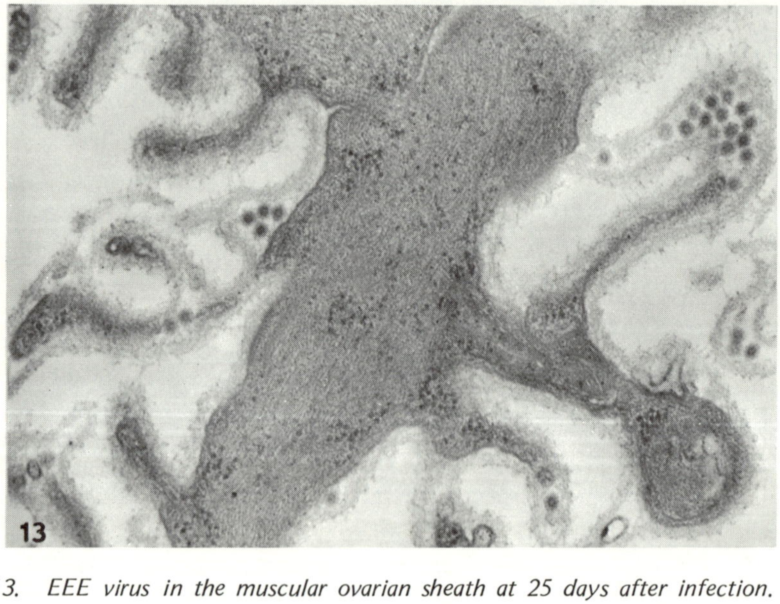

Fig. 13. EEE virus in the muscular ovarian sheath at 25 days after infection. Virus did not penetrate into deep ovarian tissues or eggs. X47,000.

the cells available. It is possible, however, that the ganglion cells may follow a different pattern. The proportion of ganglion cells which becomes infected has not been determined by electron microscopy, but immunofluorescence often indicates an overwhelming involvement (e.g., dengue virus in *Aedes spp.*; Rosen, 1973). With the possible exception of the ganglia, the limitation of arbovirus infections to relatively few cells in any organ appears to operate throughout the mosquito body.

ARBOVIRUSES IN TICKS AND OTHER ARTHROPODS

Far less is known about the development of arboviruses in ticks and biting flies than in mosquitoes. Early work by Pavolovsky and Solovier (1940, 1941) on infection of *Ixodes* and *Haemaphysalis spp.* ticks with Group B tick-borne encephalitis (TBE) virus indicated a rapid dispersal of virus from an infected blood meal to gut, salivary glands, ovaries, Malpighian tubules and brain, virus persisted in some cases for up to 53 days. These observations were later confirmed by electron microscopy (Nosek, 1972). Chernesky and McLean (1969), in a comprehensive study, followed the course of Powassan virus (a Group B virus of the TBE complex) transstadially in *Dermacentor andersoni* ticks by sequential organ titration and immunofluorescence. In nymphs and larvae infected by ingestion of viremic blood, virus was first demonstrated in gut diverticula; and in nymphs the infection progressed to Malpighian tubules and salivary glands. In adult ticks virus grew initially in the epithelial cells of gut diverticula and subsequently gained access to other organs including Malpighian tubules, brain, Gene's organ and salivary glands. Powassan virus concentration in salivary glands was particularly high (a single gland contained $10^{8.0}$ adult mouse i.c. LD_{50} of virus). Virus titers reached plateau levels in all of these organs by day 24 and continued, usually without diminution, until the end of the experiment at 61 days. Specific immunofluorescence in organs, including salivary glands, was usually focal, but some salivary gland alveoli fluoresced throughout their entirety. Many other organs of infected adult ticks did not yield virus; these included heart, trachea, nerve mass, ovary, testis, and fat body. The kinetics of virus growth correlated with metamorphic stages--virus titers dropped at the beginning and increased at the end of molts in parallel with degeneration and replacement of potential host cells. No viral cytopathic effects were seen.

Dermacentor andersoni has also been studied in relation to Colorado tick fever (CTF) virus, for which it is a prime vector. Although very different in physical characteristics from the Group B viruses of the TBE complex (Murphy et al., 1968), CTF virus was shown by Rozeboom and Burgdorfer (1959) to behave rather similarly in its arthropod host. Larvae, nymphs and adults could be infected and virus could be maintained and/or increased transstadially. No significant loss of CTF virus titer occurred in ticks kept for 10 to 13 months after an infectious blood feeding (Rozeboom and Burgdorfer, 1959), but transovarial transmission could not be demonstrated (Eklund et al., 1960). Without ultrastructural or immunofluorescent data, little more can be said about the real nature of CTF persistence in its tick vector. This shortcoming of titration data in explaining biologic processes within arthropod hosts is characteristic of the state of our knowledge of the rest of the more than 50 tick-borne arboviruses.

The least amount of information on viral natural history in arthropods concerns those agents borne by biting flies, some of which are quite important medically. Despite much effort and even some electron microscopic data on bluetongue virus in its vector, *Culicoides spp*. (Bowne and Jones, 1966), there is no kinetic, quantitative evidence of the extent and character of the arthropod phase of bluetongue infection. Finally, studies of vesicular stomatitis virus (VSV) in phlebotomine sandflies (*Lutzomyia spp*.) by Tesh and his colleagues (1972) showed, in yet another example, the need for study of mechanisms of viral maintenance and transmission. Their clear proof of transovarial transmission of a vertebrate virus by a dipteran insect prompted a collaborative study of possible mechanisms (Tesh, Johnson, Whitfield and Murphy, unpublished studies, 1972), but because of the insensitivity of microscopic approaches, we were not successful. Certainly, more experimental work on viruses in biting flies is warranted.

CONCLUSIONS

Limitations to Viral Proliferation in Arthropods. A considerable degree of invertebrate host specificity is common among the arthropod-borne animal viruses; the degree of this specificity in relation to species, genus, family, order or class level of arthropod hosts has recently been reviewed by Varma (1972). There are many examples of complete refractoriness of particular arthropods when challenged with a given arbovirus (e.g., Murray Valley encephalitis virus in *Anopheles annulipes*, McLean, 1955) and there are many examples of extremely high threshold phenomena acting to prevent arthropod infection except when the virus concentration in the blood meal is unnaturally high (e.g., EEE virus in *Culex quinquefasciatus*, Chamberlain et al., 1959). Such refractoriness has generally been considered "intrinsic" and the expression of active destruction of virus by some means unknown or a gradual dying off of ingested virus in the absence of favorable environmental requirements. Factors such as toxicity of digestive fluids to viruses, impermeability of a peritropic membrane, physical or chemical deficiencies of gut cell membranes, failure of adsorption and/or penetration of gut cells, and activities of arthropod "surface defense" mechanisms of unknown nature have been suggested (Chamberlain and Sudia, 1961). The expression of any of these "intrinsic" resistance mechanisms might be genetically determined. It is this sort of refractoriness which has defined the precise vector-borne disease problems of mankind; without it precise vector-virus host associations would not be delineated and we might be overrun by arboviruses using as invertebrate hosts any biting arthropods at hand ! Schlesinger (1971) has argued that in arbovirus evolution, as molded by intense selective pressures in vertebrate and arthropod phases of the virus life cycle, the key absolute restrictions have taken place in arthropods. In fact, the narrow range of competent arthropod vectors, the lifelong, harmless viral persistence and replication, and the transovarial transmission of some viruses have led Schlesinger to favor the propostion that arboviruses originated in arthropods.

Since it has been concluded that mechanisms operative in extreme arthropod refractoriness must be at least quantitatively different from those affecting the "quantal balance," "limitation", or "homeostasis" of arthropod-virus systems which do successfully cycle in nature, it becomes clear that understanding the latter may provide a key to human intervention and disease control. This paper has primarily dealt with evidence that viral infection in vector species is usually, perhaps always, modulated. Some degree of modulation may offer survival advantage to the arthropod host and virus alike. Experiments to prove this modulation, mostly based upon titration and microscopic techniques, have not favored any hypothetical mechanism to explain it, nor have they served to localize the mechanism to the arthropod, the virus, or the arthropod-virus interaction. In the ultrastructural studies carried out in this laboratory and in the published reports of others, there has been no evidence of hemocyte migration, proliferation, or accumulation in infected tissues, although such a cellular defense has been demonstrated in the bee-wax moth, *Galleria mellonella*, infected with *Salmonella typhimurium* (Kurstak et al., 1969). Neither has there been evidence of degradation of progeny virus particles in arthropod tissues despite their formation over long periods of time. Modulation seems to act in vivo primarily by limiting the proportion of potential host cells which actually become productively infected. The single exception to this conclusion comes from in vitro studies of Raghow and his colleagues (1973). In *A. albopictus* cells infected in culture with Ross River virus there was a fall in viral infectivity after an initial rise; this fall was associated with digestion of progeny virus within multivesicular bodies and vacuoles of infected cells. This process progressed to a "cure" of infected cells in the absence of cytopathic changes. Thus, persistent infection resulted from slow cycles of infection and digestion-mediated recovery. This phenomenon has not been seen in vertebrate cells infected with this (Murphy et al., 1973) or related (Murphy and Whitfield, 1970) viruses nor in vivo in arthropods, but it must be studied further.

Possible Mechanisms Active in Arthropod Resistance. In animal virology, several general virus-limiting mechanisms are known which act by restricting viral spread from those cells initially infected; those may all be considered "primitive" mechanisms. First, a defense elicited by the host itself, namely, interferon synthesis, acts without viral specificity in preventing continuing replication cycles. There is preliminary evidence of interferon-like activity in mosquitoes and mosquito cells. Bergold and Ramirez (1972) used the known interferon inducer, Poly I - Poly C, to inhibit VSV synthesis in *Aedes aegypti* and *A. albopictus* cells in culture, and to inhibit virus growth in whole mosquitoes. Enzmann (1973), likewise, showed the presence of an interferon-like substance active in suppressing Sindbis virus in *A. albopictus* cells. These authors recognized the need for rigorous physical-chemical characterization of the active principle before concluding that arthropods may effectively make and use interferon.

A second mechanism which is very effective in restricting viral spread is non-interferon mediated interference, or auto-interference (Huang and Baltimore, 1970; Mims, 1956). This modulation is a property of viruses themselves, or more properly, a property of defective virus particles. Interference is most commonly seen in influenza infections where particles defective in RNA content are responsible (von Magnus, 1954), and in rhabdovirus infections (e.g., the arbovirus, VSV) where abnormally short particles (T particles), also defective in RNA and transcriptase activity, are responsible (Howatson, 1970). In both cases defective particles interfere by competing for replication of viral RNA. In effect, cyclic changes in the ratio of defective to infectious virus particles may modulate the viral yield and even produce titer plateaus over a long time course. In an infected arthropod, the yield of defective particles from the initial infection in any target organ could seed other cells in these organs (or elsewhere?) and competitively inhibit viral RNA synthesis. Recently, VSV defective interference was shown to be quite effective in vivo in mice (Doyle and Holland, 1973), but at this time nothing is known of the infectious: defective ratio of viruses in arthropods themselves. Defective interfering particles (with abnormal RNA) have been found in two Group A arboviruses in cell culture; these are Sindbis virus (Schlesinger et al., 1972; Shenk and Stollar, 1972; Eaton and Faulkner, 1973) and Semliki Forest virus (Levin et al., 1973). In these newly described systems the degree of interference has been substantial, just as it is in VSV infections. It is important that these in vitro findings now be tested in intact mosquitoes by sequential assays of infectious: defective particle ratios.

Another mechanism active in restricting viral yield and thereby viral spread is defectiveness caused by temperature-sensitive mutation. Several arboviruses, isolated from mosquito cell cultures months after initiation of infection, have been shown to differ from parental strains by variation in plaque morphology, virus particle surface antigenicity, and temperature sensitivity (Shenk et al., 1974; Sinarachatanant and Olsen, 1973; Singh, 1971; Peleg, 1971; Simizu and Takayama, 1971). Such mutations are often associated with decrease in vertebrate virulence and yield. If such viral modifications are shown to occur in vivo in arthropods, this mechanism could help explain the modulation of virus yield often seen late in the course of arthropod infection. The mutant particles could act as interfering species (e.g., the RNA^+ mutants of Sindbis virus, Shenk et al., 1974), or a low viremia of such mutants in vertebrate hosts could abort the viral life cycle effectively by failing to infect arthropods feeding upon these hosts.

There are merits in setting forth the above phenomena as the modulating mechanisms in arthropod infections. Each may be effective in the absence of real immune mechanisms; each might even mimic the latter in the degree of reduction of virus titers and transmissibility in nature. Homologous interference could also mimic the specificity of a real immune response.

The progress made thus far to elucidate virus-limiting mechanisms in arthropod vectors has been encouraging if not definitive; some of the answers, while presumptive, only await further confirmatory efforts. At least we can now appreciate the complexity of the mechanisms involved. Not long ago the gut barrier alone stood out as a vulnerable point in vector infection; now we have reason to suspect that a virus is modulated at many sites throughout the body, ostensibly to keep it within bounds that are physiologically tolerable to the vector species involved. The survival advantage of this phenomenon for the arthropod is obvious. It is possible that the indicated mechanisms may lend themselves to intentional genetic manipulation by man. Hopefully, with more understanding of their precise mode of action, we may be able to introduce certain of these regulative characteristics selectively into wild vector populations and, in doing so, make poor vectors out of good ones.

We must keep in mind, however, that living organisms seem to have an infinite capacity to circumvent barriers placed in the way of their continued propagation. We have seen this repeatedly in insects, which have developed resistance to many kinds of insecticides. It occurs in viruses also; selective pressures could eventually permit emergence of virus strains relatively unaffected by the reinforced regulatory mechanisms we introduce into the vector gene pool. Regardless of the type of control--chemical or genetic--we will probably always be challenged to stay one step ahead.

REFERENCES

Altman, R. M. (1963). *Am. J. Trop. Med. Hyg., 12*:425-434.

Bergold, G. H., and Ramirez, N. (1972). In "Moving Frontiers in Invertebrate Virology." (J. L. Melnick, ed.) pp. 56-59. S. Karger, Basel.

Bergold, G. H., and Weibel, J. (1962). *Virology, 17*:554-562.

Boorman, J. (1960). *Trans. Roy. Soc. Trop. Med. Hyg., 54*:362-365.

Bowne, J. G., and Jones, R. H. (1966). *Virology, 30*:127-133.

Casals, J. (1971). In "Comparative Virology." (K. Maramorosch and E. Kurstak, eds.) pp. 307-333. Academic Press, New York.

Chamberlain, R. W., Nelson, D. B., and Sudia, W. D. (1954). *Amer. J. Hyg., 60*:278.

Chamberlain, R. W., and Sudia, W. D. (1957). *Proc. Soc. Exp. Biol. Med., 95*:261.

Chamberlain, R. W., and Sudia, W. D. (1961). *Annual Rev. Entomol., 6*:371-390.

Chamberlain, R. W., Sudia, W. D., and Gillett, J. D. (1959). *Am. J. Hyg., 70*:221-236.

Chernesky, M.A., and McLean, D. M. (1969). *Can. J. Microbiol. 15*:1399-1408.

Doi, R. (1970). *Japan J. Exp. Med., 40*:101-115.

Doyle, M., and Holland, J. J. (1973). *Proc. Nat. Acad. Sci. USA., 70*:2105-2108.

Eaton, B. T., and Faulkner, P. (1973). *Virology, 51*:85-93.

Eklund, C. M., Kohls, G. M., Kennedy, R. C. (1960). In "Biology of Viruses of the tick-borne encephalitis complex." (H. Libikova, ed.). pp. 401-406. Czechoslovak Academy of Sciences, Prague.

Eldridge, B. F., Suyemoto, W., LeDuc, J. W., Keefe, T. J., and Burger, J. F., and Russell, P. K. (1974). Personal communication.

Enzmann, P.-J. (1973). *Arch. ges. Virusforsch., 41*:382-389.

Gaidamovich, S. Y., Khutorestskaya, N. V., Lvova, A. I., and Sveshnikova, N. J. (1973). *Intervirology, 1*:193-200.

Howatson, A. F. (1970). *Advan. Virus Res., 16*:195-256.

Huang, A., and Baltimore, D. (1970). *Nature, 226*:325-327.

Janzen, H. G., Rhodes, A. J., and Doane, F. W. (1970). *Can. J. Microbiol., 16*:581-586.

Janzen, H. G., and Wright, K. A. (1971). *Can. J. Zoology 49*:1343-1345.

Kurstak, E., Goring, I., and Vega, C. (1969). *Antonie van Leeuwenhoek, 35*:45-51.

Lam, K. S. K., and Marshall, I. D. (1968a). *Am. J. Trop. Med. Hyg., 17*:637-644.

Lam, K. S. K., and Marshall, I. D. (1968b). *Am. J. Trop. Med. Hyg., 17*:625-636.

LaMotte, L. C. (1960). *Am. J. Hyg., 72*:73-87.

Larsen, J. R., and Ashley, R. F. (1971). *Am. J. Trop. Med. Hyg., 20*:754-760.

Levin, J. G., and Ramseur, J. M., and Grimley, P. M. (1973). *J. Virol., 12*:1401-1406.

Maguire, T. (1973). Personal communication.

McLean, D. M. (1955). *Austral. J. Exptl. Biol. Med. Sci., 33*:53-66.

Miles, J. A. R., Pillai, J. S., and Maguire, T. (1973). *J. Med. Entomology, 10*:176-185.

Mims, C. A. (1956). *Brit. J. Exp. Pathol., 37*:129-142.

Murphy, F. A., Coleman, P. H., Harrison, A. K., and Gary, G. W. Jr., (1968). *Virology, 35*:28-40.

Murphy, F. A., Harrison, A. K., Gary, G. W., Jr., Whitfield, S. G., and Forrester, F. T. (1968). *Lab. Invest. 19*:652-662.

Murphy, F. A., Taylor, W. P., Mims, C. A., and Marshall, I. D. (1973). *J. Infect. Dis.*, *127*:129.

Murphy, F. A., and Whitfield, S. G. (1970). *Exp. Mol. Pathol.*, *13*:131-146.

Nosek, J., Ciampor, F., Kozuch, O., and Rajcani, J. (1972). *Acta Virologica*, *16*:493-497.

Pavlovsky, E. N., and Soloviev, V. D. (1940). *Arh. Biol. Nauk.*, *59*:111-117.

Pavlovsky, E. N., and Soloviev, V. D. (1941). *Trudy Voj. Med. Inst. Akad. Krasnoiarsk.*, *25*:9-18.

Peers, R. R. (1972). *Can. J. Microbiol.*, *18*:741-745.

Peleg, J. (1971). *Current Topics in Microbiol & Immunol.*, *55*:155.

Raghow, R. S., Grace, T. D. C., Filshie, B. K., Bartley, W., and Dalgarno, L. (1973). *J. Gen. Virol.*, *21*:109.

Rosen, L. (1973). Personal communication.

Rozeboom, L. E., and Burgdofer, W. (1959). *Am. J. Hyg.*, *69*:138-145.

Rozeboom, L. E., and Kassira, E. N. (1969). *J. Med. Entomol.*, *6*:(4), 407-411.

Schaeffer, M., and Arnold, E. H. (1954). *Am. J. Hyg.*, *60*:231-236.

Schlesinger, R. W., (1971). *Current Topics in Microbiol & Immunol.*, *55*:241.

Schlesinger, S., Schlesinger, M., and Burge, B. W. (1972). *Virology*, *48*:615-617.

Shenk, T. E., Koshelnyk, K. A., and Stollar, V. (1974). *J. Virol.*, *13*:439.

Shenk, T. E., and Stollar, V. (1972). *Biochem. Biophys. Res. Commun.*, *49*:60-67.

Simizu, B., and Takayama, N. (1971). *Arch ges. Virusforsch.*, *34*:242.

Sinarachatanant, P., and Olsen, L. C. (1973). *J. Virol.*, *12*:275.

Singh, K. R. P. (1971). *Current Topics in Microbiol. and Immunol.*, *55*:131.

Tesh, R. B., Chaniotis, B. N., and Johnson, K. M. (1972). *Science*, *175*:1477-1478

Thomas, L. A. (1963). *Am. J. Hyg.*, *78*:150-165.

Varma, M. G. R. (1972). In "Moving Frontiers in Invertebrate Virology." (J. L. Melnick, ed.) pp. 49-56. S. Karger, Basel.

vonMagnus, P. (1954). *Advan. Virus Res., 2*:59-79.

Watts, D. M., Pantuwatana, S., DeFoliart, G. R., Yuill, T. M., and Thompson, W. H. (1973). *Science, 182*:1140-1141.

Whitfield, S. G., Murphy, F. A., and Sudia, W. D. (1971). *Virology, 43*:110-122.

Whitfield, S. G., Murphy, F. A., and Sudia, W. D. (1973). *Virology, 56*:70-87.

W.H.O. Scientific Group. (1967). Arbovirus and Human Disease. *W.H.O. Techn. Rept. Series., 369*:1-84.

Wright, K. A. (1969). *Can. J. Zoology., 47*:579-587.

INFECTION OF ARTHROPOD VECTORS BY PLANT PATHOGENS

Karl Maramorosch
Institute of Microbiology, Rutgers University
New Brunswick, New Jersey

Many plant pathogenic agents, such as viruses and mycoplasma-resembling mollicutes, require alternating plant and arthropod hosts for their survival in nature (Maramorosch, 1963). I shall discuss briefly observations concerning interactions between leafhoppers (Cicadellidae) transmitting a virus, a spiroplasma, and an as yet poorly defined plant pathogenic mollicute that causes the aster yellows disease.

The wound tumor virus, which causes tumors in several plant species, systemically invades its leafhopper vector, *Agallia constricta* Van Duzee, when insects acquire the virus by feeding on diseased plants (Maramorosch, 1970). After a few days virions appear in the filter chamber of the gut. Subsequently the virus multiplies in the gut, fatbody, muscles, Malpighian tubules, trachea, salivary gland, and even epidermal cells and the central nervous system. The viroplasmic foci, where virions are assembled, resemble foci of virus multiplication described from human cells infected with reovirus in tissue culture. When leafhoppers are artificially injected with wound tumor-containing inocula by piercing their abdomens with fine glass capillaries, virions are detected after a few days within lamellar structures of fatbody cell cytoplasm (Fig. 1). These virions might represent the original inoculum, or the progeny of virions that multiplied in other tissues of the insect. After a week, the first viroplasms form, accompanied by severe cytopathic changes. As the number of viroplasms gradually increases, virions fill the lamellar structures and crystalline virus inclusions are formed. Abdominal inoculation results in systemic infection, similar to that which follows natural acquisition of wound tumor virus by feeding, except for the absence of virions in the gut. Thus the gut barrier is eliminated in injected leafhoppers, but other possible mechanisms of resistance seem to be comparable in the two systems.

The sequential changes in assembly sites of the plant pathogenic virus in the insect vector are reminiscent of the assembly of reovirus. The original inoculum is taken up by the hemolymph and transported to various organs and tissues, or taken up by some part of the fatbody cells. In the fatbody the original inoculum apparently becomes engulfed in defined structures, from which the viral genome is released, reaching viroplasmic matrices. Virions may accumulate in vacuolated areas of the cytoplasm, gradually becoming regularly packed, forming microcrystals (Fig. 2).

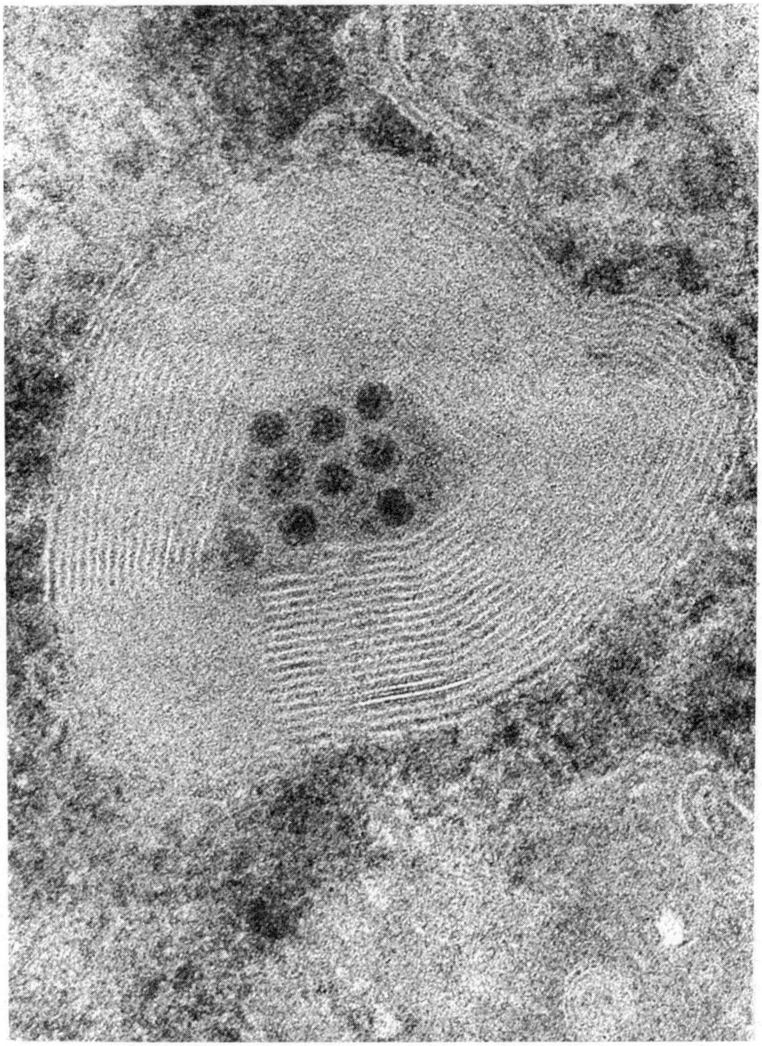

Fig. 1. Virions of wound tumor virus within a lamellar structure in the fatbody cell cytoplasm of a leafhopper vector (Agallia constricta) Magnification x 140,000.

The role of the hemolymph is not only that of a vehicle, transporting the virions within the insect body, but also that of a site of multiplication and storage. When hemocytes, attached to various internal organs, were sectioned and examined by electron microscopy, large quantities of empty virus shells were detected. These protein coats gave an insight into the significance of vector hemolymph in the infection process and in the transmission of propagative plant pathogenic viruses. The early abundance of such coats may indicate that the viral RNA slipped out so as to infect new sites in

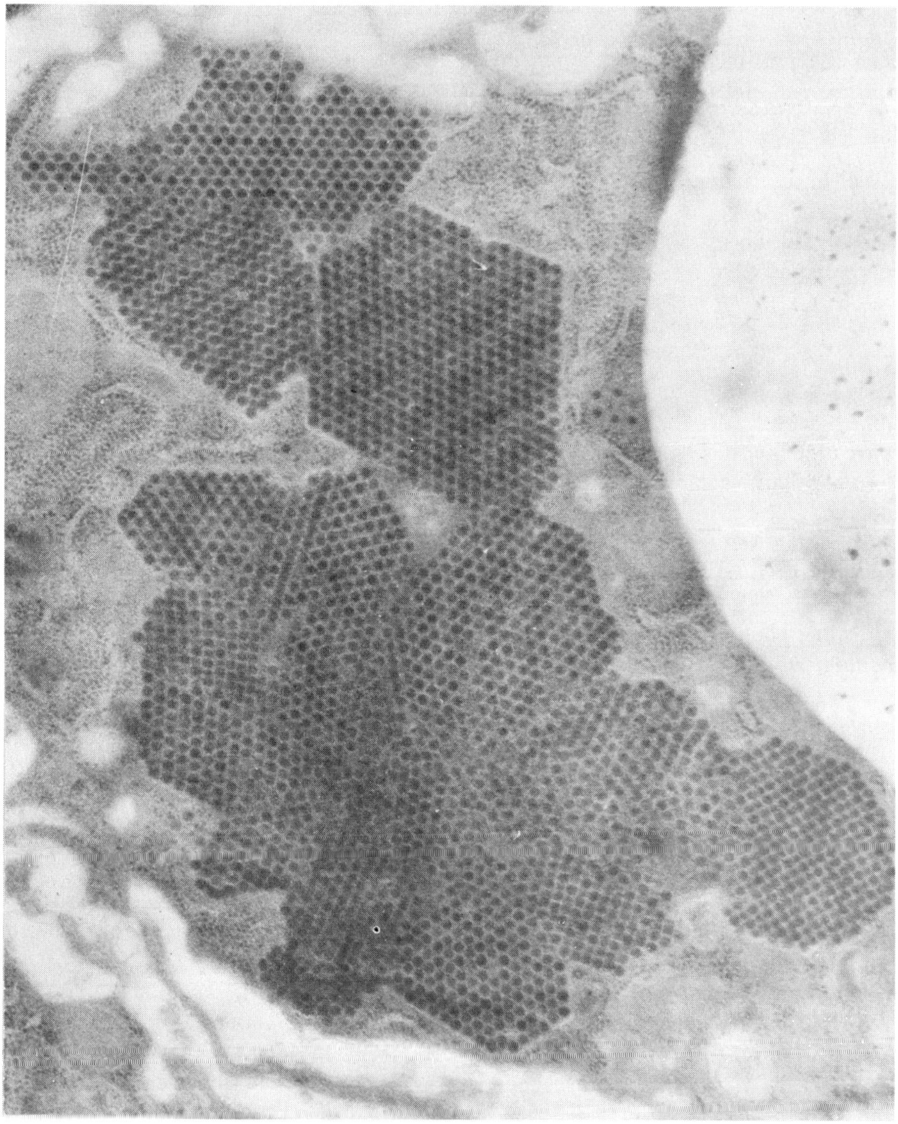

Fig. 2. Microcrystals of wound tumor virus in the fatbody tissues of an insect vector (Agallia constricta) Magnification x 60,000.

various tissues and organs of the vector. The hemocytes are the medium through which the virus multiplies and travels from one organ to another, until it reaches the salivary gland. After the penetration through the gut membrane, viremia seems to play an important role in rendering the insect infectious.

Despite cytopathic lesions in the brain and despite the formation of huge virus crystals in the fatbody cell cytoplasm and other locations, insect vectors seem to remain practically unaffected by the viremia. Their life span and reproduction appear normal. It is not well understood what, if any, mechanism protects the vectors from deleterious effects of the systemic virus infection. The huge accumulations of virions of wound tumor virus, as well as similar crystalline arrays of arboviruses in various organs and tissues of invertebrate hosts, might represent a means of protection rather than storage. One could speculate that virions, once assembled in crystalline arrays, become immobilized and can no longer participate in the developmental cycle. The high concentration of virions in the form of microcrystals is characteristic for certain viruses multiplying in invertebrate hosts and in plants, but not for viruses within vertebrate tissues.

If the wound tumor virus were highly pathogenic to its invertebrate host, one would expect that insects could cope effectively with exposure to low virus concentrations but succumb to high concentrations. Yet, despite unusually high concentrations manifested by the microcrystal formation, the leafhopper vectors apparently live and propagate normally. The explanation proposed here is that the microcrystals serve as a blind alley for the virus--it can no longer proliferate once it forms a crystal--and thus microcrystal formation might serve as a protective mechanism.

One could also speculate that the lamellar structures, illustrated in Fig. 1, trap the original virus inoculum and thus constitute a protective device to prevent the foreign material from reaching other areas. This hypothesis could be tested experimentally, by using colloidal gold and checking whether the engulfing and sequential infection by virions is being altered.

A different phenomenon, concerning the mollicute organism causing the aster yellows disease, has been described nearly twenty years ago by the late L. O. Kunkel (1955, 1957). At the time, it was believed that the aster yellows disease was caused by a virus, and the phenomenon was described as "cross-protection" by related virus strains in insect vectors, as well as in plants. The term "cross-immunity" has also been used, but the observation actually related to interference between two strains of the pathogen. When six-spotted aster leafhoppers, *Macrosteles fascrifrons* Ståhl, were first confined on China aster plants, *Callistephus chinensis* Nees, infected with the eastern strain of aster yellows, and the insects were transferred after two weeks to plants infected with a California strain of aster yellows, these leafhoppers transmitted only the eastern strain for the remainder of their lives. When the sequence of acquisition was reversed and the insects were first given access to plants with the California strain, only that strain was transmitted later. When additional strains were tested, the results were no longer fitting the simple "cross-immunity" explanation (Freitag, 1969). The phenomenon indicated various degrees of interference, but the mechanism of the phenomenon remained an enigma.

The interaction between related strains of the wall-less microorganism of corn stunt disease, tentatively classified as a spiroplasma (Maramorosch, 1974), has been termed "cross-protection" or "unilateral protection" when first reported (Maramorosch, 1958). The Rio Grande strain of corn stunt was transmitted by *Dalbulus maidis* (D. & W.) leafhoppers throughout their life even though they were subsequently fed on plants infected with the Mesa Central strain. When insects were first exposed to Mesa Central-infected plants and later to Rio Grande, they transmitted Mesa Central initially, but then transmitted the Rio Grande strain. There is no information available concerning the mechanism involved in the interference between strains of mollicutes in insect vectors, but it is known that related mollicutes, such as the agent of corn stunt and that of citrus stubborn disease, share antigenic properties (Tuly et al., 1974; McIntosh et al., 1974).

REFERENCES

Freitag, J. H. (1969). In: *Viruses, Vectors, and Vegetation* (K. Maramorosch, Ed.). Interscience-John Wiley & Sons, New York : 303-325.

Kunkel, L. O. (1955). *Advances Virus Research, 3*:251-273.

Kunkel, L. O. (1957). *Science, 126*:1233.

McIntosh, A. H., Skowronski, B. S., and Maramorosch, K. (1974). *Phytopathologische Zeitschrift* (in press).

Maramorosch, K. (1958). *Virology, 6*:448-459.

Maramorosch, K. (1963). *Ann. Rev. Entomology, 8*:369-414.

Maramorosch, K. (1970). *World Review of Pest Control, 9*:29-41.

Maramorosch, K. (1974). *Ann. Rev. Microbiol., 28*: (in press).

Tuly, J. G., Whitcomb, R. F., Bove, J. M., and Saglio, P. (1973). *Science, 182*:827-829.

FACTORS AFFECTING VIRUS INFECTION OF INSECT GUT TISSUE

T. W. Tinsley
Natural Environment Research Council
Unit of Invertebrate Virology
Oxford, England

The case for the existence of defensive reactions in insects has been clearly stated by Salt (1970) in these terms: "on present knowledge it is clear that the cellular defence reactions of insects are simpler than those of vertebrates. That conclusion, however, does not include the implication that they are in any way makeshift. Evidence that is overwhelming testifies that they are suitable for their purpose, in spite of their economy of means and their helplessness against a few parasites. That evidence is the undoubted success of insects, as shown by the enormous number of species, their countless individuals and their ubiquity. Insects could not have become so dominant if their defence reactions had been anything but fully effective." This is an important concept and it stresses the need to evaluate the nature and mode of action of these defensive reactions. Salt also pointed out that insects live in almost all terrestrial and fresh water habitats and consume almost every type of organic matter as food. He suggested that as a result insects are exposed to a greater range of diseases than any other group of animals and that they possess certain structural and metabolic qualities that are useful in reducing the possibilities of infection taking place. The exo-skeleton, made up of chitin and tanned proteins, provides a tough, chemically resistant and, in most individuals, a waterproof covering for the softer and more vulnerable tissues of the body. Further, the chemical composition of the gut fluids is such that many micro-organisms are quickly and efficiently destroyed after ingestion. However, such factors would not be adequate in themselves to protect insects against infectious virus taken in by mouth and therefore the existence of additional systems must be considered.

The search for immune responses in insects analagous to the vertebrate antigen/antibody reactions has not met with any success (Bernheimer et al, 1952; Briggs, 1958; Stephens, 1959, Stephens Chadwick, 1967a; Kamon and Shulov, 1965; Phillips, 1966). The fact that insects do not produce vertebrate type immunoglobulins is not too surprising as they do not possess the requisite lymphatic or glandular organisation. The Phylum Arthropoda represents a climax group and, in the middle Paleozoic era, vertebrates and arthropods were already so widely divergent as to provide ample opportunity for the development of widely different but nonetheless effective systems of defence. The existence of systems in arthropods either not connected with or widely divergent from those present in vertebrates, necessarily demands new techniques for their elucidation and a new terminology for their description (Stephens Chadwick, 1967b; Salt, 1970).

The mouth is the significant portal of entry for viruses into insects and so initial defensive mechanisms could be expected at any point along the route from the mouth to the gut tissues. Aizawa (1962) investigated the inhibitory or antiviral properties of gut juices on the nuclear polyhedrosis virus of the silk-worm *Bombyx mori.* Untreated material and gut juices adjusted to pH 7.0 with buffer both reduced virus infectivity by about 50% after incubation at 28°C. for 60 minutes. Aizawa concluded from these experiments that an antiviral substance existed in the gut juices. However, this problem was studied again by Aruga and Watanabe (1964) who took the precaution of including a control treatment whereby virus was incubated with pH 10.0 buffer to simulate the pH of lepidopteran gut juice. These authors could only find slight antiviral activity in the gut juice and they considered that it could constitute only a minor factor in the general system of resistance. Rehac̃ek (1965) could find no evidence of digestive enzymes inhibiting or destroying the infectivity of arboviruses when the gut juices of ticks were incubated with virus for 30 minutes at 22° C. More precise experimentation is clearly needed to resolve these differing opinions, probably by incubating gut juices with purified virus preparations and using insect cell cultures as a test system.

The permeability of the peritrophic membrane has frequently been suggested as a mediating influence on the penetration of virus particles. Chamberlain and Sudia (1961) were of the opinion that with two viruses of equal size, equal rates of either retention or diffusion across the peritrophic membrane would be expected but, as they pointed out, experience with arboviruses in particular provided no support for this idea. However, it is also possible that the nature of electrical charge on this membrane when related to the charge on the virus particles may have some relevance to permeability, but little information is available on this aspect.

The mid-gut epithelial cells seem to be the focal entry point for all virus groups associated with insects (Chamberlain and Sudia, 1961; Sinha, 1967; Harrap and Robertson, 1968; Harrap, 1970). Chamberlain and Sudia (1961) discussed the possibility of specific receptor sites on the mid-gut epithelial cells acting as determinants of virus attachment. However, when one considers the range of virus types capable of infecting insects, it would be necessary to postulate the existence of a wide range of specific receptor sites. Moreover, the mode of entry of virions into the epithelial cells might well determine the type and specificity of such receptor sites. Very little is known about the way virions or possibly nuclei acids enter the gut cells. Chamberlain and Sudia suggested that the receptor site hypothesis could be tested by first feeding "killed" virus to occupy the essential receptor sites and then to follow this at a suitable interval with infectious virus. These very interesting and relevant experiments seem not to have been done and could not perhaps be refined by using U.V. irradiated virus and radio-actively labelled infectious virus. The possibility that the "gut barrier" might be simply a manifestation of a surface inhibitory substance secreted by or associated with the gut epithelial cells was put forward as an alternative hypothesis by Chamberlain and Sudia (1961). This idea would also lend itself to critical experimentation using inactivated and infectious virus preparations with the interesting possiblity that the activity of the inhibitor could be deflected by attachment to the non-infectious virus, leaving the way clear for infectious virus. Some support for this hypothesis is provided by the numerous examples in the literature of successful infections being brought about in apparently resistant hosts simply by increasing the ingested dose. However, the "infection threshold" of arboviruses in mosquitoes varies widely with different viruses and very often critical dosages are required for systemic infections to occur (Chamberlain and Sudia, 1961, Chamberlain 1968). Therefore, if a non-specific surface inhibitor exists then it must either be a quantal response or function with varying degrees of efficiency, depending on the type of ingested virus. On the other hand, if the inhibitors are virus specific then it follows that a wide range of types exists. This would be a very complex situation and, at the present time, would be very difficult to test with existing techniques. Therefore, it would be more meaningful to investigate the presence or absence of pre-formed non-specific inhibitors on the surface of the mid-gut epithelium.

On the other hand, it is equally possible that the critical stages in the determination of successful virus infection occur after the entry of virus particles into the mid-gut epithelium. Therefore, the "gut barrier" may not just reflect the functions of the processes already discussed, but result from a tissue-mediated response manifested after the entry of infectious virus into such tissues. McLean (1955) suggested that the factor determining whether or not Murray Valley encephalitis was capable of undergoing a "biological" cycle in the vector was the capacity of the mid-gut to support virus replication. If conditions for the liberation of nucleic acid after virus entry were unfavourable, then obviously no infection would take place and so the gut-cells would represent, to all intents and purposes, a barrier of non-susceptible tissue. Whether this state could be changed through stress factors or changes in host metabolism is not known. Alternatively, the activity of the liberated viral nucleic acid could be repressed to such an extent that only very low levels of virus replication occurred, resulting in sub-clinical infections, and again these would be very difficult to detect and measure with present insensitive methods. However, assuming that conditions for virus replication were optimal in the gut-cells if the progeny virus could be contained within the gut tissue in some way, then movement of virions into the haemocoele would not occur, preventing systemic invasion of the host. The effects of by-passing the gut through intrahaemocoelic infection of virus is relevant to these considerations. The original concept of the "gut barrier" developed from comparisons of the magnitude of dosages necessary to cause infection by ingestion in relation to the minimal quantities required by injection into the haemocoele (McLean, 1955; Bailey and Gibbs, 1964; Peers, 1972). Non-vectors of both plant viruses and arboviruses can be converted into efficient vectors by simply puncturing the gut after ingestion of the virus, thus allowing systemic invasion and passage into or infection of the salivary glands (Storey, 1932, 1939; Merrill and ten Broeck, 1935; McLean, 1955). General susceptibility of mosquitoes to all biologically transmitted arboviruses was demonstrable after injection (Hurlbut, 1951, 1953, 1956; McLean, 1955). Further, non-haemophagous insects such as various members of the Coleoptera and Lepidoptera can be infected successfully with a wide range of arboviruses by intra-haemocoelic injection (Hurlbut, 1956; Hurlbut and Thomas, 1960, 1969). Scherer and Hurlbut (1967) have shown that Nodamura virus, an insect-borne picornavirus, will replicate in larvae of the Indian meal moth, *Plodia interpunctella,* and with fatal results in the wax-moth *Galleria mellonella L.*, and in honey-bees (Bailey and Scott, 1973) following injection into the haemocoele. Whether the Nodamura virus can infect these hosts after ingestion is not known. Viruses that are pathogenic to their insect host certainly need only to be injected in very small doses to produce fatal results, whereas very large doses are needed *per os* to produce the same level of mortality. There is the implication that if defensive mechanisms do exist in the haemocoele, i.e. haemolymph factors, then they are not immediately operative. These observations lead one to suggest that the defensive reactions of insects are so organised as to prevent entry of virus particles from the mid-gut tissue into the haemocoele.

An observation of significance was made by Whitfield et al (1973) who found that following successful infection of the mid-gut an equilibrium was reached in that the number of infected cells and number of virus particles appeared to be reasonably constant. This was in agreement with an earlier suggestion of Chamberlain and Sudia (1961) that an "infectivity plateau" occurred in the mid-gut before other tissues were or could be invaded. Failure in the rate of virus replication to reach the requisite infectivity plateau and the consequent mobility to invade the blood system would constitute a containment of the infection. Rehaček (1965) was of the opinion that the threshold phenomenon played a major role in the development of infection in arthropods and was very likely to be influenced by receptors or inhibitors in the gut cells, less by the action of digestive enzymes and least by the permeability of the peritrophic membrane.

Insect pathogenic viruses often occur as latent or sub-clinical infections in natural host populations. These can be activated to become frank infections, necessarily involving invasion of the haemocoele, by a wide range of physical and chemical stimuli which put the host under stress (Aruga, 1963). The mode of action of these stress factors is not understood, but it is possible that they either remove the barriers against systemic infection, or stimulate a higher rate of virus replication, so that the infectivity plateau is reached. A possible explanation of both the gut barrier reaction and of latent infections could be the activities of a tissue-mediated viral inhibitor which depressed virus replication. This inhibitor could be a normal constituent of the mid-gut and related tissues, because production of an inhibitor as a response to virus particles would not be swift enough to be effective. Varma (1972) suggested that the disappearance or elimination of virus from gut cells could be the result of destruction of the virus by antiviral substances already present in the cells. However, there is no evidence in the literature of the occurrence of such inhibitors in mid-gut tissues, but this could be investigated by incubating highly purified virus with mid-gut cell extracts and then determining the degree of inhibition of virus infectivity, preferably in an insect cell culture system. The attachment of inhibitors to the virus surface, possibly leading to aggregation, could be tested by either incorporating radio-active materials into the gut extract, or injection of radio-active labelled markers into the living host some time before excision of the mid-gut tissue.

Preliminary work in Oxford may have some relevance to this problem. Aggregates of virus particles can be obtained when extracts of healthy insects, treated only with low speed centrifugation, are incubated with highly purified virus particles at 4° C. for 12 hours. These observations were made during the course of a wider study of the growth curves of the densonucleosis virus ($^{R}/1$: $^{4}/35$: $^{S}/S$: $^{1}/0$) in *Galleria mellonella L.* When the mixture of virus plus healthy insect extracts was run in a 2.5% polyacrylamide gel system (PAGE), two bands could be seen in the stained gel columns. One band (B) represented normal virus particles and corresponded in rate of migration to the purified virus sample; the other band (A) migrated much more slowly into the gel, indicating that a change in overall charge had occurred, possibly by the incorporation of host material. This slow-running band was excised from a comparable unstained gel and examined with the electron microscope and was shown to contain virus particles surrounded by an amorphous substance. When the densonucleosis virus was treated with 1% sodium dodecyl sulphate (SDS) and run in a 10% PAGE system incorporating β mercaptoethanol, four polypeptide bands are consistently obtained. When band (A) excised from the 2.5% gels was treated in a similar manner and run on the 10% gel system, an extra band was observed in addition to four virus polypeptide bands. When comparable gel columns were stained with the Schiff reagent, specific for carbohydrate moieties, this fifth band showed faint red colouration indicating the presence of carbohydrate. A logical extension of these investigations was to determine if similar bands corresponding to coated particles could be found when extracts from infected insects were tested by PAGE. *G. mellonella* larvae infected *per os* with the densonucleosis virus were sampled at daily intervals after infection feeding. Samples were run on 2.5% gels and on day 5 a strongly staining band in the position of band (A) was visible close to the top of the column, while purified particles showed the normal position for virus much further down the gel. Excision of the band (A) revealed coated virus particles with the electron microscope and treatment of this material with SDS followed by electrophoresis in 10% acrylamide gels, also showed the additional band in the normal virus profile. When virus incubated with healthy insect extract was co-run on acrylamide gels with the 5-day sample from infected insects, band (A) was clearly visible at the top of the column and stained more intensely than bands in the (A) position produced by either of the separate treatments. These preliminary experiments indicate that substances found in both healthy and infected insects can be associated in some way with virus particles, though it is not known yet if the infectivity of such coated virus particles is affected. However, these attached substances were demonstrated in whole insect extracts, and it remains now to determine if they are present in the mid-gut tissue. The chemical nature of the additional substances has not been fully established, but certainly protein and probably carbohydrate are involved, which is suggestive of a glycoprotein.

Longworth et al (1968) noted a 17S contaminant during the initial purification of the densonucleosis virus from *G. mellonella.* This substance was invariably associated with virus particles and was only eliminated from virus preparations by fairly extensive treatment. The 17S fraction had a very different amino-acid composition from virus protein, was highly immunogenic and showed no relationship with virus protein in serological tests. The effect of the 17S substance on the infectivity of the densonucleosis virus is not known. Bailey, Gibbs and Woods (1970) criticised the methods used by other workers in the purification of insect viruses and urged the use of a less rigorous system: the data provided by these authors from experiments with the densonucleosis virus, albeit for other purposes, are relevant to a consideration of the nature and role of these contaminating fractions. Extracts made from younger (5th instar) *Galleria* larvae had less of the contaminants, particularly the 17S component, than those derived from older (8th instar) larvae used by Longworth et al (1968). It would be of interest to follow the development of the small S components in the course of insect development. More importantly, when virus was extracted from a mixture of healthy larvae labelled with ^{14}C, and larvae infected with the densonucleosis virus, Bailey et al found that both the small S material and the virus preparation had residual radio-activity. Indeed, the virus preparation possessed over 60% of the total residual activity of the whole mixed preparation. This is suggestive of the "contaminant" taking up a radio-active label and then attaching to or being associated with the purified or possibly semi-purified preparation. The nature and role of the 17S host protein component would repay further study, particularly in relation to virus infection and replication.

Salt (1970) was impressed by the key role played by mucopolysaccharides in the reactions of the insect host following the deposition of parasites' eggs. He thought that if immunoglobulins could not be invoked in defence reactions, then mucopolysaccharides would head the list of possible candidates. The preliminary results reported here suggest that glycoproteins other than immunoglobulins may also have a role to play, very probably in contributing to the gut barrier reactions. Salt also suggested that the involvement of polysaccharides, especially mucopolysaccharides, in both mammalian and insect cellular defence reactions, seemed too frequent to be a mere coincidence. It was possible that an original invertebrate system based on mucopolysaccharides was taken over by insects and vertebrates; the latter, however, found the system unsatisfactory either because it was insufficiently specific or perhaps too easily mimicked and so developed a parallel system based on immunoglobulins. A similar case can be made for the retention and development of glycoproteins other than immunoglobulins as a defence reaction in the invertebrate, which in time became more sophisticated in their action and probably more specific. The insect has no requirement for a system of acquired immunity, principally because its life span is short, and yet, in relative terms, the virus infection can occupy a large proportion of the insect's life. Therefore, whatever defence system may be in operation, it is clear that it must be swift in appearance and is very probably a normal host constituent. There could also be a finite quantity of inhibitor which would explain why the insect hosts can deal effectively with low virus dosages commonly met in nature, but are in serious difficulties when exposed to high doses. It is significant that some vertebrate animals do have a preformed inhibitor system based on glycoproteins that are quite different from immunoglobulins. The best known of these is the α inhibitor which is effective against influenza virus and is found in various organs, including the sub-maxilliary glands

(Cohen, 1963).

REFERENCES

Aizawa, K. (1962). Antiviral substance in the gut-juice of the silkworm *Bombyx mori L.*, *J. Insect. Path. 4*, pp. 72 - 76.

Aruga, H. (1963). Induction of virus infections. In "Insect Pathology", Vol. I, pp. 499 - 530. Ed., E. A. Steinhaus, Academic Press, New York.

Aruga, H. and Watanabe, H. (1964). Resistance to *per os* infection with cytoplasmic polyhedrosis virus in the silkworm *Bombyx mori L.*, *J. Insect. Path., 6,* pp. 387 - 394.

Bailey, L. and Gibbs, A. J. (1964). Acute infection of bees with paralysis virus. *J. Insect. Path., 6,* pp. 395 - 407.

Bailey, L., Gibbs, A. J. and Woods, R. R. (1970). A simple way of purifying several insect viruses. *J. gen. Virol., 6,* pp. 175 - 177.

Bailey, L. and Scott, H. A. (1973). The pathogenicity of Nodamura virus for insects. *Nature, 241,* p. 545.

Bernheimer, A. W., Caspari, E., Kaiser, A. D. (1952). Studies on antibody formation in caterpillars. *J. Exp. Zool., 119,* pp. 23 - 25.

Briggs, J. D. (1958). Humoral immunity in Lepidopterous larvae. *J. Exp. Zool., 138,* pp. 155 - 188.

Chamberlain, R. W. (1968). Arboviruses, the arthropodborne animal viruses. *Current Topics Microbiology and Immunology, 42,* pp. 38 - 58.

Chamberlain, R. W. and Sudia, W. D. (1961). Mechanism of transmission of viruses by mosquitoes. *Ann. Rev. Ent., 6,* pp. 371 - 390.

Cohen, A. (1963). Mechanisms of cell infection. I. Virus attachment and penetration, Chapter 4, pp. 153 - 196. In "Mechanisms of Virus Infection". (W. Smith, Ed.) Academic Press, London and New York.

Harrap, K. A. (1970). Cell infection by a nuclear polyhedrosis virus. *Virology, 42,* pp. 311 - 318.

Harrap, K. A. and Robertson, J. S. (1968). A possible infection pathway in the development of nuclear polyhedrosis virus. *J. gen. Virol., 3,* pp. 221 - 225.

Hurlbut, H. S. (1951). The propagation of Japanese encephalitis virus in the mosquito by parenteral introduction and serial passage. *Am. J. Trop. Med., 31,* pp. 448 - 451.

Hurlbut, H. S. (1953). The experimental transmission of a Coxsachi-like virus by mosquitoes. *J. Egypt. Med. Assoc., 36,* pp. 495 - 498.

Hurlbut, H. S. (1956). West Nile virus infection in arthropods. *Am. J. Trop. Med. Hyg., 5,* pp. 76 - 85.

Hurlbut, H. S. and Thomas, J. I. (1960). The experimental host range of the arthropod-borne animal viruses in arthropods. *Virology, 12,* pp. 381 - 407.

Hurlbut, H. S. and Thomas, J. I. (1969). Further studies on the arthropod host range of arboviruses. *J. med. Ent., 6,* pp. 423 - 427.

Kamon, E. and Shulov, A. (1965). Immune response of locusts to venom of the scorpion. *J. Invertebrate Path., 7,* pp. 192 - 198.

Longworth, J. F., Tinsley, T. W., Barwise, A. H., and Walker, I. O. (1968). Purification of a non-occluded virus of *Galleria mellonella, J. Gen. Virol., 3,* pp. 167 - 174.

Merrill, M. H. and ten Broeck, C. (1935). The transmission of equine encephalomyelitis virus by *Aedes aegypti, Jour. Exp. Med., 62,* pp. 687 - 695.

McLean, D. M. (1955). Multiplication of viruses in mosquitoes following feeding and injection into the cavity. *Austr. J. Exp. Biol. Med. Soc., 33,* pp. 53 - 66.

Peers, R. R. (1972). Bunyamwera virus replication in mosquitoes. *Can. J. Microbiol. 18,* pp. 741 - 745.

Phillips, J. H. (1960). Immunological processes and recognition of foreigness in the invertebrates. In "Phylogeny of Immunity", (R. T. Smith, R. A. Meiseler and R. A. Good, eds.) pp. 133 - 140. University of Florida Press, Gainesville.

Rehac̃ek, J. (1965). Development of animal viruses and Rickettsiae in ticks and mites. *Ann. Rev. Ent., 10,* pp. 1 - 24.

Salt, G. (1970). "The cellular defence reactions of insects". Cambridge Monographs in Experimental Biology, No. 16, Cambridge University Press.

Scherer, W. F. and Hurlbut, H. S. (1967). Nodamura virus from Japan: A new and unusual arbovirus resistant to diethyl ether and chloroform. *Am. J. Epidemiology, 86,* pp. 271 - 285.

Sinha, R. C. (1967). Response of wound-tumor virus infection in insects to vector age and temperature. *Virology, 31,* pp. 746 - 748.

Stephens, J. M. (1959). Immune responses of some insects to some bacterial antigens. *Can. Jour. Microbiol. 5,* pp. 203 - 228.

Stephens Chadwick, J. (1967a). Serological responses of insects. *Federation Proceedings, 26,* pp. 1675 - 1679.

Stephens Chadwick, J. (1967b). Some aspects of immune responses in insects. In "Differentiation and defence mechanisms in lower organisms", *3,* "In vitro" pp. 120 - 128. (M. M. Segel, ed.). Symposium, Tissue Culture Association. Distr. Williams and Wilkins Co., Baltimore.

Storey, H. H. (1932). The inheritance by an insect vector of the ability to transmit a plant virus. *Proc. Roy. Soc.,* London (B), *112,* pp. 46 - 60.

Storey, H. H. (1939). Transmission of plant viruses by insects. *Bot. Rev., 5,* pp. 240 - 272.

Varma, M. G. R. (1972). Invertebrate host specificity in arthropod borne animal viruses. In "Moving Frontiers in Invertebrate Virology". (T. W. Tinsley and K. A. Harrap, eds.). Monographs in Virology, No. 6, S. Karger, Basel.

Whitfield, S. G., Murphy, F. A., and Sudia, D. W. (1973). St. Louis encephalitis virus: an ultrastructural study of infection in a mosquito vector. *Virology, 56,* pp. 70 - 87.

THE PERITROPHIC MEMBRANE: ITS ROLE AS A BARRIER TO INFECTION OF THE ARTHROPOD HOST

Thomas C. Orihel
Tulane University
School of Public Health and Tropical Medicine
New Orleans, Louisiana

The epithelium of the insect midgut, unlike that of the fore- and hindgut does not have a cuticular lining. However, in many insects there is a membrane called the *peritrophic membrane* which apparently functions to protect cells of the midgut from injury by hard or sharp particles of food performing much the same function as mucus in the mammalian alimentary tract. Although this view is supported to a large extent by the absence of a PM in many insects having a fluid diet, it occurs nevertheless in a wide variety of blood sucking species as well (Day and Waterhouse, 1953). Apparently the membrane is freely permeable to digestive enzymes and to the products of digestion (Wigglesworth, 1956). Experiments have demonstrated that it is permeable to most dyes except those with the largest colloidal particles.

Aside from its role in physiological functions, the question is raised as to the significance of the PM as a barrier to the infection of arthropod species which may be biological vectors of parasites. To what extent does the peritrophic membrane block or interfere with the migration of parasites (such as microfilariae, trypanosomes, etc. ingested with blood or tissue fluids) from the midgut to other tissues outside the gut?

The report which follows is an attempt to review what is known about the membrane and its role, if any, as a barrier to infection in arthropod vectors of parasites.

First, a few words about the formation of the peritrophic membrane. In some insects it may consist of a series of thin lamellae, either independent or loosely attached to one another secreted by cells of the midgut. In each case, the delamination is said to be brought about by the pressure of secretion from underlying cells. In spite of numerous experimental studies and observations, there is no general agreement that delamination is ever the main source of the PM (Day and Waterhouse, 1953). Stohler (1957) described a variation in this type of membrane formation which appears to be characteristic for a variety of blood sucking arthropods. He observed in working with *Aedes aegypti*, that the PM is secreted in viscous form by all or part of the midgut epithelium in response to a blood meal. Freyvogel and Jaquet (1965) expressed the opinion that (in the mosquito) the secretion is provoked by a mechanical factor i.e., by distention of the midgut epithelium rather than associated with the ingestion of food per se. Bertram and Bird (1961) in their study of the structure of the midgut epithelium in *Aedes aegypti* and its modification following a blood meal observed that the secretion forming the PM is derived from the cytoplasmic matrix of the epithelial cells. Initially, it appears as a finely particulate, amorphous material which after 15 to 24 hours consists simply of condensed granular material, loosely layered, rather than a well-formed envelope of compact laminate structure.

The PM may be produced as a viscous secretion by special cells at the junction of the fore- and midgut. As the viscous secretion passes back posteriorly it makes its way through the narrow space between the invaginated esophagus and midgut which acts as an annular press and molds the membrane into a continous tube that extends from the proventriculus through the midgut and into the hindgut. This mode of formation was first fully described in the tsetse fly (*Glossina*) by Wigglesworth (1929). Preliminary studies on the PM by electron microscopy by Willett (1966) indicated that it was a relatively uniform and structureless mass.

In the different groups of blood-sucking arthropods which serve as vectors of parasites, the role, if any, of the peritrophic membrane in relation to infection may differ from group to group.

Mosquitoes. A peritrophic membrane has been demonstrated in a variety of mosquito species (Christophers, 1960; Clements, 1963; Stohler, 1957; Waterhouse, 1953; Freyvogel and Staubli, 1965; Richardson and Romoser, 1972). Many are vectors of filariasis, malaria, and other parasites, etc. Usually, within a few minutes after engorgement with blood, an amorphous layer of material secreted by cells of the midgut epithelium separates the blood bolus from the epithelium. During the first few hours, the membrane appears as a thick, viscous structure which, as time progresses, becomes thinner and solidifies forming a hard, inelastic membrane. By the time digestion of the blood is well advanced, the membrane has begun to soften. In most cases, however, it persists as long as, or usually longer than, the blood meal.

In the mosquito, the PM does not appear to be a formidable barrier to the migration of microfilariae out of the midgut and into the thoracic muscles. This may be attributed in part to the rapidity with which microfilariae move away from the blood meal. Esslinger (1962) found that in *Anopheles quadrimaculatus*, a PM was already present, though very soft and delicate, within 5 minutes after engorgement. He also noted that microfilariae of *Brugia pahangi* ingested with the blood had by this time already escaped the blood mass and penetrated both the PM and midgut wall. Migration out of the midgut was at its peak within a period of 15 minutes after engorgement, and after 30 minutes few microfilariae remained in the gut contents. Ewert (1965) working with the same system (*B. pahangi* in *Anopheles quadrimaculatus*) similarly noted that 90% of the *B. pahangi* microfilariae escaped the midgut within the first hour. Wharton (1957) observed that microfilariae of a related species i.e., *Brugia malayi* began their migration from midgut to haemocoele of *Mansonia longipalpis* within ten minutes after engorgement. Other investigators (Schacher, 1960; Laurence and Pester, 1961 a, b; Ramachandran, 1966; Ramachandran and Zaini, 1967) recorded similar observations. It is evident that microfilariae can and do penetrate the PM in its early stage of formation while it is still quite soft. Esslinger (loc. cit) also observed that penetration of the PM by the microfilaria is accomplished through the use of its cephalic hook. Whether this is the principal or only mechanism for penetration has not been determined.

Iyengar (1936) expressed the view that microfilariae did not enter the midgut of the mosquito but remained in the tubular or cardiac region and penetrated the epithelium at this point. O'Connor and Beatty (1936) held the view that microfilariae of *Wuchereria bancrofti* in *Culex fatigans* migrated anteriorly from the expanded midgut to the cardiac region and at that point escaped into the hemocoel and thence into the thoracic muscles. In general, however, it is widely accepted that penetration of the PM and midgut can be accomplished at almost any point, most often in the expanded region of the midgut.

Lavoipierre (1958) made the interesting observation while working with *Dirofilaria immitis* in *Aedes aegypti* that large numbers of microfilariae impeded or actually prevented the formation of a PM which rendered the digestion of the blood virtually impossible and caused the death of the mosquito. It should be recalled that not all microfilariae penetrate the gut wall in order to accomplish their migration to other sites in the body. Some species of *Dirofilaria* e.g., *D. immitis*, *D. tenuis*, develop in the malpighian tubules and simply migrate directly from midgut posteriorly to the open ends of the tubules, often very rapidly. Kartman (1953) showed in controlled experiments that clot formation could be a significant mechanical barrier to migration. Other observations suggest that digestive enzymes or some other substances in the midgut might be lethal to microfilariae and prevent their migration.

Stohler (1957) made some interesting observations on the role of the PM in malaria infection. Using *Aedes aegypti* as a vector for *Plasmodium gallinaceum* a malaria parasite of fowl, he found that the PM secreted in fluid form could be readily penetrated by the ookinetes as long as it was soft. After 20 to 30 hours however, when the membrane became hardened, the malaria parasites were no longer able to penetrate it, were trapped within it and died. Certainly, in this particular system the PM may exert some regulatory function on the degree of infection in the mosquito.

Simulids. Species of *Simulium* are vectors of *Onchocerca volvulus* in man and of other onchocercas in animals. Lewis (1950) first noted that PM in *Simulium damnosum*, and from the very first believed it played an important part in limiting infection in the arthropod host. He observed that the membrane formed rapidly from the posterior part of the midgut after ingestion of blood but not after a meal of sugar water. The PM completely enveloped the blood meal within 30 seconds after engorgement. Though very delicate and difficult to see initially, within 30 minutes it had a well defined outline. Lewis described it as a tough, inelastic membrane composed of a series of laminae which persisted for more than 24 hours. He suggested that only the microfilariae (*O. volvulus*) that remained in the tubular or cardiac portion of the midgut during the period of PM formation were destined to reach the thoracic muscles and develop to the infective stage. However, he did recognize the possibility that microfilariae might passs through the PM while it was being formed and was still soft. Later studies by Duke and Lewis (1964) demonstrated that less than half of the ingested microfilariae of *O. volvulus* made their way out of the midgut during the first 6 to 10 hours following the blood meal; the remainder were trapped within the PM and died. However, almost all of the microfilariae that managed to escape the membrane barrier developed to the infective stage. Laurence (1966) also studied *O. volvulus* in *Simulium damnosum* and paid special attention to the route used by microfilariae to escape from the blood meal into the tissues. There was little evidence of migration from the cardiac portion of the gut as suggested by Lewis (1953) but strong evidence of a rapid penetration out of the blood meal into the tissues immediately after engorgement. Migration out of the midgut and into the abdomen and tissues reached a peak at 30 minutes after engorgement and ceased entirely after 2 to 4 hours. Roughly 40 to 50% of the microfilariae were found outside the midgut after the first 60 minutes. In contrast to the observations of Duke and Lewis, cited above, Laurence found that only about 25% of the microfilariae were trapped in the blood meal. Eichler (1973) working with *O. gutturosa* in *Simulium ornatum* made similar observations. In this system however, microfilariae migration was severely limited. Approximately 25% of the ingested microfilariae migrated out of the midgut, most within one hour of feeding; the remainder were trapped in the blood meal.

Form the observations made thus far, it is evident that in *Simulium* spp. the PM does play an important role in limiting infection with *Onchocerca*. Microfilariae evidently are capable of penetrating the PM only while it is still very soft, i.e., during the first hour or so after formation. As compared with mocrofilarial migration in mosquitoes, the number of *O. volvulus* that manage to escape during the first hour or two is comparatively small. Perhaps the environment of the midgut does not favor survival of microfilariae for much more than a few hours and may even render them incapable of migration within minutes after their entry into the midgut.

Glossina. The peritrophic membrane is of special interest in the tsetse flies, *Glossina* spp., the vectors of African trypanosomes (*T. rhodesiense, T. gambiense, T. brucei* etc.). One of the perplexing problems associated with African trypanosomiasis is the variable and usually low susceptibility of tsetse flies to infection by trypanosomes either in the field or under controlled conditions in the laboratory. In the most general terms it seems that the passage from vertebrate to fly is less of a hurdle to the parasite than is survival in the fly. So far, little is known of the factors, whether they are inherent in the fly or the trypanosomes or both, which permit development of the parasite in the fly. Of course, the role of the PM has been the subject of numerous studies in this regard dating back to those of Hoare (1931).

In contrast to the arthropods already discussed, the peritrophic membrane formed in *Glossina* is a tubular structure secreted by cells in the proventriculus. The details of its formation have been described by Wigglesworth (1929) and Hoare (1931). More recently, Willett (1966) observed that the PM begins to form in the unfed fly soon after eclosion and grows at a rate of about 1 mm/hour during the first 24 hours and, thereafter, more slowly, reaching the hindgut at about 72 to 96 hours if the fly remains unfed. The posterior end of the membrane, still intact, is broken up in the hindgut by posteriorly-directed spines. Thus far, PMs have been observed in at least eight species of *Glossina* (Willett, 1966).

Early studies (Hoare, 1931; Taylor, 1932; Yorke et al., 1933) suggested that the trypanosomes escaped the midgut and PM by passing posteriorly to the hindgut through the open end of the membrane and then moved anteriorly making their way through the ectoperitrophic spaces. In order to reach the salivary glands, the trypanosome proventricular forms penetrated the soft PM (ectoperitrophic to endoperitorphic) in the region of its formation in the proventriculus. This latter aspect of the migration was conclusively demonstrated by Fairbairn (1958) who showed that the trypanosomes were capable of penetration of the fluid secretion as well as the formed membrane in the roof of the proventriculus.

Willett's studies clearly indicate that the blood meal, especially the first, is contained within the intact membrane and that trypanosomes probably reach the ectoperitrophic spaces by penetration of the soft membrane. He hypothesized that very young tsetse flies taking their first blood meal probably produce additional membrane through which the trypanosomes can pass. There was no indication that ingesting a large quantity of blood could cause rupture of the PM and thereby provide another avenue of escape. Still others have attempted to clarify the routes by which trypanosomes reach the ectoperitrophic spaces. Freeman (1973) recently reviewed the possible routes of migration to ectoperitrophic sites. These include those mentioned above i.e., circumnavigation of the membrane, crossing a ruptured membrane or penetrating the soft membrane. Her observations supported the concept that the trypanosomes actually penetrate the PM in its anterior (soft) region.

Harmsen (1973) is unwilling to believe that the membrane itself can be such a major obstacle to infection and suggests that, to quote him, "...The barrier mechanism involves a much more basic adaptation of the trypanosome to life inside the tsetse (fly's) alimentary system." It is Harmsen's view that trypanosomes may undergo a physiological transformation while in the crop of young flies which makes them capable of survival in the fly's midgut. The destruction of non-transformed trypanosomes in the midgut (after leaving the crop) of mature tsetse flies is considered to be the main establishment barrier. Still unanswered by Harmsen is how ectoperitrophic infection is established. Penetration of the proximal zone of the membrane still seems to be the most plausible answer. It is clear that in the tsetse fly-trypanosome system, many questions regarding barriers to infection including the role of the PM, are still to be resolved.

Sandflies. From the work of Feng (1951) we get some idea of the role of the

PM in the establishment of *Leishmania* and trypanosome infections in the sandflies, *Phlebotomus* spp. Feng observed that the PM played a significant role in the determination of the vector capacity of *Phlebotomus* spp. for *L. donovani* in China. Apparently the flagellate forms of *L. donovani* are unable to penetrate the peritrophic membrane. For example, in *P. mongolensis* an intact membrane is formed around the blood meal and ingested parasites. Even though digestion of the blood may take 5 to 6 days (depending on its size) the PM remains intact, though after 48 to 72 hours it becomes softened and shrunken. Eventually, it along with any undigested blood residue is passed *en toto* into the hindgut. During the entire process of digestion, the flagellate forms of *L. donovani* were seen within the membrane, never in the ectoperitrophic spaces. Flagellates were never observed to establish themselves in the proventriculus or hindgut. In contrast, in *P. chinensis*, the PM is an intact sac too, but it looses its integrity by the 3rd or 4th day, breaks up, and the partially digested blood meal and any flagellate forms of *L. donovani* that may be present are liberated and continue their development in the proventriculus. This sandfly species is a good vector of *L. donovani*.

Feng also made some interesting observations on the development of the toad trypanosome, *Trypanosoma bocagei* in *Phlebotomus squamirostris*. Soon after the fly feeds on an infected toad a PM forms around the entire blood meal. However, the posterior end of the membrane is not tightly closed. After 24 to 48 hours, blood begins to leak out from the posterior end of the sac. The crithidial forms of the parasite pass down into the hindgut to the wall of which they attach and continue to multiply and develop to the infective stage. A peritrophic membrane has been described in other sandflies, including *P. papatasii*, *P. minutus* and *P. perniciosus*. In these species to it appears to be a sealed tube.

Culicoids. Species of *Culicoides* are vectors of filariae and viruses in animals and man. However, in spite of their potential medical and veterinary importance little is known about their biology and interactions with their parasites. This is due in large part to the fact that they are difficult to colonize and maintain in the laboratory. Megahed (1956) made a histological study of the alimentary tract of *Culicoides nubeculosus* and in the process described the formation and structure of the PM. He observed that it forms in response to the ingestion of blood, is a very delicate membrane secreted by cells of the midgut and is composed of several successive layers of lamellae. The posterior end of the sac apparently is perforated to allow fecal matter to pass into the hindgut. It remains for about 72 hours at which time it is discharged into the hindgut. Although species of *Culicoides* are vectors of certain species of *Onchocerca* of animals and of *Dipetalonema perstans* and *D. streptocerca* of man and possibly of *Mansonella ozzardi* of man, little has been learned about the interactions of arthropod host and parasite. This is an especially fertile field for study and one which might be expected to attract the attention of parasitologists and entomologists as techniques for colonization of these arthropods continue to be explored and refined.

In summary, it would be accurate to state that in most arthropod vectors of the parasites mentioned above, a peritrophic membrane is present. In most cases it is an impenetrable barrier to infection during a substantial period of its existence. However, it does not prevent infection of the arthropod but, at best, may limit the degree of infection. On the other hand as suggested by Stohler (1961), the possibility that the PM may prevent transmission in some forms cannot be excluded. Most of the observations made in regard to the role of the membrane as a barrier to infection have been in systems where the arthropod under study is a vector of variable potential. In many instances, observations on the interaction of the parasites and membrane reflect only casual or incidental observations and were not the principal objectives of the studies. There have been few studies of the PM in arthropods that are known to be incapable of serving the vector role for a given parasite. Medical entomologists, parasitologists and others are very much interested in the factors responsible for susceptibility and refractoriness in arthropods which are potential vectors of parasites especially the filariae but the role of the PM is seldom mentioned in this regard. Certainly it is an aspect of host-parasite interaction deserving of further study.

REFERENCES

Bertram, D. S., and Bird, R. G. (1961). Trans. Roy. Soc. Trop. Med. *Hyg.*, *55*:404-423.

Christophers, Sir R. (1960). Cambridge: University Press.

Clements, A. N. (1963). International Series of Monographs on Pure and Applied Biology. Div. *Zoo.*, *17*:128-150.

Day, M. F., and Waterhouse, D. F. (1953). In Roeder, K. D.: Insect Physiology. J. Wiley, New York, 273-310.

Duke, B. O. L., and Lewis, D. J. (1964). Ann. Trop. Med. *Parasit.*, *58*:83-88.

Eichler, D. A. (1973). *J. Helminth.*, *47*:73-88.

Esslinger, J. H. (1962). Am. J. Trop. Med. *Hyg.*, *11*:749-758.

Ewert, A. (1965). Am. J. Trop. Med. *Hyg.*, *14*:254-259.

Fairbairn, H. (1958). Ann. Trop. Med. *Parasit.*, *52*:18-19.

Feng, L. C. (1951). Pek. Nat. Hist. *Bull.*, *19*:327-334.

Freeman, J. C. (1973). *Acta Trop.*, *30*:347-355.

Freyvogel, T. A., and Jaquet, C. (1965). *Acta Trop.*, *22*:148-154.

Freyvogel, T. A., and Staubli, W. (1965). *Acta Trop.*, *22*:119-147.

Harmsen, R. (1973). Trans. Roy. Soc. Trop. Med. *Hyg., 67*:364-373.

Hoare, C. A. (1931). Trans. Roy. Soc. Trop. Med. *Hyg., 25*:57-64.

Iyengar, M. O. T. (1936). *Parasitology, 28*:190-194.

Kartman, L. (1953). Exper. *Parasit., 2*:27-78.

Laurence, B. R. (1966). *J. Helminth., 40*:337-342.

Laurence, B. R., and Pester, F. R. N. (1961a). J. Trop. Med. *Hyg., 64*:169-171.

Laurence, B. R., and Pester, F. R. N. (1961b). *J. Helminth., 35*:285-300.

Lavoipierre, M. M. J. (1958). Ann. Trop. Med. *Parasit., 52*:326-345.

Lewis, D. J. (1950). *Nature, 165*:978.

Lewis, D. J. (1953). *Bull. Ent. Res., 43*:597-644.

Megahed, M. M. (1956). *Parasitology, 46*:22-47.

O'Connor, F. W., and Beatty, H. (1936). Trans. Roy. Soc. Trop. Med. *Hyg., 30*:125-127.

Ramachandran, C. P. (1966). *J. Med. Ent., 3*:239-252.

Ramachandran, C. P., and Zaini, M. A. (1967). Med. *J. Malaya, 22*:136-144.

Richardson, M. W., and Romoser, W. S. (1972). *J. Med. Ent., 9*:495-500.

Schacher, J. F. (1962). *J. Parasit., 48*:679-693.

Stohler, H. R. (1957). *Acta Trop., 14*:302-352.

Stohler, H. R. (1961). *Acta Trop., 18*:236-266.

Taylor, A. W. (1932). *Parasitology, 24*:401.

Waterhouse, D. F. (1953). Aust. J. *Zool., 1*:299-318.

Wharton, R. H. (1957). Ann. Trop. Med. *Parasit., 51*:278-296.

Wigglesworth, V. B. (1929). *Parasitology, 21*:288-321.

Wigglesworth, V. B. (1956). Methuen's Monographs on Biological Subjects, Methuen

& Co. Ltd., London, 43-45.

Willett, K. C. (1966). Exper. *Parasit.*, *18*:290-295.

Yorke, W., Murgatroyd, F., and Hawking, F. (1933). Ann. Trop. Med. *Parasit.*, *27*:347-354.

EARLY EVENTS IN THE INFECTION OF THE ARTHROPOD GUT BY PATHOGENIC INSECT VIRUSES

J. D. Paschke and Max D. Summers

Department of Entomology
Laboratory
Purdue University
West Lafayette
and
The Cell Research Institute
Department of Botany
The University of Texas
Austin, Texas

In light of the advances recently made in the study of infectious processes in virus infection of insects, it seems timely to present a review of our current knowledge of the phenomena involved. Particularly important are the events occurring early during the infection, a heretofore little studied aspect of the problem. By necessity we will draw upon a limited number of well documented *in vivo* and *in vitro* examples to elucidate mechanisms of infection.

We will depart from the generally employed terminology used to describe the various forms and structures associated with the insect viruses, especially the inclusion body viruses, and wherever possible employ the terminology of Caspar et al. (1962).

It is hoped that attention will be drawn to the insect-virus model as a system to be used for a better understanding of viral infections. Before and during the preparation of this review, it became more increasingly evident that many gross phenomena of viral infectious processes are not well understood and dictate a need for further study at the cellular level.

BARRIERS

It is an accepted axiom that the natural portal of entry of the pathogenic insect virus is via the digestive tract (Heimpel and Harshbarger, 1965) and specifically the midgut region. The generalized alimentary tract of the insect is composed of three distinct areas: 1) the foregut; 2) the midgut; and 3) the hindgut. The fore and hindgut are of ectodermal origin and are lined with cuticle, thus forming a significant barrier to infection.

Only the recent studies of Cunningham (1971), Granados (1973a), Harrap (1969, 1970), Harrap and Robertson (1968), Kawanishi et al. (1972b), Summers (1969, 1971), and Tanada and Leutenegger (1970) have partially explained the infection of the midgut epithelial cells. The above studies utilized granulosis or nuclear polyhedrosis viruses and an insect poxvirus of lepidoptera species, and the conclusions reached may not be generalized for all insect-virus relationships.

In the cited studies, the investigators did not direct attention to the role of the peritrophic membrane (PM) of the host midgut since it was obvious in these preliminary studies that this did not appear to constitute an absolute or significantly restrictive barrier to infection. However, if one considers all possible barriers to infection, the PM, as well as the biochemical environment of the midgut lumen may act significantly to limit or restrict infection.

The above mentioned barriers require serious consideration when one examines infectious processes in cases involving a host-virus system which exhibits infection at low levels, or in the case of negative results by *per os* techniques in cross infectivity studies.

BIOCHEMICAL ENVIRONMENT OF THE MIDGUT LUMEN

Virus Degradation In our (J.D.P.) laboratory, we are investigating icosahedral cytoplasmic deoxyriboviurses (ICDV) which are only slightly pathogenic to mosquitoes (Paschke el al., 1972; Stoltz and Summers, 1971; Woodard and Chapman, 1968). Infection of laboratory reared mosquitoes approximates 10% apparently irrespective of the dose of virus administered. This phenomenon of susceptibility is being investigated with regard to the genetics of the host as well as the *in vivo* kinetics of infection.

Stoltz and Summers (1971) conducted studies to determine the pathway of infection of mosquito iridescent virus (MIV). They utilized fixed and unfixed MIV as inocula and incorporated marker viruses to better trace the movement of virus in the midgut. The latter were fixed and unfixed inclusions of the *Trichoplusia ni* granulosis and polyhedra of the *Spodoptera frugiperda* nuclear polyhedrosis. Their results indicated a significant, if not complete, degradation of ingested MIV particles shortly after they entered the midgut.

Similar to the results of Stoltz and Summers (1971), we have found electron microscopic evidence suggesting degradation of MIV in the midgut of the mosquito (unpublished). However, the results indicated that there was no significant degradation of MIV at the fore end of the midgut, and it appeared that the degradation which occurred was more intense as one progressed from the anterior to the posterior end of the midgut. Intact virus was observed in the approximate region of the juncture of the fifth and sixth abdominal segments of the mosquito larvae. Intact virus was not observed in the hindgut. These studies were conducted in a fashion similar to that described by Stoltz and Summers (1971) and incorporated the granulosis virus capsules of *Spodoptera frugiperda* as a marker. The results indicated only minimal degradation of the granulosis capsules as compared to the complete breakdown of capsules and polyhedra in the studies conducted by Stoltz and Summers (1971). Furthermore, even though significant degradation of the inclusions was observed, the enveloped virions were observed to remain intact. In our study, the pH of the midgut of mosquito larvae (first, second, and fourth instar) was determined to be 6.6-7.1, and it was shown that purified MIV solutions appeared to be stable at pH 3 to 8 in electrophoretic mobility studies (Wagner et al., 1973). In addition, the midgut appears to be very effectively buffered since larvae exposed to the high and low ends of the pH indicator range rapidly returned the pH to its near-neutral point (W. R. Campbell, unpublished).

In our study (W. R. Campbell, unpublished), the gut pH apparently did not significantly alter the MIV virion, thus certain active factors produced by the mosquito larvae were investigated as to their importance in degradative phenomena. While midgut (with PM and ingested food) homogenate incubated with purified RMIV for 12 hours at 30°C, pH 7.0 showed (as observed in electron microscope) no degradation, at pH 3.0 the virus in whole gut homogenate was degraded, as in the control. Under the conditions of these degradative studies, it was not possible to detect significant breakdown of RMIV at a pH near that of the normal midgut pH and the tests suggest that an active factor in the midgut is either lacking or undetectable under the conditions of the experiment (W. R. Campbell, unpublished).

Rachiplusia ou multiploid polyhedrosis virus (RONPV), *Trichoplusia ni* granulosis virus (TNGV), and *Spodoptera frugiperda* granulosis virus (SFGV) enveloped virions apparently survive the gut environment of their respective hosts to enter susceptible midgut cells (Kawanishi et al., 1972b). Unfortunately electron microscopic observations of TNGV interactions with *Spodoptera* midgut cell microvilli have not been demonstrated, even though *T. ni* virus particles have been observed in the microvilli of *T. ni* and *S. frugiperda* virus in the microvilli of *S. frugiperda* (Tables I and II). In the vicinity of the microvilli, virus degradation is only rarely observed if at all in *T. ni* larvae with TNGV. This is in contrast to extensive degradation of virus in the gut lumen of both *T. ni* and *S. frugiperda* larvae. Present estimates are that virus degradation is at least greater than 50-60% (Summers, unpublished).

To better understand degradative phenomena, studies on the effects of pH and ionic strength (I) have been conducted on nuclear polyhedrosis viruses (NPV) and granulosis viruses (GV). Shapiro and Ignoffo (1969), Kawanishi and Paschke (1970), Summers and Paschke (1970), showed that infectious virions were obtainable *in vitro* by alkali dissolution of the inclusion bodies. Bioassay of these preparations was either by *per os* routes or by intrahaemocoelic injection. Summers and Paschke (1970) were the first to lay the basis of quantified bioassays using protein and nucleic acid based on optical density (O.D.).

Summers and Paschke (1970) studying the TNGV demonstrated that virus held in 0.01 M phosphate buffer, pH 7.8, fractionated into three sedimenting forms on 10-40% (wt/vol) sucrose gradients. Two of these fractions were of the enveloped form, and no physical differences other than sedimentation could be attributed to the two forms. The slowest sedimenting of the three forms was determined by electron microscopy to be the nucleocapsid. When purified virions were exposed to 0.05 M Na_2CO_3, pH 11.2 for 14-18 hrs and then dialyzed against 0.1 M $NaCO_3$, pH 11.2

TABLE I. ELECTRON MICROSCOPE OBSERVATION OF OCCLUDED AND ICOSAHEDRAL VIRUSES IN THE GUT LUMEN OF 1st AND 2nd INSTAR *AEDES TAENIORHYNCHUS*[a]

	Icosahedral Virus		Occluded Virus: Protein Crystal		Occluded Virus: Non-occluded Virion		Peritrophic Membrane a Physical Barrier	Observed Virus Entry
Virus[b]	Degraded	Not Degraded	Degraded	Not Degraded	Degraded	Not Degraded		
MIV	+	–					+	–
FMIV	–	+					+	–
TIV	+	–					+	–
FTIV	–	+					+	–
TnGV			+	–	–	+	+	–
FTnGV			–	+	–	+	+	–
RONPV			+	–	–	+	+	–

[a]Summers unpublished; the results presented in this table do not represent data of completed experiments, only an up-to-date statement of present studies in progress.

[b]F = Virus fixed in 2.0% glutaraldehyde and subsequently in 2.0% OsO_4.

MIV = Mosquito Iridescent Virus of *Aedes taeniorhynchus.*

TIV = Tipula Iridescent Virus.

TnGV = Granulosis Virus of *Trichoplusia ni.*

RONPV = Nuclear Polyhedrosis Virus (multiploid virions) of *Rachiplusia ou.*

TABLE II. VIRUS INVASION AND INFECTION: MIDGUT CELLS 0-2 HOURS POSTINFECTION[a]

	Protein Crystal Removed	Virus Not Degraded	Virus Degraded	Virus Entry: Fusion	Virus Entry: Phago-cytosis	Virus Uncoating: Cyto-plasm	Virus Uncoating: Nuclear Pore	Other	Peritrophic Membrane as a Barrier
NUCLEAR POLYHEDROSIS									
Aglais urticae[1]	0[b]	0	0	+?	—	+?	—	0	0
Trichoplusia ni[2]	+[c]	+	+	+	—	—	?	0	±
GRANULOSIS									
Trichoplusia ni									
Summers (1969, 1971)[3]	+	+	+	+?	—	—	+	—	±
Tanada (1970)[4]	0	0	0	0	0	0	0	0	0
Spodoptera frugiperda[5]	+	+	+	?	?	0	0	0	±
TnGV in *S. frugiperda*[5]	+	+	+	?	?	0	0	0	±
SfGV in *Trichoplusia ni*[5]	+	+	+	?	?	0	0	0	±
FIXED VIRUS PREPARATIONS									
Trichoplusia ni GV[5]	—[d]	+	—	—	—				±
Spodoptera frugiperda GV[5]	—	+	—	—	—				±
Tipula Iridescent Virus[5]		+	?	—	—				±

[1] Harrap, K.A., Virology 42, 311 (1970).
[2] Kawanishi, C.Y., Summers, M.D., Stoltz, D., and Arnott, H.J., J. Invertebr. Pathol. 20:104-108 (1972).
[3] Summers, M.D., J. Ultrastruct. Res. *35*, 606 (1971).
[4] Tanada, Y., and Leutenegger, R., J. Ultrastruct. Res. *30*, 589 (1970).
[5] Summers, M.D. (unpublished data).
[a] The results in this table represent observations up to the present time and do not constitute data from completed studies.
[b] 0 — No data available or provided by investigator for designated virus-host cell interaction.
[c] + — Virus-host cell interaction observed.
[d] — —Virus-host cell interaction not observed.

for 24 hrs the treatment resulted in a top fraction sedimenting at approximately 1 cm from the top of 10-40% sucrose gradients and a faster sedimenting single major peak of virions. This top fraction was infective in *per os* and intrahaemocoelic bioassays against *T. ni* fourth stage larvae, but survey by electron microscopy did not detect virions, nucleocapsids, or significant particular components associated with the virus. Biological characterization indicated the top fraction to be sensitive to DNase I. The ionic conditions of the buffering systems were not considered in any of the described studies.

Kawanishi and Paschke (1970) found glycine-NaOH-NaCl (GNN) buffers superior to standard carbonate solution (Bergold, 1963) in the dissolution of purified polyhedra from RONPV, and that a pH 11.5 buffer resulted in the best yield of virions.

They determined that buffering was important in obtaining higher yields as suggested by the fact that I=0.1, pH 12.0 glycine buffer mean yield of virions was higher than a corresponding NaOH-NaCl solution with the same pH and I.

Except for Bergold's findings (1947), the effect of ionic strength on virion yield, integrity, or infectivity has not been considered. Virions are rapidly released from polyhedra at I values much above 0.1 but they are soon disrupted by these same conditions (Kawanishi and Paschke, 1970). Gudauskas and Cannerday (1968) found a decrease in infectivity of preparations of *Heliothis* and *Trichiplusia* NPV dissolution mixtures between 0.5 and 24 hrs. The data also showed a sharp rise in infectivity when I of their buffer decreased below 0.1.

The appearance of nucleocapsids in dissolution mixtures of RONPV after 24 hrs exposure is apparently independent of the buffer components but dependent on the buffering capacity, quantity of polyhedra dissolved, final pH of the dissolution mixture of RONPV after 24 hrs exposure and ionic strength (Kawanishi and Paschke, 1970). Dialysis of RONPV at pH 7.8 with a 0.01 M phosphate buffer did not result in the appearance of nucleocapsids as Summers and Paschke (1970) found with purified *T. ni* granulosis virions. Nucleocapsids from RONPV did appear at a pH 9.3, I=0.08. The disruptive effects of Ca^{++} and Mg^{++} on the virion envelope as described by Bergold could not be reproduced by Summers and Paschke (1970) or Kawanishi and Paschke (1970).

Dissolution of Occluded Protein Crystal During the natural infection of the insect host by polyhedrosis or granulosis virus the protein crystal which occludes the virions is solubilized to release the contained virions into the midgut lumen. The proteinaceous crystal is and/or biological role is not known. Numerous investigations of solubilization processes have been conducted, to either better understand the nature and structure of occluding protein, or in connection with the study of the occluded virions (Aizawa, 1953; Bergold, 1947, 1948, 1959; Day et al., 1953; Estes and Faust, 1966; Faust and Adams, 1966; Harrap, 1972; Ignoffo and Dutky, 1963; Kawanishi and Paschke, 1970; Longworth et al., 1972; Tanada et al., 1973; Vago and Croissant, 1959). More recently Egawa and Summers (1972) and Kawanishi et al., (1972a) have conducted detailed studies with regard to the kinetics of the protein solubilization and accompanying ultrastructural changes.

In the gut lumen of the respective and non-susceptible host of the granulosis virus of *Trichoplusia ni* and *Spodoptera frugiperda* both TNGV and SFGV readily lose the proteinic crystal and release virus (Tables I and II). This apparently is a non-specific process in the *in vivo* invasion and infection sequence and from this aspect supports the results of an earlier *in vitro* study by Faust and Adams, (1966). The protein crystal dissolution process has been studied *in vivo* by electron microscopy (Summers, 1971). Contrary to previous reports solubilization does not require the carbonate ion or high pH for crystal dissociation (Egawa and Summers, 1972). Dissociation in a biological system at lower pH conditions has been confirmed *in vivo* by observations (Stoltz and Summers, 1971) on the fate of the virus crystal in the gut lumen of mosquito larvae, where the pH is known to be neutral (W. R. Campbell, unpublished).

Summers (1971) has shown that in the *T. ni* gut lumen the crystal of the TNGV is first dissociated in the regions adjacent to the envelope of the virus and solubilization proceeds from the interior to the surface. In an attempt to study the effect of specific agents on the crystal, observations on solubilization correlated with the *in vivo* studies have shown that various solvents produce alterations in crystal structure which were characteristic for a given solvent (Kawanishi et al., 1972a). Dilute sodium carbonate, *7M* urea, as well as the presence of n-propanol caused longitudinal fragmentation of the protein structure early in the solubilization process. Solubilization in n-propanol apparently paralleled the crystal lattice and produced fragments with fibrous surfaces. The pattern of carbonate solubilization of the crystal and release of free virions most closely resembled the *in vivo* process in the midgut lumen of larval *T. ni*. Guanidine hydrochloride, in contrast, appeared capable of penetrating and solubilizing small cavities within the crystal to give the crystal structure a vacuolate appearance.

Tanada et al. (1973) isolated by centrifugation and purification Sephadex-G200 a synergistic factor from solubilized capsules of the granulosis of the armyworm, *Pseudaletia unipuncta*. The ID_{50} of the synergistic factor corresponded to 0.0015 OD_{280}. Its optimum pH was 8.5.

Synergism was most evident when the "factor" was fed to larvae together with polyhedra or 24 hrs prior to ingestion of polyhedra. Unfortunately, no information is available to explain the significance of the synergistic factor as it influences the infection by nuclear polyhedrosis virus.

From the foregoing review of the conditions relating to the stability of virions and solubilization *in vivo* and/or *in vitro* of crystalline protein, it becomes apparent that the gut lumen provides a suitable enough environment for those viruses which have been studied even though some degradation takes place. The TNGV "top fraction" described by Summers and Paschke (1970) remains infective in the rather extreme hydrogen ion concentration of the *Trichoplusia ni* midgut. As judged on the basis of *in vitro* studies on degradation of the virions and based on the studies of the pH conditions of the midgut, it now seems that one should examine more closely the effect of the various enzymes which are commonly associated with the midgut lumen of susceptible insect species.

We know from the well studied examples a number of important parameters involved in the initial phases of infection: 1) the pH of the midgut; 2) the effect of the ionic conditions and buffering conditions on virions; 3) relative stability of occluding protein crystal (NPV, CPV, or GV); and 4) the stability of the virion and/or components under varied conditions. It is obvious that in those cases where the virus is highly infectious, we have optimal conditions with regard to the gut lumen environment for a full expression of virulence.

PERITROPHIC MEMBRANE

The second barrier which may interfere or inhibit infection is the peritrophic membrane (PM). It has been established (Schildmacher, 1950; Zhuzhikov, 1964; Waterhouse, 1953; Wigglesworth, 1930, 1965; Peters, 1969) that the PM functions to protect the midgut epithelial cells and to allow only the passage of small diffusible molecules. The passage of the relatively large virion of the rod type viruses across the PM has been the subject of much discussion but with a paucity of experimentation. The PM is of two types (Wigglesworth, 1965) depending upon the way in which they are formed. In the first type, which is found in caterpillars as well as other representative groups, the membrane is made up of concentric lamellae, independent or loosely attached to one another. In this type, formation is by separation of thin sheets from the surface of the cell throughout the length of the midgut. Each new sheet appears as a limiting membrane at the surface of the striated border. In the second type, which is found in larval and adult Diptera and earwigs, the PM is secreted by a group of cells at the anterior end of the midgut and forms a homogenous tube the length of the midgut. The origin and composition of the peritrophic membrane of mosquito larvae has been investigated by Richards and Richards (1971), and they describe in detail the progressive changes taking place in the formation of the membrane during morphogenesis. Ultrastructural studies of the PM in *Aedes taeniorhynchus* denotes structural layers in the PM of early instar larvae that correspond to those described by Richards and Richards (1971), Stoltz and Summers (1971), and W. R. Campbell (unpublished). Schildmacher (1950) has shown that particles 20 nm in diameter are retained by the PM of the mosquito.

The question of how the viruses are transported across such a structure with the integrity thus far demonstrated is indeed an intriguing one. Harrap (1970) suggests that there may be holes in the PM that allow transport of virions across the barrier in lepidopterous larvae and Summers (unpublished) and C. Y. Kawanishi (personal communication) have observed large discontinuities in the PM of *T. ni*.

In the MIV studies conducted by W. R. Campbell (unpublished), the structure of the PM of *Aedes taeniorhynchus* was investigated utilizing transmission (TEM) and scanning electron microscopy (SEM). Additional information on the PM of this mosquito was gained by observing ultrathin sections of the midgut. Figs. 1a and b are TEM and SEM micrographs of the peritrophic membranes which were removed from first instar *A. taeniorhynchus* larvae. The fine structure of the PM is of layers of a fibrous network as described by Waterhouse (1953) and Richards and Richards (1971). It is not believed that an MIV virion with a hydrated diameter of 300 nm (Wagner et al., 1973) could passively diffuse across this structure.

In the case of NPV or CPV infections of lepidoptera, it is possible that the PM is of such a "loose" network that the virus may be passively transported to come into immediate contact with the midgut epithelial cell membrane. However, viral enzymes may play a significant role in penetration of the PM and only future studies designed to study the earliest events of infection will provide answers.

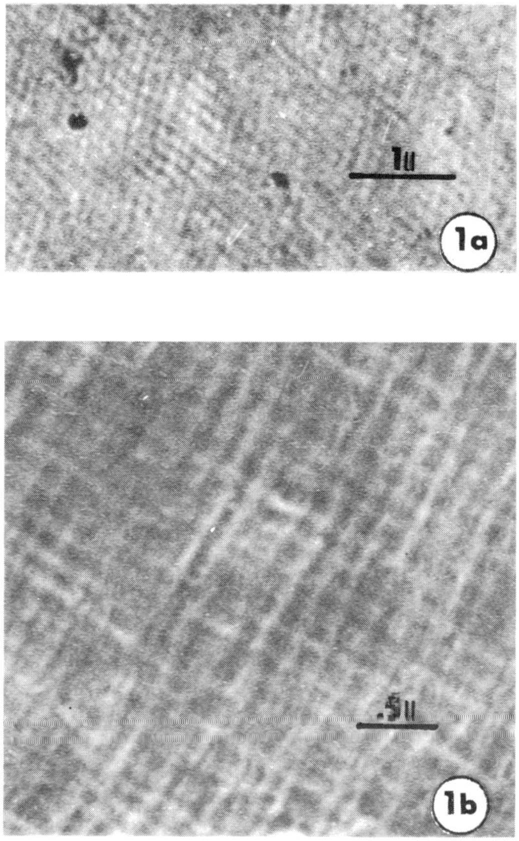

Figure 1. Transmission electron micrograph (a) and scanning electron micrograph (b) of the peritrophic membrane removed from first stage Aedes taeniorhynchus larvae. Note the predominant network and limiting integrity of the structure.

MIDGUT EPITHELIUM

The aforementioned classical studies of Cunningham (1971), Granados (1973a), Harrap (1970), Harrap and Robertson (1968), Kawanishi et al., (1972b), and Summers (1969, 1971), Tanada and Leutenegger (1970) have given us preliminary but valuable information concerning attachment, penetration, and replication during early events of the virus infections processes.

It seems advisable to consider the terminology employed by various authors in their description of the process of penetration of animal viruses into cells before we launch into a review of studies of the phenomenon.

Viruses and cells probably come together by the forces of simple diffusion. Initial attachment of virus to cell apparently involves reversible electrostatic forces (reviewed by Dales, 1965, and others). There is some evidence for specific cell receptors.

Animal viruses are often internalized by a process called *viropexis* (Fazekas de St. Groth, 1948). This process is similar to pinocytosis, the mechanism by which a cell "drinks" many of its nutrients. Some authors have used pinocytosis in place of viropexis, but in the strict sense pinocytosis should be reserved for internalizing liquid. Viropexis and phagocytosis are sometimes used interchangeably.

Entry of viruses by a process of fusion will be discussed. In this case, the envelope of the virion or outer structure of nonenveloped particles fuses with the cell membrane forming a single structure and with a subsequent release of the infective unit within the cell.

Harrap and Robertson (1968) showed virus replication in the columnar midgut cells of *A. urticae* NPV 24 hrs after inoculation but without apparent infection of the goblet cells. Virus particles with and without "membranes" ("developmental membrane", (Bergold, 1963) = envelope) were distributed throughout the nucleus but fewer in number than in infected fat cells. Crystallized protein was observed in all nuclei but without occluding progeny virions. Examination of the cytoplasm of these infected cells showed enveloped virus particles lying between the nucleus and the basal cell membrane. Virus was also observed in the basal membrane. Forty-eight hours after inoculation feeding, the tracheolar cells were found infected. In these infected cells, polyhedra were formed occluding progeny virus. Harrap and Robertson concluded that progeny virus produced in the midgut columnar cells was responsible for further infection of the host such as the cells of the tracheal epithelium adjacent to the basal lamina of gut cells which showed various stages of infection. They did not elucidate the exact nature of the initial infection process of columnar cells.

Harrap (1969, 1970) further elucidated the mechanisms of infection of midgut cells of *A. urticae* NPV by demonstrating at the ultrastructural level the presence of enveloped virus at or near the plasma membrane at the apex of columnar cells. He further demonstrated (Fig. 2) the nucleocapsid of the NPV within the confines of columnar cell microvilli. Complete fusion of viral envelopes with the plasma membrane was not observed, but microvilli were often seen bearing balloon-like projections (Fig. 3) from the plasma membrane. Harrap suggested that these projections could be residual virus envelopes seen after the nucleocapsid had passed into the cell cytoplasm. Occasionally, a nucleocapsid was observed at the base of microvilli where it appeared to be uncoating.

Harrap (1970) hypothesized the infectious processes for NPV based on the observations he had made on the NPV of *A. urticae*. The main features of his hypothesis were the adsorption of the virion (enveloped nucleocapsid) at the plasma membrane of the microvillus with subsequent fusion of viral envelope and plasma membrane followed by release of the nucleocapsid into the cell cytoplasm. The lack of incorporation of progeny virions into polyhedron protein is a logical step if one assumes, as did Harrap, that these virions will further infect susceptible tissues. If bound in polyhedron protein, the virions would be for all purposes noninfective for

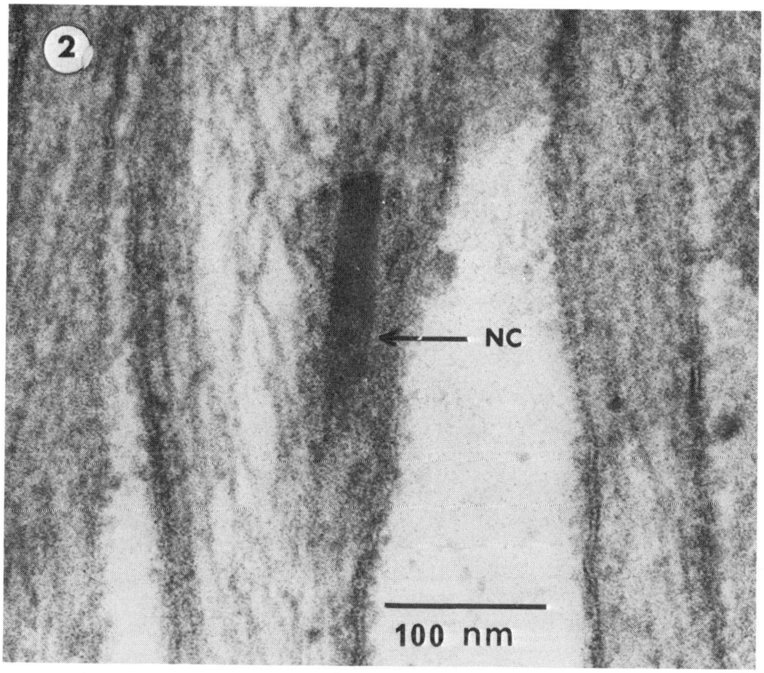

Figure 2. A virus particle (NC) within a columnar cell microvillus. (Courtesy, K. A. Harrap)

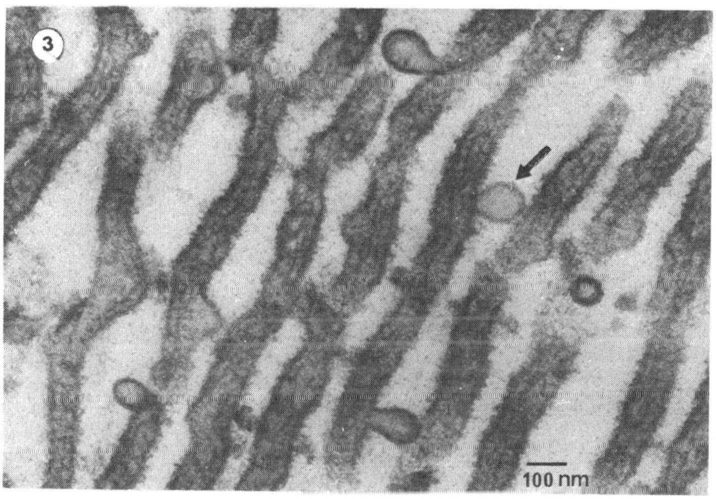

Figure 3. Balloon-like projections (arrowed) on columnar cell microvilli. (Courtesy K. A. Harrap)

other tissues. Harrap's electron micrographs (Figs. 4 and 5) give convincing evidence of the movement of progeny virions from infected nuclei and presumably to cross the plasma membrane (though not demonstrated) to infect other tissues. It was concluded by Harrap that such a mechanism of cell infection supports the view that nuclear polyhedrosis and granulosis virus "particles" are enveloped viruses, the nucleocapsid equivalent to the "intimate membrane" form (Bergold, 1963) and its contents. Harrap believed the enveloped form to be the infective form of the virus and therefore it should be considered as the *virion*.

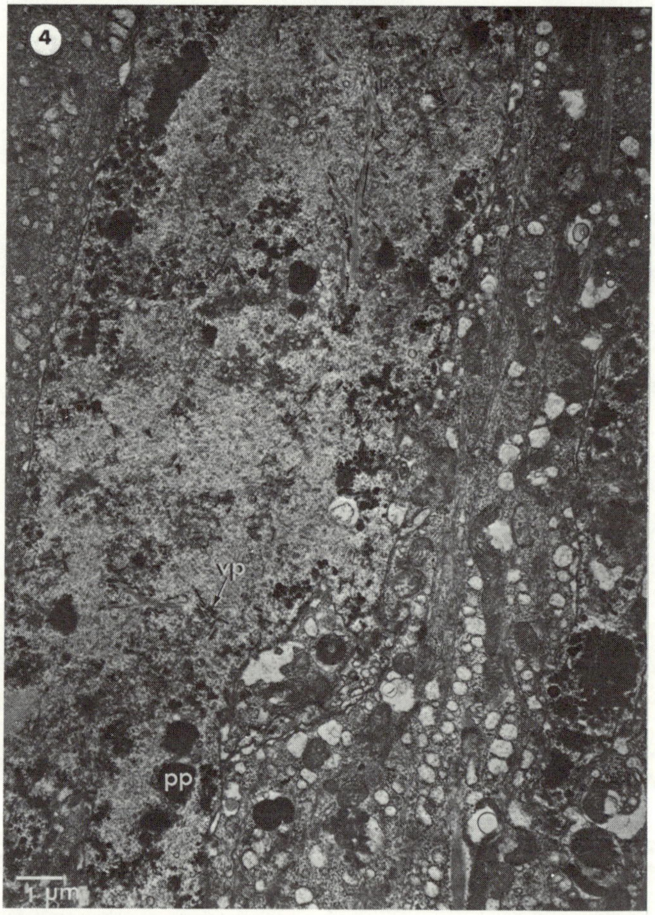

Figure 4. Virus replication in a columnar cell nucleus vp, virus particles; pp, polyhedron protein. Note the lack of virion occlusion by polyhedron protein. (Courtesy K. A. Harrap)

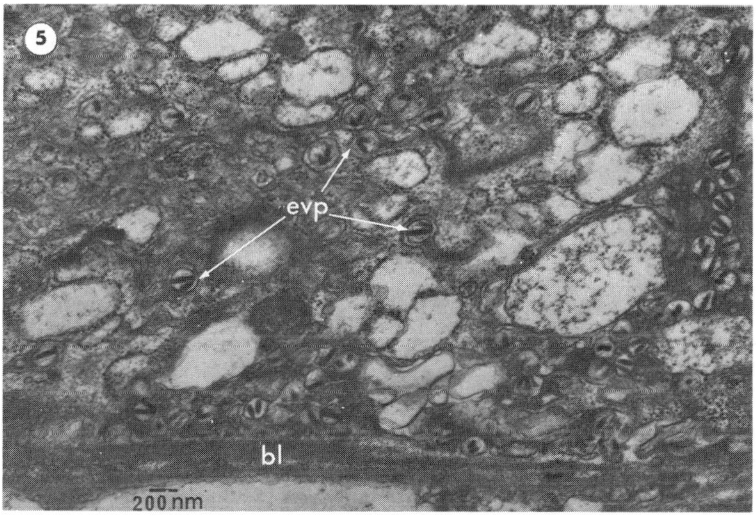

Figure 5. Enveloped virus particle (evp) in the cytoplasm of an infected columnar cell. Note the close proximity of evp to basal lamina (bl). (Courtesy K. A. Harrap)

For a comparison to the conclusions reached by Harrap (1970), Stairs and Ellis (1971), and Khosaka et al. (1971) suggested that the nucleocapsid form of the virus is the infectious factor; however, the virus preparations in these studies were poorly characterized and both used intrahaemocoelic injection instead of *per os* routes as a means of bioassay.

Summers (1969) found that virions of the TNGV were released into the midgut lumen from the occluding protein by dissolution of the protein adjacent to the virion, and in general the crystal separated at each end. Presumably the high pH conditions bring about the dissociation, a condition which may at times cause dissociation of the virion envelope. Summers judged the latter to be rare in the study he conducted. Two hrs post-inoculation, he noted many released virions associated with midgut cell microvilli; phagocytosis was not observed. Summers (1971) believed, even though the evidence was obscure (Fig. 6), that the virus envelope was lost at the cell surface since nucleocapsid was observed in the microvilli (Fig. 7).

Two to six hrs after infection, many nucleocapsids were observed in nonspecific association with the nuclear envelope. However, some appeared to be associated end-on with the nuclear pores (Fig. 8). Structures similar to empty capsids were also associated in a like manner with nuclear pores (Figs. 9-10).

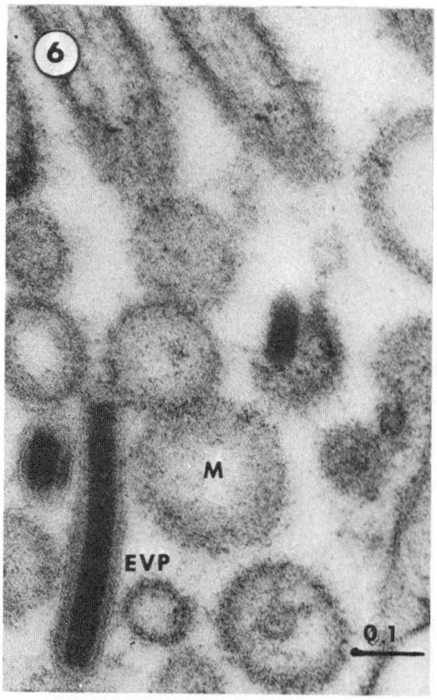

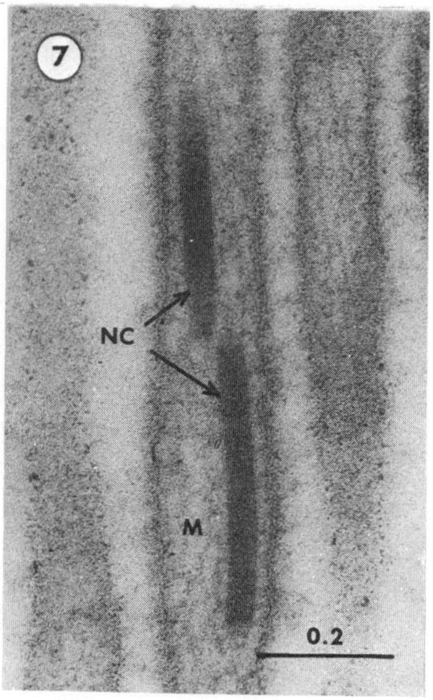

Figure 6. Possible interaction of TNGV enveloped virions (EVP) with the microvilli (M) of the midgut columnar cell of T. ni.

Figure 7. Longitudinal section of midgut microvillus (M) containing the nucleocapsids (NC) of TNGV. Figs. 6-10, scale equals 1.0 = or portions thereof as indicated. (Courtesy M. D. Summers)

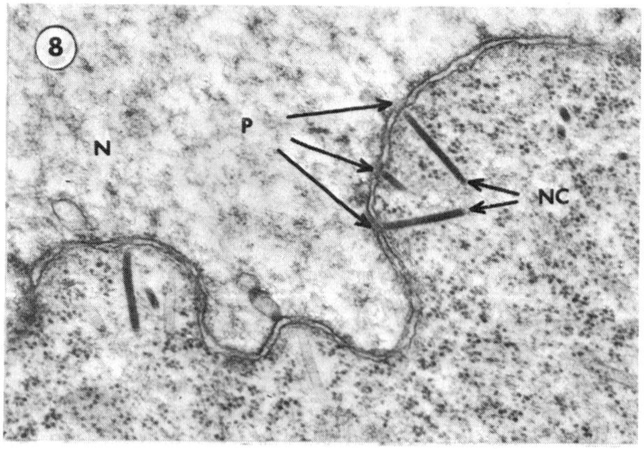

Figure 8. End-on association of intact TNGV nucleocapsids (NC) with the nuclear pores (P) of midgut columnar cells of T. ni. N = nucleus.

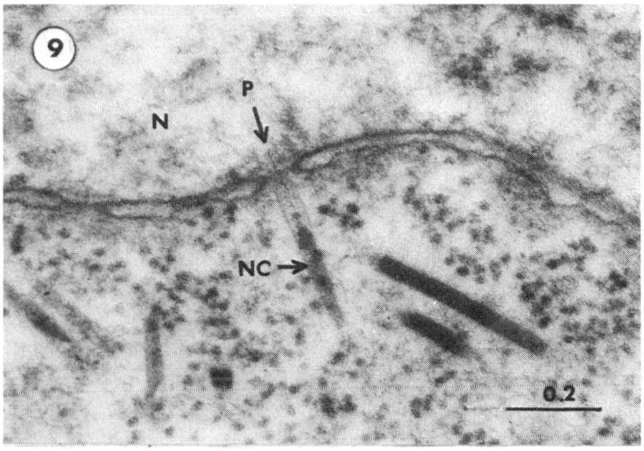

Figure 9. End-on association of partially filled TNGV nucleocapsids (NC) with the nuclear pore (P) of midgut columnar cells of T. ni. N = nucleus. (Courtesy M. D. Summers)

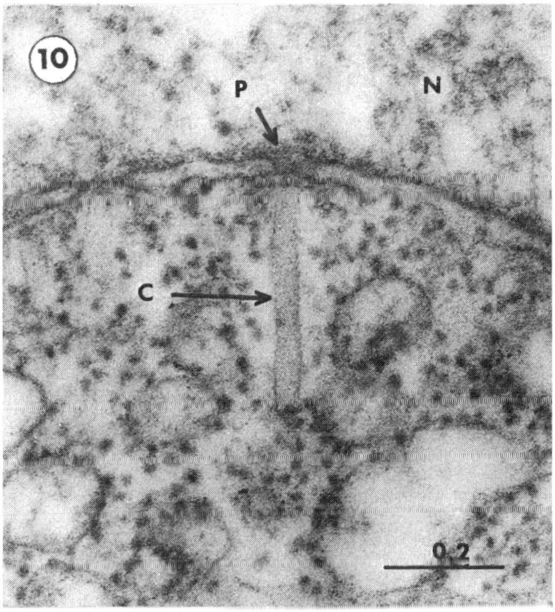

Figure 10. Empty capsid (C) of TNGV associated with the nuclear pore (P) of the midgut epithelial cells of T. ni. N = nucleus. (Courtesy M. D. Summers)

From 6-20 hrs after inoculation, changes in the nucleoplasm suggested the eclipse phase, and by 24 hrs progeny virus was evident. Some progeny virus was contained in vesicles and Summers concluded that virus transport to the basement membrane was accomplished by this means. Presumably these vesicles are passed through the basement membrane and into the haemocoel. Alternatives to this process are: 1) the *de novo* synthesis of membrane components; 2) acquisition of a membrane structure of cytoplasmic origin; or 3) nucleocapsids could be released into the cytoplasm and bud through the plasma membrane of the basal infolding (Smith, 1968) of the midgut cells. "After budding through the infolding, the enveloped particle would be 'sandwiched' between the membrane of the basal infolding, thus appearing to be in the cell when, in fact, it is actually outside of the cell." Summers concluded this latter supposition as the most likely based on the observations he made during the study.

Summers found no ultrastructural evidence of occluded virions or evidence for the biosynthesis of the protein crystal. These observations are in general agreement with those of Harrap and Robertson (1968) for the NPV of *A. urticae*.

Tanada and Leuteneggar (1970) suggested *two* possible pathways for the invasion of the granulosis virus of *Trichoplusia ni*. On the basis of their electron microscope study conducted 24 hrs post-inoculation, they postulated that the occluded virus (TNGV) was released into the midgut lumen after dissolution of the protein surrounding the virions, followed by invasion by movement intercellularly between two columnar cells to the basement membrane, and from there to the haemocoel. In the second process, it was postulated, but without electron microscopic evidence, that virions entered the cell by pinocytosis where they subsequently "multiplied" in the midgut columnar cell nuclei. Infected cell nuclear membranes disintegrated at an early stage and developing virus rods passed into the cytoplasm, especially near the base of the cell. Completed virus rods, singly or as aggregates, passed through the basement membrane into the haemocoel where they infected susceptible tissues. In either process, it was postulated that the virion enters the haemocoel to infect other tissues. Tanada and Leutenegger found capsules (protein crystal occluding granulosis virions) in only one instance and suggested that formation of the protein body was rare in midgut infection by granulosis virus.

Cunningham (1971) in studying the NPV of *Lambdina fiscellaria* noted the lack of polyhedron protein occlusion of mature progeny virions in infected midgut cells. Virions singly or in bundles, in vesicles were found in the cytoplasm of the cells whether or not nuclei were infected. It is unclear from Cunningham's description what the origin of the virions might be. Cunningham presumed that these virions "escaped" into the haemocoel from the midgut epithelium to infect other susceptible tissues.

The evidence for the viral envelope fusion to the plasma membrane of the midgut cells was recently provided by Kawanishi et al. (1972b). He convincingly demonstrated the process in electron micrographs (Figs. 11-14) of RONPV virions fused with the midgut microvillar membrane of *T. ni*, enveloped either as single or multiple nucleocapsids. Nucleocapsids were observed within the microvillus (Fig. 15) as well as in the cytoplasm of these cells (Fig. 16). Enveloped virus was not observed in the cytoplasm during the infection process. RONPV capsids were observed attached to the nuclear envelope (Fig. 17) in much the same manner as described by Summers (1971). Membrane fusion was the only pathway observed in this study and what was presumed to be progeny nucleocapsids were present in the nuclei as early as 4 hrs after exposure to the virus.

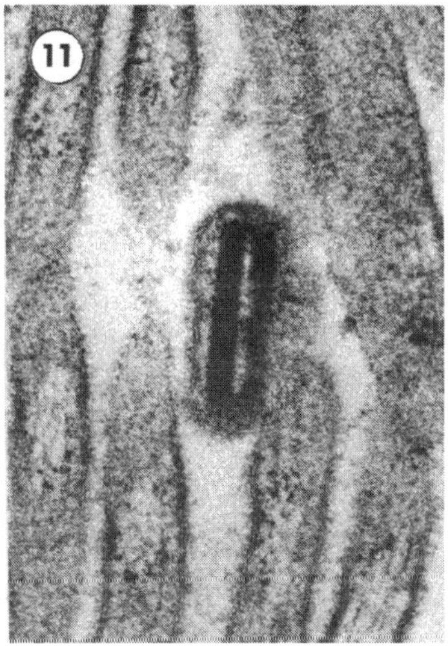

Figure 11. A Rachiplusia ou nuclear polyhedrosis virus (RONPV) particle adjacent to a columnar cell microvillus of the larval midgut of Trichoplusia ni. X 130,000.

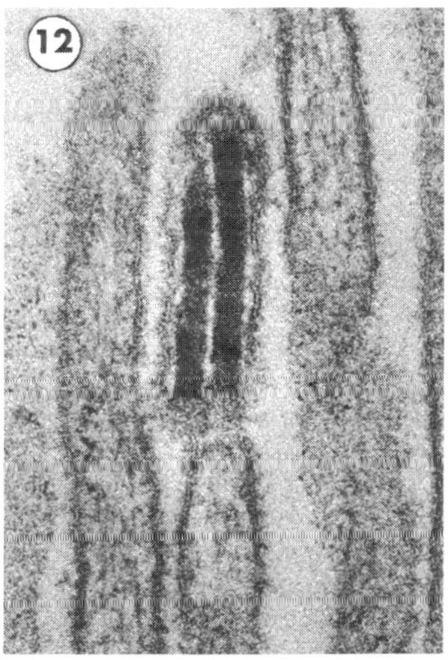

Figure 12. A diffuse juncture between RONPV viral envelope and microvillus membrane resulting from a putative tip-to-tip interaction. X 130,000.

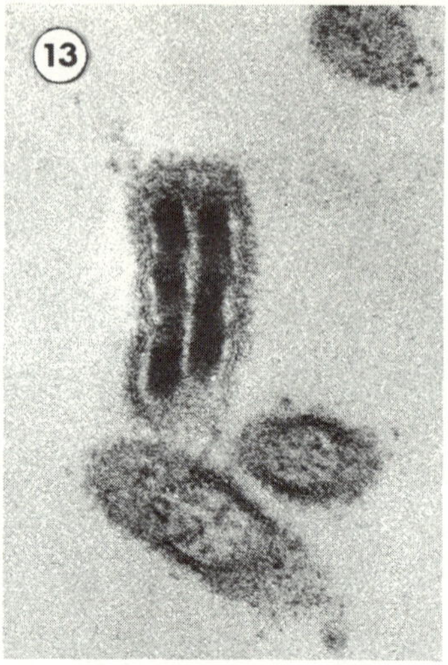

Figure 13. Electron micrograph suggestive of confluence between viral envelope and microvillus membrane. X 130,000.

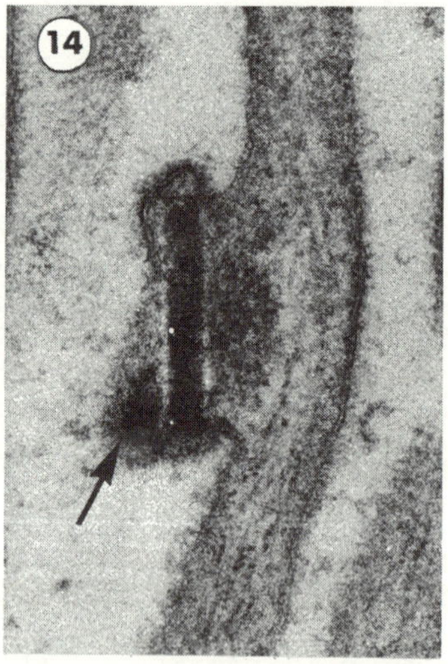

Figure 14. RONPV nucleocapsids in the process of entering microvillus after side-to-side fusion of viral envelope and microvillus membrane. Arrow indicates part of a possible second nucleocapsid present in this section. X 130,000. (Courtesy C. Y. Kawanishi)

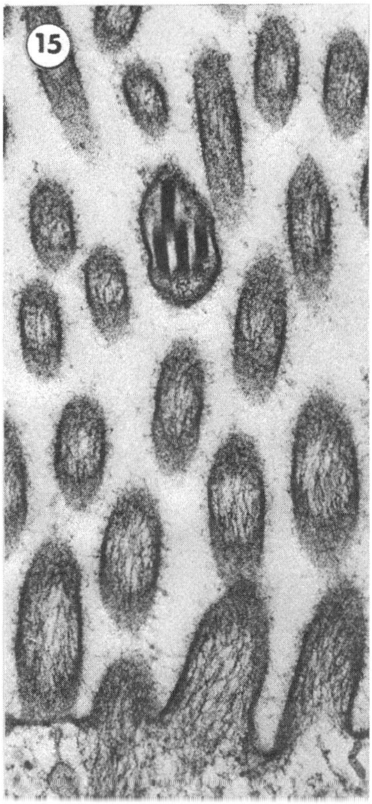

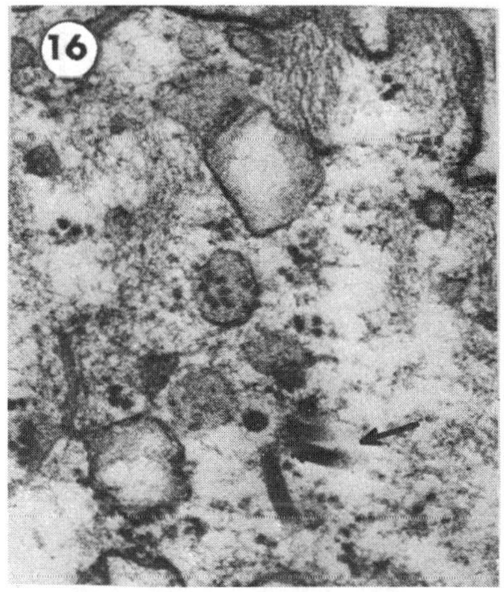

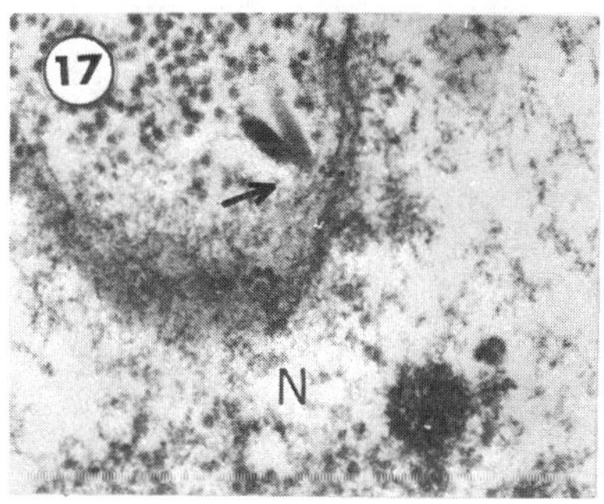

Figure 15. Oblique section of RONPV nucleocapsid assemblage within a microvillus. Apex of columnar cell body is observable at the bottom of the micrograph. X 66,000.

Figure 16. RONPV nucleocapsid assemblage within the cytoplasm of a columnar cell (arrow). Bases of the microvilli are present at the top of the micrograph. X 66,000.

Figure 17. RONPV nucleocapsids (arrow) attached to the basal region of a colmnar cell nucleus (N) within the infoldings of the nuclear envelope. X 66,000. (Courtesy C. Y. Kawanishi)

In other studies, C. Y. Kawanishi and H. J. Arnott (personal communication) found that when late third or early fourth instar *T. ni* larvae were fed RONV and held at 78°F, virogenic stroma were visible in the midgut columnar cell nuclei 18 hrs postinoculation (Fig. 18), and polyhedra at approximately 24 hrs postinoculation. Few virus particles were observed occluded by crystalline protein. Most particles remained unenveloped and many passed from the nuclei into the cell cytoplasm by some unobserved process. Nucleocapsids then "bud" through the plasma membrane or its basal infoldings into the haemocoel (Fig. 19). The particles presumably are responsible for the subsequent infection of haemocoelic tissues.

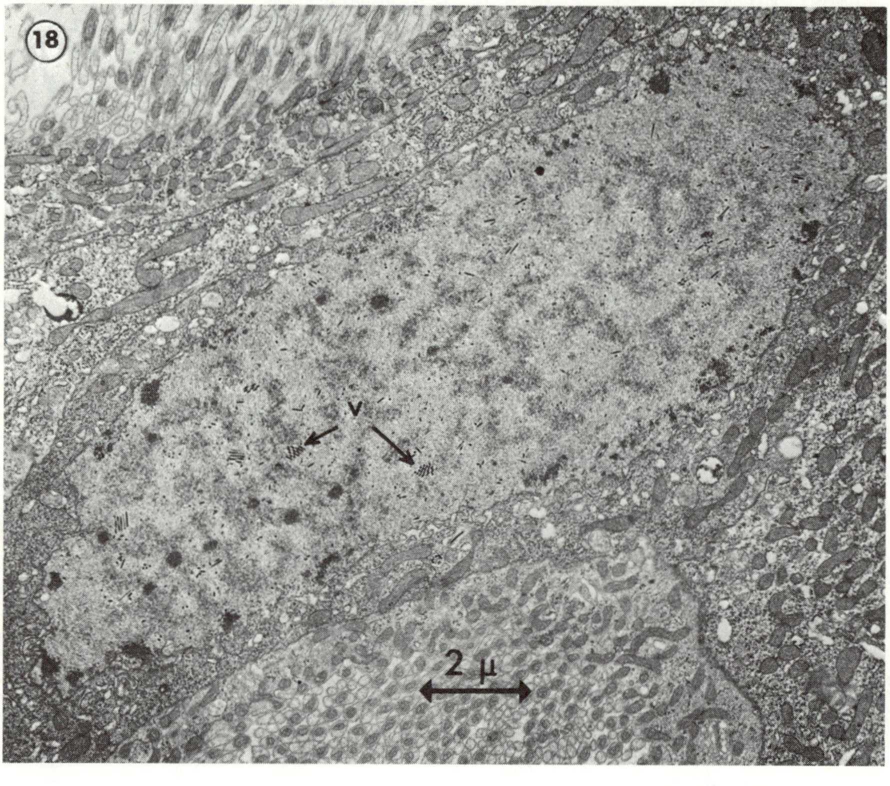

Figure 18. T. ni larvae midgut columnar cell with progeny RONPV (V) 18 hours after inoculation. (Courtesy C. Y. Kawanishi and H. J. Arnott)

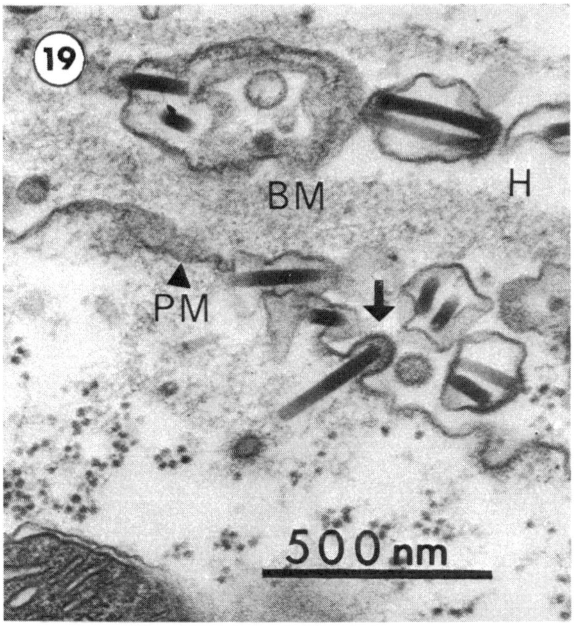

Figure 19. Infected midgut columnar cell with nucleocapsid (arrow) (NC) "budding" through the basal area. BM = basement membrane or basal lamina; H = haemocoel; PM = plasma membrane. (Courtesy C. Y. Kawanishi and A. J. Arnott)

Entry of an insect poxvirus (IPV) into the cells of the midgut of the host by fusion of the virus envelope with the host cell membrane has been demonstrated by Granados (1973a). His study on the entry (Figs. 20-26) of *Amsacta moorei* poxvirus, an insect poxvirus, into the intestinal epithelium of *Estigmene acrea* larvae by fusion of the virus envelope and the microvillar membrane were found to be similar in process as the study conducted by Kawanishi et al., (1972b) with RONPV.

Granados (1973a) inoculated (*per os*) 1-2 μl of a suspension containing 5 X 10^7 inclusion bodies per ml into second and third stage larvae of *E. acrea*. Larvae were sacrificed periodically and the intestines were fixed for electron microscopy. Intact virions were first detected in the gut lumen in the posterior region of the midgut one to two hrs after feeding. The virions were observed adjacent to and closely associated with microvilli of columnar cells (Fig. 20). The region of junction between cell membrane and the virus envelope was not always clear but is easily discernible in Figure 21. Following fusion of the envelope with the membrane of the microvillus, the virus core and portions of the lateral bodies entered into the microvillus (Fig. 22). After fusion, it was apparent that the virus envelope was lost at the time of entry process. Virus entry appeared to take place near the tip of the microvilli.

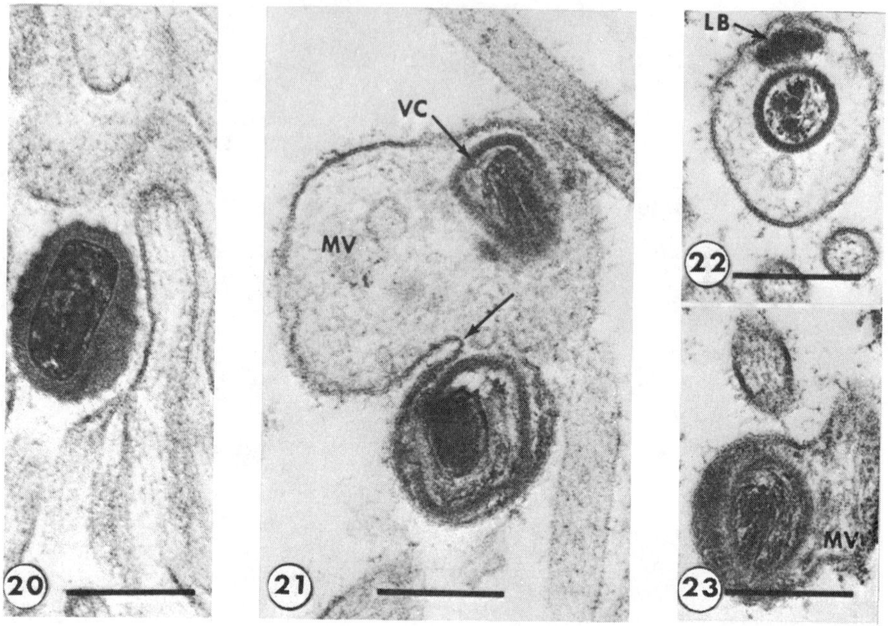

Abbreviations: VC = virus core; MV = microvillus; LB = lateral body. The scale markings are 2500 Å in length.

Figure 20. An Amsacta poxvirus particle among microvilli at the apex of a columnar cell. X 82,000.

Figures 21-26. Virions at various stages of entry into microvilli. All micrographs (Figs. 21-26) were taken from larvae fixed 1-2 hr after inoculation.

Figure 21. Fusion of a virus particle with the microvillus. Note confluence between the viral envelope and the microvillus membrane (arrow). This enlarged portion of the microvillus contains a core from another virus particle. X 82,000.

Figure 22. Transverse section of a microvillus containing a virus core and portion of the lateral body. X 82,000.

Figure 23. Virus core in the process of entering a microvillus after side-to-side fusion of viral envelope and microvillus membrane. X 82,000.

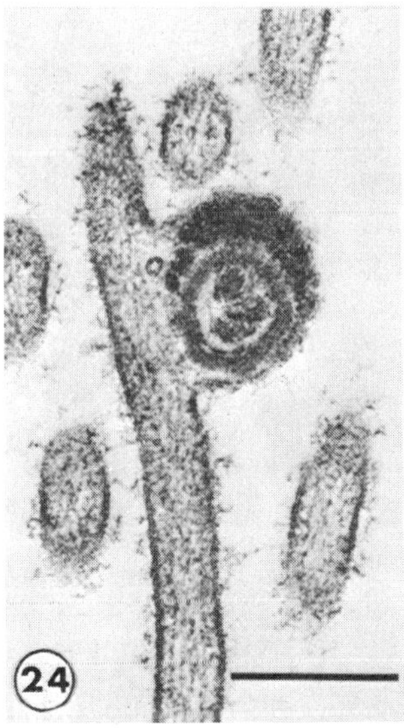

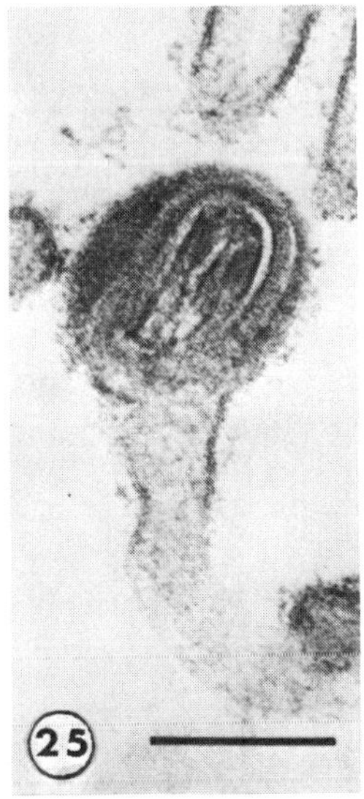

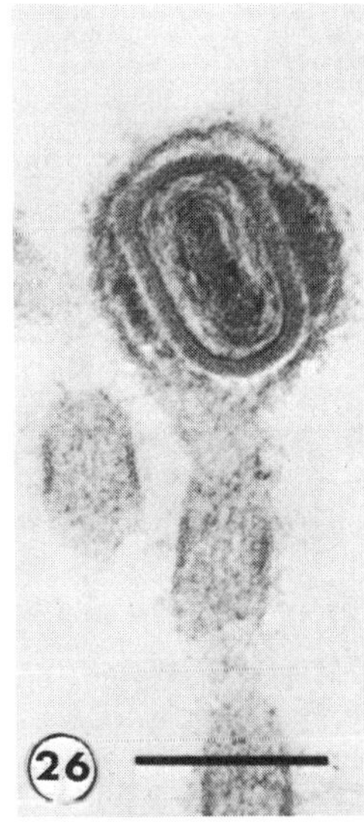

Figure 24. Microvillus-associated virus. The virion appears to be attached laterally to the microvillus. X 82,000.

Figure 25. Virus particle which appears to be fused with the tip of a microvillus. X 82,000.

Figure 26. Another virus particle which appears to be associated with the tip of a microvillus. X 82,000. (Courtesy R. R. Granados)

As many as three viral cores were found in an individual microvillus. Microvilli containing cores were greatly enlarged in the immediate area surrounding these viral structures. This is apparently the result of the relatively large viral "core" (300 nm in length and 150 nm in diameter) entering into the microvillus of only 110 nm in diameter. In contrast to insect poxvirus, vertebrate poxviruses are, in general, only known to enter cells by viropexis (Dales, 1965). In the study conducted by Granados (1973a) viropexis did not occur in the penetration of the virus in the midgut cells. However, Granados (1973b) has observed viropexis (Figs. 27-30) by hemocytes of purified insect poxvirus injected into the haemocoel of *E. acrea*. Similar observations were made by Devauchelle et al., (1971) in their study of the poxvirus of *Melolontha melolontha*. This suggests two mechanisms of entry depending upon the nature of the tissue site and perhaps the form of virus at the site of adsorption.

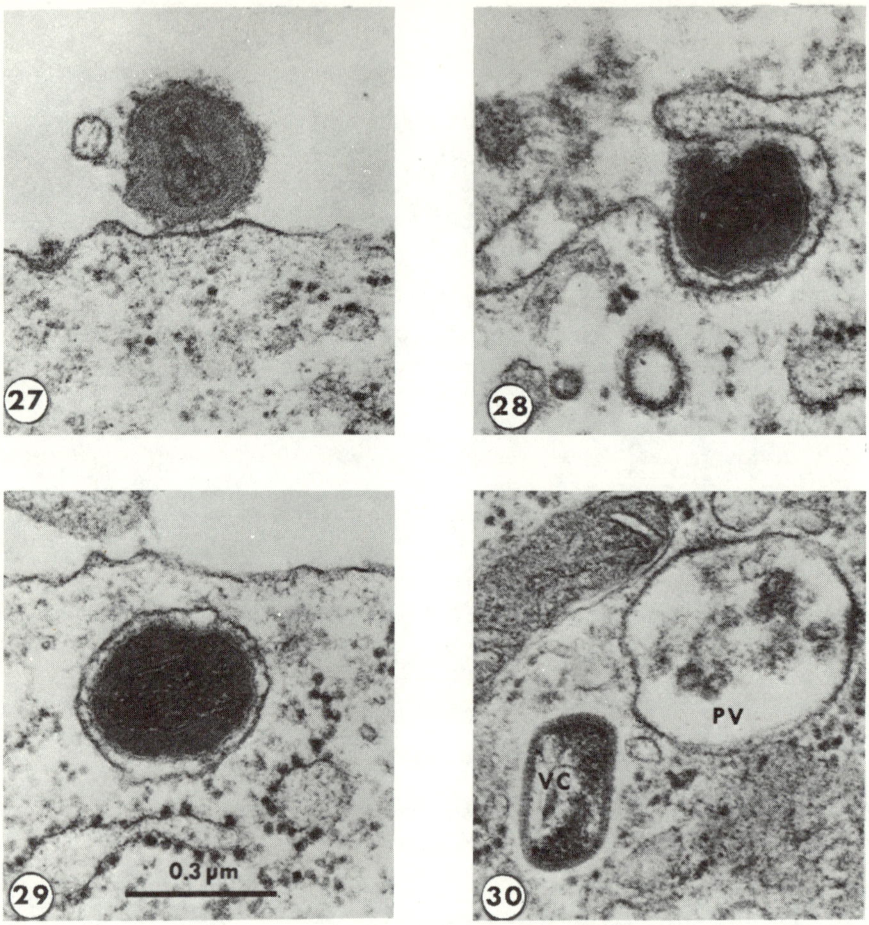

Figures 27-30. Viropexis of Amsacta poxvirus by E. acrea hemocytes. X 82,000.

Figure 27. Attachment of virion to the plasma membrane.

Figure 28. Invagination of the plasma membrane and engulfment of a virus particle.

Figure 29. Virus particle contained within a phagocytic vacuole in the cell cytoplasm.

Figure 30. Virus core (VC) presumably released from an adjacent phagocytic vacuole (PV). (Courtesy R. R. Granados)

Little information is available concerning the adsorption, penetration, and early replication of the cytoplasmic-polyhedrosis viruses in those insects which are infected by these agents. Aruga et al. (1963) (cited by Kobayashi, 1971) studied the invasion route of the cytoplasmic-polyhedrosis viruses (CPV) in the silkworm and noted that the ingested virus penetrated into cells of the midgut in a region posterior to the seventh segment and from this initial location the virus spread from cell to cell towards the cephalic portion of the midgut.

IN VITRO STUDIES

Research in insect virology has been distinctly hampered by the lack of a suitable method of assay as used by the mammalian virologist. The use of tissue culture for the study of mammalian viruses has allowed the rapid advances in virology that have taken place during the past two decades. Even though tissue culture has been available to the insect virologist utilizing established cell lines and/or primary cultures, the ability to infect the cell lines has been discouraging. The CPV's (Grace, 1962; Kawarabata and Hayashi, 1972) and the NPV's (Goodwin et al., 1970; Grace, 1967) have been demonstrated to replicate in established cell lines from insects. Some success has been reported in establishing replication of iridescent viruses in lepidopterous cell lines (Hukuhara and Hashimoto, 1967). The fact that these lines have been inoculated and infected now permits further study of the development of the plaque assay and will allow the study of the mechanisms involved in the infection of cells by the insect viruses.

Insect tissue culture and its use in invertebrate virus research have been recently reviewed by Vago (1971), and it is our intention to review only the advances which have taken place since the publication of those reviews, with the exception of those pertinent papers which were published prior to 1970 and which have bearing on our present discussion.

Studies of insect virus attachment, penetration, uncoating and replication in tissue culture are incomplete for comparison, and we have been awaiting the results of such studies from several laboratories. Vaughn et al. (1972) shows virus particles present in tissue culture cells in what appear to be phagocytic vacuoles. Vaughn used a cell line from *Spodoptera frugiperda* which had been tested for virus susceptibility. In this case, as in the primary cell system, virions from polyhedra of the NPV of the fall armyworm were not typically infectious (no polyhedra formation) for the cells, but hemolymph preparations were (Goodwin et al., 1970). A series of studies was conducted on the sequence of events from virus attachment to virus release for the fall armyworm NPV and *S. frugiperda* cell line. The preliminary studies were designed only to give some reference to the timing and to lay a groundwork for detailed studies planned for the future. In the first series of tests, virions in suspension, prepared by dissolving polyhedra in dilute sodium carbonate, were used and "as expected the typical sequence of events leading to the formation of polyhedra did not occur." However, Vaughn did note interaction between the cells and the virions. Within 15 to 20 minutes after inoculation, virions were seen entering the cell through the cell membrane. The process resembled phagocytosis but too few observations were made to be certain that this phenomenon does occur or that this is how the cells in reality, are infected. Twelve hours after inoculation, numerous rods were found enclosed in vacuoles within the cytoplasm of these cells. The vacuoles contained one to several virions and in addition some virions were found free in the cytoplasm. A few cells were seen which appeared to have virions in the nuclei, but there was not clear evidence of viral replication. No polyhedra were ever seen in any cells when the cultures were inoculated with virion suspensions.

When Vaughn inoculated cultures with hemolymph collected from a diseased larva, none of the above early stages were observed. The EM sections did not contain extracellular virions although the hemolymph had an LD_{50} titre of $10^{-3.5}$ which indicated a large amount of infectious material. No attachment to the cell membranes was observed, nor were virus particles identified in the cytoplasm. Within 20 hrs after inoculation, replication of virions was noted in some cell nuclei, and by 44 hrs polyhedra were seen in the nuclei of infected cells.

It is not clear from Vaughn's discussion whether he was dealing with purified virions from the fall armyworm NPV or whether the particles referred to in this case as virions were the nucleocapsid form of the virus. If Vaughn were dealing with the nucleocapsid of purified preparations of NPV then perhaps this might explain the low infectivity of his viral suspension; on the other hand, when dealing with hemolymph, he undoubtedly had the enveloped form of the virus, that is, the virion, which could account for the higher infectivity since it may be that envelope of the virion is necessary in attachment at the cell membrane to promote infection.

Vaughn considers that the improved culture medium used for the propagation of the *S. frugiperda* cells is responsible for the considerably improved viral replication observed in this cell line. A number of factors as pointed out above, in addition to the improved culture medium, may have been responsible.

During the past two years, attempts have been made to infect mosquito cell lines with MIV, and only recently has infection of the cell lines been accomplished (Webb et al., 1974).

Singh's and Peleg's *A. aegypti* cells were maintained at 25° on M and M medium containing 100 units of penicillin, 100 micrograms streptomycin, and 0.25 micrograms of Fungizone per ml. Approximately 5 X 10^5 cells were explanted to 30 milliliter T-flats and allowed to grow for 2-3 days. Flats were used for inoculation before there was appreciable contact inhibition. Purified virus of either the T or R strains was filtered through a 0.45 Millipore filter after which the concentration of virus was adjusted to 0.1 OD unit per ml. Aliquots of 0.1 ml of the suspension, or a decimal dilution, were added directly to the existing medium in the flats and the cells were incubated at 21, 25, 31, and 35°C.

Inspection of the cells by light microscopy revealed that those incubated at the elevated temperatures elicited a dose dependent cytopathological effect (CPE) within three days after inoculation, while those incubated at lower temperatures appeared normal. No CPE was evident at 21°C and evidence for a reproducible CPE at 25°C was inconclusive. The CPE at 31°C was initially manifested as a slight granular appearance and increased vacuolization of the cytoplasm, followed by a tendency for the adjacent cells to form small clusters. Incubation beyond 3 days resulted in formation of tenuous cytoplasmic filaments, frequently containing refractile bodies, that appeared to adhere to surrounding clusters or individual cells. Subsequently the cells became refractile and rounded and detached from the substrate. Destruction of the cell monolayer was complete by 7-8 days.

Electron microscope observation of cells which had been fixed and prepared for ultrathin sectioning denoted the presence of MIV virions in cells which were inoculated and incubated for 2.5 days at 31°. These virions were found in an area of the cell cytoplasm which was relatively void of cytoplasmic organelles and which can be considered as the viroplasm. In addition to these foci of infection, it was noted that the virions in many cases were found in close proximity to the plasma membrane, whether at the external surface of the cell, or near one of the numerous vacuoles present in many of the cells. Protrusions of the plasma membrane, probably homologous to the cytoplasmic filaments noted by light microscopy, appeared to be a significant characteristic of the CPE. It was also evident that chromatin material concentrated near the periphery of the nuclei at the nuclear envelope and formed a dense ring. Whether or not this is a characteristic of the CPE has not been conclusively determined. Similar results by light and electron microscopy were obtained when the T strain was inoculated in cultures of Singh's *A. aegypti* cells.

Peleg's *A. aegypti* cells exhibited similar CPE characteristics as Singh's cells and with both strains of MIV. However, the earlier appearances of cytoplasmic filaments could not be consistently used for a criterion for infection of this more elongated cell type. Electron microscopic observation of cells incubated for 2.5 days at 31° with either a T or R strain of MIV was similar.

Preliminary investigation (S. R. Webb, unpublished) of the early events of viral attachment, penetration, uncoating, and replication have been initiated. Five minutes after inoculation virions can be detected at the cell membrane and in some cases these virions appeared to be undergoing a process of viropexis. Virions enclosed within membranes, that is, within a vacuole, were also observed in the cell cytoplasm 5 minutes postinoculation. At 15 minutes postinoculation, single or groups of virions in vacuoles were observed in the cytoplasm as well as some virions which were not surrounded by a membrane. These virions had either lost the membrane or had been absorbed without acquiring a membrane. It appears, therefore, that the process of attachment and penetration to susceptible cell sites takes place very rapidly and that viropexis is the main mechanism by which the virions of MIV enter into the cell. The observations made to date are in many respects much like those described by Leutenegger (1967) in her *in vivo* study of Sericesthis Iridescent Virus (SIV) infection in hemocytes of *Galleria mellonella*. Similar results were reported by Younghusband and Lee (1969) in their studies of Tipula Iridescent Virus (TIV) and its interaction with the haemocytes of *Galleria mellonella*.

Additional references to the infection of tissue culture cells, without information regarding infectious processes, can be found in the recent literature. Some of these are Goodwin et al., 1970; Ignoffo et al., 1971; Faulkner and Henderson, 1972;

Kawarabata and Hayashi, 1972; Sohi and Cunningham, 1972; and Vail et al., 1973. None of these listed references gives a detailed account of the early events of infection nor do they go into the detailed accounts of replication processes.

Kobayashi (1971) conducted a series of studies on the entry of the CPV into silkworm midgut epithelial cells cultured *in vitro* (organ culture). Purified virus particles at a concentration that corresponded to 10^{10} polyhedra per ml were added to the culture medium. Cultures were rocked gently at 20°C and at intervals of 10, 20, 30 and 60 minutes, the cells were harvested for electron microscopy. Electron micrographs of the samples showed many virus particles attached to the cell and indicated the various stages of the process of entry, the first stage of which appeared to be the attachment of the viral projection to the surface of a host cell. In the second stage, the virus particles seemed to attach more closely to the cell surface and appeared to penetrate through the projection into the cytoplasm. The final stage, the beginning of which was evident within 10 minutes after the inoculation of the virus, was characterized by the release of a core substance of the cell. This was suggested by the presence of empty particles consisting of intact outer shells on the cell surface. Kobayashi concluded that the core substance appeared to be released as a filament and injected into the cell, and he also concluded that phagocytosis did not play a part in virus penetration. The projection through which the core material entered the cell did not contact during the penetration stage and Kobayashi drew an analogy between the CPV and the injection of nucleic acid and that of the bacteriophages which do not necessarily inject nucleic acid by contraction of the tail sheath.

Hosaka and Aizawa (1964) have shown the "empty particles" have only an outer shell in contrast to intact particles of CPV which have two protein coats, an outer and inner shell, and that the release of the core substance from the intact virion results in an empty shell (Hosaka, 1964, as cited by Kobayashi, 1971). Based on these observations, Kobayashi (1971) believed that the inner shell of the CPV plays an important role in the transfer of the core substance from virus to cell. He made no attempt to determine the mechanism of ejection but assumed that the inner shell of the CPV to be an elastic "membrane" and that its contraction brings about the ejection of the core substance as shown in the case of the T25 phage (Cummings and Kozloff, 1962).

On this basis, Kobayashi suggests that CPV infects the *in vitro* cell in a manner analogous to the process in bateriophages. However, he raised the question of how CPV's pass through the peritrophic membrane, through which cells, at what specific cell site, and how the mechanism of cell entry is controlled.

DISCUSSION

In the foregoing sections, we have considered the natural barriers which can possibly inhibit the *per os* infection by viruses of their respective hosts. We have examined the effect of the environment of the midgut lumen on the occluding protein crystal of the polyhedrosis and granulosis viruses and also the effects of this environment on the virions freed from their protective coating of inclusion body protein. The virions are susceptible to the drastic conditions found in insect midgut and Summers (unpublished) estimated a 50-60% reduction in infectivity based on his observations of some of the NPV's and GV's of lepidoptra. Campbell (W. R. Campbell, unpublished) did not find that MIV degraded to the same extent as that observed by Stoltz and Summers (1971) nor did Campbell note intact virions only in the forepart of the midgut of *Aedes taeniorhynchus.* Further more, capsules of the SFGV were not degraded to the same extent as observed by Stoltz and Summers (1971). The need for quantitative studies to elucidate the amount of degradation of the virions, and the mechanisms by which degradation takes place, is obvious.

The peritrophic membrane (PM) presents a further barrier to infection of the midgut cells. In lepidopteran species (as well as other insects with the same type of PM), the PM probably does not play as significant a role in inhibition of virion transport as in those cases where the PM is a very tight laminar network as is illustrated in the mosquitoes. It seems unlikely that a virion such as MIV could passively cross the fully formed PM of the mosquito host; there is no unequivical evidence to date, that transport across the PM takes place either passively or by some active mechanism involving the virus.

However, it has been observed (Paschke, unpublished; Richards and Richards, 1971; Stoltz and Summers, 1971) that in the forepart of the mosquito midgut, near the junction with the foregut, the PM is secreted by the microvillate midgut cells in the form of granular secretory material between the microvilli. Slightly further posteriorly the thickness of this layer increases and covers the microvilli. Further to the posterior the newly forming PM becomes more granular but with no visible fibers in the first instar larvae.

MIV has been observed in the anterior area of the midgut near the region described above (Paschke, unpublished; Stoltz and Summers, 1971). It seems possible that contact between exposed microvilli and the infectious virions could take place in this region of the gut. Virions have been detected in the cell side of the PM in a number of observations made using the electron microscope (W. R. Campbell, unpublished).

Once the PM barrier has been surmounted, the virions may come into contact with a third natural barrier to infection, the microvillar plasma membrane into the midgut epithelial cells. It appears that transport across the cell membrane may take place via two and possibly three mechanisms.

The first, suggested by Harrap, 1970, and Summers, 1971, and which has been demonstrated with an NPV and also with an insect poxvirus (Kawanishi et al., 1972b; Granados, 1973a), is fusion of the viral envelope with the cell membrane after which the microvillus "takes up" the capsidform of the virus. The nucleocapsid is subsequently transported within the cell to sites of replication. In the case of the GV's, it has been demonstrated (Summers, 1971) that the nucleocapsid aligns itself with the nuclear pores for uncoating (Dales, 1973).

Vectorial movement of the nucleocapsids has not been studied thoroughly in the case of insect viruses. The lack of suitable tissue cultures for the growth of these viruses has hampered such studies. It has been shown by Dales and Chardonnet (1973) that vectorial movement of adenovirus in infected HeLa cells is provided by pathways of intracellular microtubules. Is it likely that such a mechanism is responsible in the virus infected insect cells?

Tanada and Leutenegger (1970) suggested, without evidence, that midgut epithelial cells became infected with TNGV by pinocytosis. They additionally suggested that the midgut barrier does not necessarily become infected and that the free "virus rods" entering the junction between two columnar epithelial cells subsequently move along the intercellular space into the basement membrane and from there to the haemocoel. The evidence for such mechanisms of infection is not convincing, particularly in the light of the more recent studies of Granados (1973) and Kawanishi et al., (1972b).

However, it should be pointed out that the midgut columnar epithelial cells exhibit pyncytosis at the base of the cell between the microvilli (Smith, 1968). Perhaps it is possible that entry of viruses into these midgut cells takes place through the process of viropexsis at this location on the cell plasma membrane.

No information is available concerning the mechanisms by which other insect viruses bring about infection of the host. Only *in vitro* studies and studies involving insect haemocytes provide us with possible mechanisms which may or may not play a role in the *in vivo* system.

Viropexis of Sericesthis Iridescent Virus (SIV) has been observed by Leutenegger (1967) after intrahaemocoelic inoculation of the virus. In similar studies, Younghusband and Lee (1969) noted the same phenomenon with the Tipula Iridescent Virus (TIV), and Granados (1973) with the insect poxviruses. Infection of haemocytes by a process of viropexis does not seem surprising since these cells are active phagocytes.

Viropexis has been observed by S. R. Webb (unpublished) as the *in vitro* process by which MIV enters the cells. The process apparently occurs in much the same manner as in the case of animal viruses as described by Dales (1973).

In Dales' (1973) opinion, viropexis is apparently the predominant, but not exclusive, mechanism for internalizing many animal agents. He cites the above referenced insect viruses, and the case of Sendai virus, one among the paramyxoviruses, as unequivocal demonstration that fusion occurs between the viral envelope and the plasma membrane. Dales (1973) further explains that other agents such as poxviruses and adenoviruses do not lack the capacity to lyse cells and that the virion-fusing factor may vary between different virus-cell interactions. "Thus determination as to whether fusion or viropexis occurs externally may be the consequence of a race between expression of the lytic principal of the virus and the phagocytic response of the host cell to the presence of a foreign macromolecular complex on its surface" (Dales, 1973). He further suggests that in the case of virus-cell interactions, the "non-de-formable" cell membrane such as microvilli will exhibit fusion with virions possessing powerful lysins. On the other hand, the same agents become phagocytized at regions where the host membrane is more plastic.

Previous studies of animal viruses using tissue culture cells have shown that the ability of viruses to gain successful entrance into cells is intimately associated with components of the virus envelope and the cell membrane surface structure and composition (Dourmashkin and Tyrell, 1970; Koch and Volpe, 1971).

Viruses of different groups are constructed and replicated in different manners, and it should be expected that they may also enter cells in different ways. It is important to know which mechanism of entry relates to each virus and whether or not the virus can initiate infection. One type of virus, or a portion of a virus, may enter cells in differing ways (Dourmashkin and Tyrell, 1970).

Thus, the infection of *Trichoplusia ni* by *per os* inoculation using an infectious DNA preparation as an inoculum may present a third means by which infection may proceed (Summers and Paschke, 1970). It is the fate and the role of these DNA's in the midgut lumen which requires attention, not by electron microscopy, but by biochemical techniques.

It is possible that we are dealing with all three processes in the *in vivo* infection of midgut cells and perhaps the conclusions drawn by Stairs and Ellis (1971) and Khosaka et al. (1971) who suggested that the infectious factor is non-enveloped forms of the virus provides us with yet a fourth alternative.

However, the evidence to date supports the hypothesis that in the case of the enveloped insect viruses, NPV's and GV's as well as the insect poxvirus, the normal *in vivo* pathway is most likely the fusion of the envelope with the plasma membrane of the midgut epithelial cells.

Unfortunately little information is available concerning the mechanisms of uncoating in the midgut cells of the insects. The observations of Summers (1969, 1971) on TNGV nucleocapsids aligning at the nuclear pore is the most significant. He showed nucleocapsids in "full" and "empty" states which suggests that the viral genome is released directly through the pore into the nucleus. Such an interaction is analogous to adenovirus 5 & 7 which uncoat at the nuclear pore (Morgan et al., 1969; Chardonnet and Dales, 1970; and Dales, 1973).

In addition, there is a paucity of information regarding the biochemistry of replication and the lack of occlusion of progeny virion even though inclusion body protein is crystalized in the infected midgut cells. Furthermore, little is known of the mechanisms involved in infection of the other tissues. Subsequent infection of other tissue cell types by progeny virus from the midgut cells may be accomplished through release of these progeny virions through a process of budding at the basal membrane where the progeny virus becomes enveloped with a portion of the cell membrane as has been suggested by Appleyard et al. (1971), Devauchelle et al. (1971), Harrap (1970), Injac et al. (1971), C. Y. Kawanishi and H. J. Arnott (unpublished), and Summers (1971).

The fate of the membrane associated virus which is budded from the basal membrane into the haemolymph of the infected insect has not been determined. Perhaps it is of no consequence since infectious nucleic acid released from the infected midgut cells may be responsible or capable of infecting other tissues. This would mean that in the insects it is the infectious nucleic acid which may contain specificity for designating a host cell type for infection. However, it is logical at this time to suggest that perhaps this specificity is designated by information encoded at the direction of the virus, in the vesicle as it is passed from the midgut cell into the haemolymph of the insect.

SPECIFICITY

Insect virology is an area which at one time was narrowly defined within its scope, and its use was directed toward the biological control of insect pests. Because of this narrow definition, and a subsequent lack of attention to careful basic studies on virus characterization and host interactions, it was generally concluded, and still is by many, that the specificity of this group of viruses was just as narrowly defined. With the development of research in the area of general virology, the term specificity now has a more far-reaching and complicated meaning from the biochemical and molecular standpoint than it did a few years ago. We can also extend this to the manifestation of disease in organisms.

In insects, we are finding more and more non-occluded viruses with structural properties similar to those of the vertebrate viruses. The crystalline array virus (Jutila, et al., 1970), and *Gonometa* virus (Longworth, et al., 1973) are similar to the picornavirus group, e.g. poliovirus. The densonucleosis virus of *Galleria mellonella* (Kurstak and Côte, 1969) and those of other insects (Tinsley and Longworth, 1973) are similar to the adeno-associated viruses and other viruses of the parvovirus group. The insect poxviruses and their structural relationship to the vertebrate poxviruses does not need additional explanation. The insect cytoplasmic-polyhedrosis viruses and their structural similarity to rice dwarf and wound tumor viruses, plant viruses, and the reovirus of vertebrates, plus the presence in an RNA polymerase nicely links all of these together (Lewandowski, et al., 1971). Mosquito Iridescent Virus and *Tipula* Iridescent Virus are remarkably similar to those icosahedral cytoplasmic deoxyriboviruses of fish, frogs, octupus, green algae, fungi, and the African swine fever virus of vertebrates (Stoltz, 1971). The occluded insect viruses characterized by the occlusion of the virion within a protein crystal were, until recently, considered in a class by themselves. However, there are some notable exceptions to this; there is a group of not widely known bacterial viruses which contain DNA, and are rod shaped with structural similarities like those of granulosis and nuclear polyhedrosis nucleocapsids of ticks and mites (Bradley, 1967). And more recently the discovery of an occluded virus in the infected cells of the hepatopancreas of the pink shrimp has been observed (Couch, 1974). This later discovery more or less removes the "uniqueness" from the occluded insect viruses that they once so advantageously enjoyed.

We believe in view of present day knowledge concerning the infectious processes by viruses that we need to carefully reconfirm the specificity of insect viruses to those previously narrowly defined hosts. Many persons interested in the utilization of insect viruses for biological control are confident that the boundaries of host specificity do not overlap, especially to the vertebrates, but let us be forewarned by such reports in the literature (Longworth et al., 1972) of the detection in vertebrates of IgM antibodies specific to an insect picornavirus, a virus structurally similar to poliovirus. The densonucleosis of Galleria is reported (Kurstak et al., 1969) to infect and perhaps transform mouse cells. Also, there has been a report of occluded insect viruses infecting *Bacillus* species of bacteria (Wells and Heimpel, 1970).

Before we can ascertain the biological potential of insect viruses for biological control, as well as "crossing over", it is imperative that we have a positive identification of these viruses before we can compare and properly describe their structural relatedness similar to those of vertebrate viruses. Unfortunately we are a long way from the extensive kind and quantity of work that needs to yet be painstakingly carried out using systems sensitive enough to give significant results.

More importantly, insect virology has a future and potential perhaps unparalleled as compared to other areas of virology. We have an even greater obligation and commitment than to limit ourselves to biological control. The role that medically important vectors have played as mediators of disease and disease organisms throughout the world, and the impact they have had on the total environment and mankind, cannot be denied. It is interesting, in view of this, that very little is yet known about the more specific mechanisms by which mosquitoes and other vectors transmit arboviruses. In addition, let us look at the opportunity to contribute to the science of virology as a whole. It is obvious that the insect viruses, and virus systems, have not been employed to the greatest extent. Because of this, insect viruses should, and can be, used to explore the more interesting and more relevant relationships between viruses and cells as they exist under the more complicated circumstances of an *in vivo* model system and compared to that of the *in vitro* system.

Many studies and comparisons of disease processes are limited to vertebrates and vertebrate tissue culture systems. It seems reasonable that by not giving attention to similar studies in the area of invertebrate virology, valuable information to be derived from the explorations of fundamental mechanisms of infectious processes and pathology common to all species of living forms are being overlooked. The numbers of insect viruses with structural, biochemical and biophysical similarities to those of the vertebrates are becoming more and more numerous as more investigations are made on the invertebrate disease processes. The phenomena of latency of viral infection has been suspended for many years in insects with some most interesting case and experimental histories presented. It seems reasonable that this might be an excellent area for comparison of basic studies on the relationships of viruses and their hosts under persistent or transformed situations. Another area which has not been investigated satisfactorily is that of transovarial transmission of virus from the female through the egg to her offspring.

Although differences are present, the epithelial cells of the midgut of the insect are remarkably similar in many aspects to the epithelial digestive cells of vertebrates, and in addition, their functions are similar to those of the vertebrates. Recent studies, as we have pointed out above, in insects, have elaborated on the processes by which some insect viruses are able to invade and infect insect hosts. It seems likely that similar or comparative processes may be involved with those in the vertebrates as they invade and infect a host. The importance of such studies should be obvious. Unfortunately, there is a paucity of information on the invertebrate side of virus research. It is most important that we investigate and compare interactions using the invertebrate model and the vertebrate systems so that we can make serious attempts to understand the mechanisms of specificity. Such a careful extension of a comparative virological approach undoubtedly will yield important information which can be utilized at the applied level, whether medical or for the use of viruses in pest control. Insect virologists must carry out their work with a degree of sophistication and sophisticated techniques utilized by the vertebrate virologist. There is a lack of attention to the area of comparative virology and therefore a lack of interest for working with, and characterizing, the invertebrate virus systems. Still within the framework of these meetings we are stimulating and providing the justified basis for additional scientific effort in this area. Without this stimulation, the productivity of insect virology is not going to progress rapidly enough to allow us to provide and utilize the information base for the progress that is badly needed.

ACKNOWLEDGEMENTS

The authors wish to express their gratitude for contribution of unpublished results to the following: W. R. Campbell and S. R. Webb, Department of Entomology, Purdue University, West Lafayette, Indiana; C. Y. Kawanishi and H. J. Arnott, formerly with the Cell Research Institute, Department of Botany, The University of Texas. In addition, we would like to thank the following for allowing us to reproduce figures from their published results: R. R. Granados, Boyce Thompson Institute; K. A. Harrap, NERC, Unit of Invertebrate Virology, Oxford, U. K. and C. Y. Kawanishi.

REFERENCES

Aizawa, K. (1953). *Jap. J. Appl. Entomol. Zool. 17,* 145.

Aizawa, K. and Iida, S. (1963). *J. Invertebr. Pathol. 5,* 344.

Appleyard, J., Hapel, A. J., and Boutler, E. A. (1971). *J. Gen. Virol. 13,* 9.

Aruga, H., Yoshitake, N., Hukuhara, T., and Owada, M. (1963). *J. Sericult. Sci.* Japan *32,* 58.

Bergold, G. (1947). *Z. f. Naturforsch. 26,* 122.

Bergold, G. (1948). *Z. f. Naturforsch. 36,* 338.

Bergold, G. (1959). *In* "The Viruses" (F. M. Burnett and W. M. Stanley, eds.), Vol. I, pp. 505-523. Academic Press, New York.

Bergold, G. (1963). *In* "Insect Pathology, an Advanced Treatise" (E. A. Steinhaus, ed.). Vol. I, pp. 413-456. Academic Press, New York.

Caspar, D. L. D., Dulbecco, R., Klug, A., Leuoff, A., Stoker, M. G. P., Tournier, P., and Wildy, P. (1962). *Cold Spring Harbor Symp. on Quant. Biol. 27*, 49.

Chardonnet, Y., and Dales, S. (1970). *Virol. 40*, 478.

Cummings, D. J., and Kozloff. (1962). *J. Mol. Biol. 5*, 50.

Cunningham, J. C. (1971). *Can. J. Microbiol. 17*, 69.

Dales, S. (1965). *Prgr. Med. Virol. 7*, 1.

Dales, S. (1973). *Bacteriol. Rev. 37*, 103.

Dales, S., and Chardonnet, Y. (1973). *Virol. 56*, 465.

Day, M. F., Common, I. F. B., Farant, J. L., and Potter, C. (1953). *Aust. J. Biol. Sci. 6*, 574.

Devauchelle, G., Bergoin, M., and Vago, C. (1961). *J. Ultrastruct. Res. 37*, 301.

Dourmashkin, R. R., and Tyrell, D. A. J. (1970). *J. Gen. Virol. 9*, 77.

Egawa, K., and Summers, M. D. (1972). *J. Invertebr. Pathol. 19*, 395.

Estes, Z. E., and Faust, R. M. (1966). *J. Invertebr. Pathol. 8*, 141.

Faulfner, P. (1962). *Virol. 16*, 479.

Faulkner, P., and Henderson, J. F. (1972). *Virol. 50*, 920.

Faust, R. M., and Adams, J. R. (1966). *J. Invertebr. Pathol. 8*, 526.

Fazekas de St. Groth, S. (1948). *Nature 162*, 294.

Goodwin, R. H., Vaughn, J. L., Adams, J. R., and Louloudes, S. J. (1970). *J. Invertebr. Pathol. 16*, 284.

Grace, T. D. C. (1962). *Virol. 18*, 33.

Grace, T. D. C. (1967). *In Vitro 3*, 104.

Granados, R. R. (1973a). *Virol. 52*, 305.

Granados, R. R. (1973b). *Misc. Publ. Ent. Soc. Amer. 9*, 73.

Gudanskas, R. T., and Cannerday, D. (1968). *J. Invertebr. Pathol. 12*, 405.

Harrap, K. A., and Robertson, J. S. (1968). *J. Gen. Virol. 3*, 221.

Harrap, K. A. (1969). *Proc. 1st Int. Congr. Virol.*, Helsinki, 1968,*1*, 281.

Harrap, K. A. (1970). *Virol. 42*, 311.

Harrap, K. A. (1972). *Virol. 50*, 114.

Heimpel, A. M., and Harshbarger, J. C. (1963). *Bacteriol. Rev. 29*, 397.

Himeno, M., and Onodera, K. (1969). *J. Invertebr. Pathol. 13*, 87.

Hosaka, Y., and Aizawa, K. (1964). *I. Insect Pathol. 6*, 53.

Huger, A. (1963). *In* "Insect Pathology, an Advanced Treatise" (E. A. Steinhaus, ed.), Vol. I, pp. 531-576. Academic Press, New York.

Hukanhara, T., and Hashimoto, Y. (1967). *J. Invertebr. Pathol. 9*, 278.

Ignoffo, C. M., and Dutky, S. R. (1963). *J. Insect Pathol. 5*, 422.

Ignoffo, C. M., Shapiro, M., and Hink, W. F. (1971). *J. Invertebr. Pathol. 18*, 131.

Injac, M., Vago, C., Duthoit, J., and Veyrunes. (1971). *C. R. Acad. Sci. Paris D273*, 439.

Jutila, J., Henry, J. E., Anacker, R. L., and Brown, W. R. (1970). *J. Invertebr. Pathol. 15*, 225.

Kawanishi, C. Y., and Paschke, J. D. (1970). *Proc. 4th Int. Colloq. Insect Pathol., College Park, Md.*, 1970, 127.

Kawanishi, C. Y., Egawa, K., and Summers, M. D. (1972a). *J. Invertebr. Pathol. 20*, 95.

Kawanishi, C. Y., Summers, M. D., Stoltz, D. B., and Arnott, H. J. (1972b). *J. Invertebr. Pathol. 20*, 104.

Kawarabata, T., and Hayashi, Y. (1972). *J. Invertebr. Pathol. 19*, 414.

Khosaka, T., Himeno, M., and Onodera, K. (1971). *J. Virol. 7*, 267.

Kobayashi, M. (1971). *In* "The Cytoplasmic-Polyhedrosis Virus of the Silkworm" (H. Aruga and Y. Tanada, eds.) pp. 103-128. Univ. Tokyo Press, Tokyo.

Koch, A. S., and Volpe, P. (1972). *Proc. 2nd Int. Congr. Virol., Budapest, 1971 2,* 277.

Kurstak, E., and Côte, J. R. (1969). *C. R. Acad. Sci. Paris 272,* 762.

Kurstak, E., Bellonick, S., and Brailovsky, C. (1969). *C. R. Acad. Sci. Paris 269,* 1716.

Leutenegger, R. (1967). *Virol. 32,* 109.

Longworth, J. F., Payne, C. C., and McLeod, R. (1973). *J. Gen. Virol. 18,* 119.

Longworth, J. F., Robertson, J. S., and Payne, C. C. (1972). *J. Invertebr. Pathol. 19,* 42.

Longworth, J. F., Robertson, J. S., Tinsley, T. W., Rowlands, D. J., and Brown, F. (1973). *Nature 242,* 314.

Morgan, C., Rosenkranz, H. S., and Mednis, B. (1969). *J. Virol. 4,* 777.

Paschke, J. D., Campbell, W. R., and Webb, S. R. (1972). *Monogr. in Virol. 6,* 20.

Peters, W. (1969). *Z. Morph. Tiere 64,* 21.

Richards, A. G., and Richards, P. A. (1971). *J. Insect Physiol. 17,* 2253.

Shildmacher, H. (1950). *Biol. Zbl. 69,* 390.

Shapiro, M., and Ignoffo, C. M. (1969). *J. Invertebr. Pathol. 14,* 130.

Smith, D. S. (1968). "Insect Cells; Their Structure and Function." Oliver and Boyd, Edinburgh.

Sohi, S. S., and Cunningham, J. C. (1972). *J. Invertebr. Pathol. 19,* 51.

Stairs, G. R., and Ellis, B. J. (1967). *J. Invertebr. Pathol. 17,* 350.

Stoltz, D. B. (1971). *J. Ultrastruct. Res. 37,* 219.

Stoltz, D. B., and Summers, M. D. (1971). *J. Virol. 8,* 900.

Summers, M. D. (1969). *J. Virol. 4,* 188.

Summers, M. D., and Arnott, H. J. (1969). *J. Ultrastruct. Res. 28*, 462.

Summers, M. D., and Paschke, J. D. (1970). *J. Invertebr. Pathol. 16*, 227.

Summers, M. D. (1971). *J. Ultrastruct. Res. 35*, 605.

Tanada, Y., and Leutenegger, R. (1970). *J. Ultrastruct. Res. 30*, 589.

Tanada, Y., Himeno, M., and Omi, E. M. (1973). *J. Invertebr. Pathol. 21*, 31.

Tinsley, T. W., and Longworth, J. F. (1973). *J. Gen. Virol. 20*, 7.

Vago, C. (1971). "Invertebrate Tissue Culture" Vol. I, 441 pp. Academic Press, New York.

Vago, C., and Croissant, O. (1959). *Ann. Epiphyt. 10*, 5.

Vail, P. V., Jay, D. L., and Hink, W. F. (1973). *J. Invertebr. Pathol. 22*, 231.

Vaughn, J. L., Adams, J. R., and Wilcox, T. (1972). *Monogr. in Virol. 6*, 27.

Wagner, G. W., Paschke, J. D., Campbell, W. R., and Webb, S. R. (1973). *Virol. 52*, 72.

Waterhouse, D. F. (1953). *Aust. J. Zool. 1*, 299.

Webb, S. R., Paschke, J. D., Wagner, G. W., and Campbell, W. R. (1974). *J. Invertebr. Pathol.* (In Press).

Wells, F. E., and Heimpel, A. M. (1970). *J. Invertebr. Pathol. 16*, 301.

Wigglesworth, V. B. (1930). *Quart. J. Micr. Sci. 73*, 593.

Wigglesworth, V. B. (1965). "The Principles of Insect Physiology", 6th ed. Methuen, London.

Woodard, D. B., and Chapman, H. C. (1968). *J. Invertebr. Pathol. 11*, 296.

Younghusband, H. B., and Lee, P. E. (1969). *Virol. 38*, 247.

Zhuzhikov, D. P. (1964). *J. Insect Physiol. 10*, 273.

THE INVERTEBRATE GUT AS BARRIER TO INVADING PARASITES

A. Ralph Barr and Robert Shope*
School of Public Health
University of California
Los Angeles, California

Dr. William Harvey discussed the insect gut anatomy, describing the goblet and columnar cells, the types of cell junctions, and mechanisms regulating the transepithelial potassium pump. The potassium level in the insect gut is high. He speculated that a change in gut cell potassium level might be a defense against parasite replication.

Dr. John Paschke described passage of insect pathogenic viruses through the barriers of the cuticle-lined foregut, the peritrophic membrane to the midgut where the alkaline environment degrades the occlusion and allows the virion envelopes to fuse with the cell membrane. The viral nucleocapsid is transported through the nuclear pore to the nucleus where uncoating presumably takes place. From midgut cells, virions bud into hemocel to infect other insect tissues. The insect offers an ideal *in vivo* system for studies of comparative virology.

Dr. Fred Murphy raised several provocative questions about arborviruses infecting mosquitoes. Why is there a rise and fall of virus titer in the midgut cells although no cytopathology is evident? Why does eastern encephalitis virus bud primarily from the base of the midgut cell and not into the lumen? Why are less than 1 in 20 salivary gland cells infected with St. Louis encephalitis virus? He suggested that some mechanism(s) must control numbers of virus particles and numbers of cells infected. These might be interferon-like substances or defective interferring virus particles.

Dr. Thomas Tinsley pointed out that we need more precise data on effects of inhibitors in gut juices on viruses of insects. The changes in virus particles may determine cell specificity. Are there nonspecific receptors on midgut cells? This could be tested by feeding killed virus, then superinfecting. As an alternative, inhibitors could act on viruses after they enter the midgut. cell. An inhibitor would have to be a normal constituent because it works immediately and the barrier must be at the gut level for

*Department of Epidemiology and Public Health, Yale University, School of Medicine, New Haven, Connecticut.

most viruses because parenteral inoculation by-passes the barrier. Polyacrylamide electrophoresis of gut cells infected with densonucleosis virus reveals a carbohydrate-containing band which by electron microscopy has virus particles surrounded by amorphous substance. This may be inhibitor.

Dr. Thomas Orihel described the peritrophic membrane formed in response to a blood meal and secreted by midgut cells. The membrane is thick and viscous, acting as a mechanical barrier to microfilariae and trypanosomes.

Dr. Barr pointed out that susceptibility of vectors to pathogens may be dominant, recessive, or partially dominant to refractoriness. In most cases, there is evidence of polygenic inheritance. The problem of getting favorable genotypes to increase in frequency in vector populations was briefly discussed.

II. Analysis of Invertebrate Immunity

ANALYSIS OF INVERTEBRATE IMMUNITY

Eli Chernin
Department of Tropical Public Health
Harvard School of Public Health
Boston, Massachusetts

I suspect that the shades of Metchnikoff, Metalnikov, and Cantacuzène, although these pioneers did their work 50 or more years ago, will be in evidence as our session's participants seek to analyze the components of invertebrate immunity. This says much for the brilliant perceptions of the Masters, but it also says that the problems they opened to science have proved extremely difficult to resolve, and that in most of the intervening time we have accumulated and elaborated data rather than achieved novel perceptions and directions. The recent few years, however, provide strong signals that a revolution is underway in our thinking about invertebrate defense mechanisms, that new methodologies and experimental models - the engines of change and progress in any science - are emerging from research on both vertebrate and invertebrate immunity, and that we can reasonably anticipate attaining higher levels of insight before we celebrate the centenary of Metchnikoff's discoveries less than two decades hence.

This NIH-sponsored workshop properly focuses its main attention on the defense mechanisms of arthropods and molluscs whose importance derives from their role as intermediaries (or "nurses," to revive Sir Patrick Manson's delightful term) for organisms that threaten human health. These intermediaries preempt attention because they are patently important, because many are easily reared and manipulated, and because work on them has generated a voluminous literature providing a matrix for further studies. I intend no denigration of investigations aimed at the intermediate hosts - indeed I have contributed to these efforts - if I hazard the opinion that the conceptual breakthroughs we seek are just as likely to spring from research on invertebrates other than the well-known vectors of human disease. One need only look to recent work on invertebrates not directly related to medicine for clues to some of the concepts now taking shape (Bang, 1973; Hildeman and Reddy, 1973).

Other clues, no less important, we'll be privileged to learn about from the participants in this workshop session - Jack Colvard Jones, Fredrik B. Bang, Robert S. Anderson, Edward H. Michelson, M. R. Tripp, G. R. Wyatt, June M. Chadwick, and A. Burns Weathersby - all of whom have distinguished themselves by making fundamental contributions to our understanding of invertebrate defense mechanisms.

REFERENCES

Bang, F. B. (1973). Immune reactions among marine and other invertebrates. *BioScience, 23*:584-589.

Hildeman, W. H., and Reddy, A. L. (1973). Phylogeny of immune responsiveness: marine invertebrates. *Federation Proc. 32*:2188-2194.

FORMS AND FUNCTIONS OF INSECT HEMOCYTES[1]

Jack Colvard Jones
Department of Entomology
University of Maryland
College Park, Maryland

Insects possess a body-vascular cavity -- a hemocoele -- which contains a lymph-like blood called the hemolymph. This fluid may have either no freely circulating hemocytes (as in some of the young larvae of the higher Diptera), or may normally contain relatively few circulating cells (as in many of the true bugs -- Hemiptera), or it may possess numerous hemocytes (for example, in *Prodenia* larvae, it is believed that most of the hemocytes are in circulation) (see Jones, 1962, for references).

Hemocytes are generally comparable to the leucocytes of other animals: from the sponges to man. While a few species of insects may have hemoglobin either in solution in the hemolymph or within special non-circulating fat-body cells, no insect has circulating red blood cells, and none of the hemocytes are concerned with transporting gases to and from the internal tissues.

The hemolymph is primarily circulated by means of a single tubular dorsal vessel. Circulation may be assisted by pulsatile membranes or muscular diaphragms in the legs, wings, and in the ventral part of the abdomen, and by small pulsating ampullae at the base of the antennae and within the thorax (see Jones, 1964, for review).

Although stationary tissues like the pericardial cells which are present in all insects and the phagocytic organs which are present only among some of the orthopteroid insects probably play an important role in the internal defense of insects (Balbiani, 1886; Metchnikoff, 1892), we still have no really detailed information on the extent of their actually protecting even a single species against a specific pathogen. Considerably more information is available on the role of hemocytes, but even so our knowledge on their role in immunity is based on only a very few species and is still quite inadequate.

In this symposium on invertebrate immunity it seems appropriate to review briefly the general status of our knowledge on the forms and functions of hemocytes in insects in general, and to refer to the most important single work on the role of hemocytes

[1]Supported by N.I.H. Career Development Award GM 21,529.

in bacterial immunity in an intact insect.

METHODS OF STUDY

Hemocytes can be studied in a variety of ways: (1) they may be observed in the intact insect directly through the cuticle in selected regions, if these areas are sufficiently transparent (*e.g.*, in the wings, anal papillae, gills), (2) they can be examined in thin whole mounts of freshly collected drops of unfixed or quickly fixed hemolymph taken from intact or pre-fixed insects, (3) they can be seen in stained smears (even though this may be a most unreliable index as to types, it is still a useful beginning to their study), (4) in favorable insects, they can be viewed with the electron microscope (in many cases, however, the hemocytes do *not* fix well, so that information relative to fine structure may be very misleading), and (5) in tissue culture (usually the specialized types die out leaving only one or two types which usually radically change in appearance and behavior).

The most useful method for studying the forms of hemocytes is to examine freshly-collected hemolymph from a chilled insect on a cold stage with phase contrast microscopy. The finest cytological details can be seen in the hemocytes and then the cold stage can be turned down or off and the changes in the hemocytes can be observed. Some hemocytes round up, some send out spike-like pseudopodia, some extend a ruffled fan-like membrane, certain types undergo partial lysis, and still other types show no change in appearance for many hours in a wet whole mount kept under a vaseline-sealed coverslip (see Jones, 1962).

Although there are a number of important studies on the fine structure of insect hemocytes, it is not fully appreciated that osmic acid and glutaraldehyde frequently do not fix the hemocytes very well and many of them are severely damaged by the fixatives, as can be easily demonstrated by flooding one or several fixatives under a coverslip containing numerous different kinds of hemocytes and examining the preparation with phase contrast microscopy. Using such a method, I have seen the phase dark inclusions of spherule cells gradually lighten and concentric finely granular material appear before the spherules vanish. Heavy plasma precipitation may prevent the easy discrimination of types. Glutaraldehyde does not prevent the rapid partial lysis of cystocytes or of some oenocytoids.

CLASSIFICATION

Nine morphologically distinctive types of hemocytes have been described for the class Insecta (Jones, 1962). Some types appear only briefly during one period of the life span. In most insects only one or two morph. types predominate, but in some species -- *e.g.*, in *Prodenia eridania* larvae (Yeager, 1945) and in the cockroach *Gromphadorhina portentosa* (Jones, unpublished) -- many different types co-exist at the same stage of the insect's life. The prolonged co-existence of many types is a direct challenge to those who believe there is only one type of hemocyte.

Those workers who believe that most if not all of the different types of hemocytes represent only stages in the development of one basic type of cell apparently would discard the present identifying names, presumably in favor of a set of numbers or lump each of the stages together into a single all-encompassing type. With a similar philosophy, the vertebrate hematologist would need to discard all of the names showing the stages of development of erythrocytes, lymphocytes, and the entire series of granulocytes, and the tasks of the pathologists would be rendered impossible. Most workers who propose that insect hemocytes are only phases in the development of a single type almost never give quantitative information and base their idea on a brief examination of hemocytes with only one method and usually from only a single stage in the development of a single species of insect. In the writer's opinion, most of the types represent highly differentiated cells (*e.g.*, plasmatocytes, spherule cells, oenocytoids, podocytes, adipohemocytes, and granular hemocytes). It seems likely that the prohemocyte is a precursor cell which can differentiate into other types. It may well be that adipohemocytes are derived from plasmatocytes, but all of the plasmatocytes do not so evolve. In some insects, granular hemocytes may also differentiate from the plasmatocyte line, but this is not always true. The cystocyte may be only a special kind of plasmatocyte: more fragile than other plasmatocytes: but it may function in a totally different way from an ordinary plasmatocyte which does not break down *in vitro*, and clearly needs to be distinguished in some way.

Although it is sometimes proposed that we should have *only* a functional classification, such a system would prove to be completely useless because one cell might be phagocytic at one period and non-phagocytic at another; or one morphologically distinctive type might engage in several quite different activities: might be both phagocytic and trephocytic; or quite different looking cells could share the same function. Phagocytic cells might well act as coagulocytes. What name would one give to cells whose functions were quite unknown? A functional classification would not distinguish some hemocytes from radically different non-hemocytic cells (*e.g.*, fat body cells, pericardial cells, midgut cells, gonadal cells). In quantitative studies it is useful to distinguish also mitotically-dividing hemocytes, degenerating hemocytes, and unidentified cells.

The nine types of hemocytes that have been found in insects can be briefly categorized as follows:

(1) *Prohemocytes* are typically small round or spindle cells which have a dense basophilic cytoplasm. They are frequently the only type to be seen undergoing mitotic division. In general they are more numerous in the early stages of an insect's development than subsequently. In some stages they may be absent from the circulating hemolymph.

(2) *Plasmatocytes* are generally ovoid or spindle cells and have a single nucleus

and a scattering of heterogeneous organelles in an abundant basophilic cytoplasm.

(3) *Granular hemocytes* are mostly ovoid cells with a small nucleus and a cytoplasm filled with small uniform round eosinophilic granules.

(4) *Cystocytes* are very fragile brilliantly hyaline cells with a spherical cytoplasmic envelope and a single sharply outlined cartwheel nucleus generally at the periphery.

(5) *Adipohemocytes* are cells with few to many relatively large spherical refringent fat-like droplets in the cytoplasm. They are at least 10 times smaller than a typical fat body cell.

(6) *Spherule cells* vary from small or very large in size and contain large phase-dark, often basophilic or dully eosinophilic cytoplasmic inclusions which generally distend the cytoplasmic envelope and tend to obscure the nucleus. Although they appear to be comparable to the morula cells seen in echinoderms and tunicates, they differ strikingly from those in the echinoderms in lacking ameboid movements. The large crescent cells of some species of cockroaches (*e.g.*, in *Gromphodarhina*) possibly belong in this category. Crescent cells contain a single enormous inclusion (Ritter, 1965) which pushes the single nucleus into an extreme excentric position. The crescent cell differs from other spherule cells which I have examined in that they tend to send out numerous fine, long filamentous cytoplasmic extensions.

(7) *Oenocytoids* are usually quite conspicuous cells with an abundant thick cytoplasm which is usually intensely basophilic (in some of the higher Diptera, however, these cells are intensely eosinophilic). The cells normally have an excentric nucleus. Some oenocytoids are very fragile *in vitro* and undergo lysis.

The cytoplasm may contain fine granules, or long complexly arranged granular threads or discrete crystals.

(8) *Podocytes* are large plasmatocyte-like cells which are known to occur among some of the Lepidoptera and in the higher Diptera, usually only at one period of life, near the time of pupation. !n the Diptera the cells are characterized by having very long filamentous extensions which rapidly pull away *in vitro* and appear free in the hemolymph as bizarrely-shaped fragments (Jones, 1956). They have been compared to thrombocytes (Zachary *et al.*, 1972) and, in fact, have an ultrastructure remarkably similar to that of a vertebrate megakaryocyte (*loc. cit.*). In *Prodenia* larvae, however, the podocytes are immense stellate cells with a single large highly punctate nucleus. The cells have three to eight pointed peripheral cytoplasmic extensions (Yeager, 1945; Jones, 1959). In some tissue cultures of lepidopteran hemocytes, plasmatocytes often transform into huge podocytes.

(9) *Vermiform cells* are extremely long threadlike cells that have been seen in only a very few insects (*e.g.*, *Prodenia* larvae, Yeager, 1945). They are intensely

basophilic, have thin elongated dense nuclei, and may occur in dense clusters. Yeager observed them only in last larvae and young pupae of *Prodenia*.

Yeager's (1945) elaborate classification scheme for *Prodenia* gives the most detailed and precise description of all of the hemocytes that may be found in heat-fixed hemolymph of any species of insect. Each of the terms he employed permits a step-by-step identification of the main types and of all of the complex intermediates that may occur between various kinds of hemocytes. Although this scheme is too unwieldy for extensive analytical work, it is still an admirable model.

A word of caution is needed with regard to the study of insect hemocytes. They must be studied with phase contrast microscopy in addition to other methods. One example of the problems that occur will suffice to show how mistakes can occur. Several different types of hemocytes may undergo quite radical changes *in vitro*. Prohemocytes may partially lyse and appear as brilliantly hyaline cells with an excentric nucleus. In some insects the oenocytoids (or crystal cells) which are normally very thick cells that may contain complex inclusions, may also undergo partial lysis (their inclusions rapidly vanish) and quickly transform into brilliantly hyaline spherical cells with an excentric nucleus. Occasionally, spherule cells will also undergo partial lysis to become hyaline cells; after losing their spherular inclusions, vacuoles and small granules may remain in the cytoplasmic envelope. Finally, a special type of granular hemocyte (cystocyte) which is involved in coagulation of the hemolymph of many insects (see Grégoire, 1964) typically undergoes rapid partial lysis *in vitro* as it produces a granular precipitate in the plasma surrounding these cells. These cells are sometimes referred to as *coagulocytes* (Grégoire, 1964). Thus, four quite different types of hemocytes may transform into hyaline cells and thus falsely appear to be related, when in fact there is no relationship between them at all. They share only their fragility *in vitro*, and after death resemble each other closely. When stained in smears, they all look alike.

FUNCTIONS

Hemocytes in the normal unstressed insect appear to be mostly or entirely passive as they float in the circulating hemolymph or reside on tissue surfaces. When they do locomote on tissue surfaces, they apparently do so only very slowly either by typical or atypical ameboid movements (Arnold, 1959, 1961). Plasmatocytes often move slowly only by the use of frontal ruffled membranes (Clark and Harvey, 1965). Hemocytes on tissue surfaces however may remain totally quiescent for many hours (Jones, unpublished). We lack detailed information on the extent of locomotion of hemocytes in normal insects over long periods of time.

Hemocytes appear to be activated by normal hormonal stress (*e.g.*, near the time of ecydsis, especially near the time of pupation) and by injuries, infections or other abnormal factors. Activation may lead to a release of cells from temporary accumulations of loose cells or from compact masses of non-proliferating cells or from phagocytic organs or by an increase in numbers of cells from mitoses of freely circulating or individually sedentary cells (Jones, 1970).

Hemocytes at particular periods may normally engage in pinocytosis: that is, they may drink microscopically visible droplets of fluid or take in submicroscopic quantities of proteinaceous substances from the hemolymph. Such activities may or may not radically change the appearance and behavior of the cells.

It has long been known that hemocytes phagocytize a variety of small particles (including inert substances, bacteria, and protozoa), when these are present in or have been injected into the hemocoele (see Jones, 1962). After engulfment of bacteria, they may destroy the micro-organisms. It is not known what hemocytes do with normally engulfed muscle tissue fragments, but presumably they digest them.

Hemocytes are known to clump around and then encapsulate inert particles, bacteria, and large parasites (Salt, 1970). Encapsulation may involve a rapid transformation of some rounded plasmatocytes into long spindle-shaped cells and may involve the lysis of some hemocytes (local formation of cystocytes). Hemocytes around foreign material may slowly turn into concentric rings of fibrous connective tissue. Encapsulated material may also be melanized. Since it can readily be demonstrated that the hemolymph without hemocytes readily melanizes and since the hemocytes in normally melanized hemolymph do not themselves melanize, it is difficult to believe that hemocytes contribute critically to this phenomenon in insects. This is not to deny that certain hemocytes may have a powerful phenoloxidase. This oxidase differs from that in the plasma and we do not now know what normally activates the hemocytic enzyme.

Hemocytes in many insects induce or participate in the coagulation of the hemolymph, by rapidly clumping together within the wound site or by shedding numerous cytoplasmic filaments or fragments which clump, or by releasing material which produces a finely granular precipitate in the surrounding plasma, or otherwise causes gelation of the hemolymph.

Since hemocytes may become densely filled with protein, fat and glycogen, and since numerous hemocytes disappear during pupal life, it has been conjectured that hemocytes have some nutritive value. We still do not know what happens to the hemocytes which disappear: it is currently *assumed* that after phagocytosis of certain larval tissues they undergo lysis within the hemolymph or on tissue surfaces, and their cellular constituents are then slowly absorbed by growing imaginal tissues. While this assumption seems reasonable, there are some other assumptions about the functions of hemocytes which seem much less probable. For example, it seems most improbable that prohemocytes become myocytes, or that hemocytes materially contribute to the formation of the basement membranes, or that such relatively small cells as hemocytes would be able to enlarge sufficiently to transform into typical fat-body cells.

A functional classification of the hemocytes is presented in Table 1.

TABLE 1. Functional Categories of Main Hemocyte Types in Insects.

Types of Hemocytes Which Participate In

Phagocytosis	Coagulation of Hemolymph	Trephocytosis
Prohemocytes ±	Cystocytes +	Granular hemocytes +?
Plasmatocytes +	Podocytes +	Adipohemocytes +?
Granular hemocytes ±		Spherule cells ?
Adipohemocytes +		Oenocytoids ?

HEMOCYTES AND IMMUNITY

By far the most detailed and important studies on hemocytes and bacterial immunity in insects are those on *Galleria mellonella* by Metalnikov who summarized his work in an important but neglected monograph published in 1927. First, he showed that *Galleria* hemocytes would rapidly ingest various dyes when these were injected into the hemocoele. He concentrated his attention on one of the dyes and showed first that a diluted emulsion of carmine (1 :80 to 1 :100) would not be phagocytized when the larvae were held at 10°C. When the larvae were placed at 15 to 25°C, however, phagocytosis of the dye began to take place one hour after its injection and reached a maximum in 4 to 5 hours, at which time 70 to 80% of the hemocytes had picked up carmine. Within 24 to 48 hours all of these phagocytic cells had fallen out of circulation and the hemolymph was clear of carmine. Metalnikov found the phagocytic hemocytes densely clumped near the heart and especially at the posterior end of the larvae. He reported that if the dilution of carmine was 1 :20, it took 24 to 48 hours for the phagocytic hemocytes to clear the hemolymph and to fall out of circulation.

Metalnikov (1927) pointed out that *Galleria* larvae have an extraordinary immunity against bacteria which are highly virulent to man. The larvae have much greater immunity to human pathogens than any vertebrate: for example, the most virulent strains of the tubercle bacilli are harmless to them.

Metalnikov found that when tubercle bacilli were injected into the hemolymph of *Galleria* larvae at 35 to 37°C, phagocytosis by hemocytes began in 15 to 30 minutes. The leprosy bacillus, on the other hand, remained in the circulating hemolymph for hours without being phagocytized, but would be slowly phagocytized within 24 hours. Non-virulent *Pneumococci* which were injected into the hemolymph were rapidly phagocytized and digested by hemocytes. The maximum level of phagocytosis of *Staphylococci* and *Sarcina* occurred within 3 hours. Intense phagocytic reactions were found when either the anthrax bacillus or *Micrococcus* were injected. When either *Bacillus galleriae II* or a certain strain of *Pneumococci* were injected, there was no phagocytosis at all. When either *B. coli* or *Proteus vulgaris* were injected, the amount of phagocytosis progressively decreased.

Metalnikov considered both natural and acquired bacterial immunity in *Galleria* to be essentially cellular immunities. He believed that the phagocytic hemocytes either phagocytized or surrounded bacteria and destroyed them intracellularly, or that hemocytes secreted bacteriolytic enzymes which caused their lysis. He thought that following vaccination of the larvae phagocytosis of bacteria and their subsequent digestion were greatly accelerated (he does not give precise data). He found that while some ingested species of bacteria were not surrounded by vacuoles (*e.g., B. coli* and *Micrococcus*), other species were surrounded by vacuoles and sometimes hemocytes became extensively vacuolated (as with *Sarcina*).[2]

Intracellular digestion of the bacteria (*e.g., Staphylococci, Sarcina, B. coli* and *Proteus*) was generally not completed 8 hours after ingestion by hemocytes. Metalnikov showed that the number of bacteria greatly influenced the speed of their digestion, and profoundly affected the whole nature of immunity, in that massive doses would overwhelm all defenses.

Metalnikov showed that the degree of immunity varied greatly with the stage of development. Thus, larvae were more resistant than pharate pupae, and pharate pupae were far more resistant than young pupae. Young pupae were considerably more resistant than old pupae (pharate adults), and adults were more resistant than pupae, but less resistant than either larvae or pharate pupae. He also showed that each of the three developmental stages could acquire immunity. He observed that immunized larvae would develop immunity very rapidly (in 20 to 24 hours): they could be immunized to withstand fatal doses of cholera vibrios in 3 hours. Furthermore, he reported that the immunity of vaccinated larvae was carried over to pupae and adults (he does not give sufficient data for this important conclusion).

In one of the most intriguing experiments of all, Metalnikov studied the immunity of *Galleria* larvae to cholera after cauterizing various ganglia. The larvae could apparently be immunized after cauterizing any one of the ganglia of the central nervous system with a major exception. After cauterizing the third thoracic ganglion he found that the larvae were paralyzed and could no longer be immunized. As far as I am aware, Metalnikov did not continue his experiments along these lines: that is, he did *not*

---

[2]It is possible that the failure of some bacteria to be surrounded by a vacuole is due to the fact that the bacteria invade the cell rather than being engulfed by a cell.

explain why this particular ganglion should have been so important.

Metalnikov's work opened up many fascinating paths but none of them seems to have been followed. Until they are, our knowledge concerning cellular immunity within the intact insect will never be adequately understood.

SUMMARY

A brief review is given relative to different methods of study of hemocytes in insects and with respect to the classification of the different cell types. Nine morphologically-distinctive types have been seen among a few insect species. During most of the life span, three to four types are the predominant forms. In some species, many complex types can co-exist in the same stage in the animal's development.

The primary functions of hemocytes are phagocytosis, encapsulation, coagulation of the hemolymph, and trephocytosis. More than one type of cell may participate in a given function. Hemocytes function primarily when activated by stress, whether this is abnormal (*e.g.*, due to mechanical injury or bacterial infection) or normal (*e.g.*, when affected by hormones or changes in the biochemistry of the hemolymph). In most insects the hemocytes do not normally form basement membranes, they do not transform into somatic muscles, and they do not enlarge to form typical fat bodies.

Metalnikov's work on phagocytic immunity in *Galleria mellonella* is reviewed in some detail because it opens up many important avenues for further research on the problem of hemocytes and their role in immunity in living insects at different developmental stages.

REFERENCES

Arnold, J. W. (1959). *Canad. J. Zool.*, *37*:371.

Arnold, J. W. (1961). *Canad. J. Zool.*, *39*:755.

Balblanl, M. (1886). *C. R. Acad. Sci.*, *103*:952.

Clark, R. M. and Harvey, W. R. (1965). *J. Insect Physiol.*, *11*:161.

Grégoire, C. (1964). In "The Physiology of Insecta" (M. Rockstein, ed.), Vol. III, pp. 153 188. Academic Press, New York.

Jones, J. C. (1956). *J. Morph.*, *99*:233.

Jones, J. C. (1959). *Quart. J. Micr. Sci.*, *100*:17.

Jones, J. C. (1962). *Amer. Zool., 2*:209.

Jones, J. C. (1964). In "The Physiology of Insecta" (M. Rockstein, ed.), Vol. III, pp. 1-107. Academic Press, New York.

Jones, J. C. (1970). In "Regulation of Hemopoiesis" (A. S. Gordon, ed.), Vol. I, pp. 7-65. Appleton-Century-Crofts, New York.

Metalnikov, S. (1927). "L'Infection Microbienne et L'Immunité chez la Mite des Abeilles *Galleria mellonella*". Monographie, Inst. Pasteur, Masson et Cie, Paris.

Metchnikoff, E. (1892). "Lecons sur la Pathologie Comparée de L'Inflammation, faites à l'Institut Pasteur en Avril et Mai 1892." G. Masson, Paris.*

Ritter, H., Jr. (1965). *Science, 147*:518.

Salt, G. (1970). "The Cellular Defence Reactions of Insects". Cambridge University Press, New York.

Yeager, J. F. (1945). *J. Agri. Res., 71*:1.

Zachary, D., Hoffmann, J. A. and Porte, A. (1972). *C. R. Acad. Sci. D, 275*:393.

*Translated into English in a 1968 edition by Dover Publ. Inc.

EVIDENCE FOR GENETICAL CONTROL OF INVERTEBRATE IMMUNITY AND ITS FIELD SIGNIFICANCE

A. Ralph Barr
School of Public Health
University of California
Los Angeles, California

Disease-producing organisms which are transmitted by biological vectors are subject to all of the defense mechanisms of the vector. If the etiological agent produces no significant pathology in the vector there might be little defensive activity on the part of the vector; this may be true of many viruses since most viruses produce little obvious pathology in the vector. There may conceivably be a great deal of pathology, however, which is not at all obvious. When, on the other hand, an etiological agent produces a great deal of pathology, as is the case in many nematode infections, any defense mechanism which ameliorates the effect of the worms will have selective value and will tend to be favored in direct relation to the intensity of selection by the worms. Such defense mechanisms should have a genetic basis so there is the interesting possibility of transforming vectors into non-vectors by increasing the efficacy of their defense mechanisms through genetic selection. If such refractory genotypes could be caused to increase in frequency in the field at the expense of susceptible genotypes, diseases could be controlled by the replacement of vector populations with non-vector populations. Manipulations of this sort might be possible if the inheritance of refractoriness is simple; if it is complex there is less chance that susceptible populations could be replaced.

Inheritance of susceptibility to filarial infection in mosquitoes. Macdonald (1962a) reported the selection of a West African strain of *Aedes aegypti* for susceptibility to a Malayan strain of the semiperiodic form of *Brugia malayi* which was maintained in cats. This mosquito is not normally susceptible to infection with this filarial worm and is not known to be a natural vector of the worm. Before selection, 17 to 31% of the mosquitoes were susceptible. After 15 generations of familial selection 85% of the mosquitoes dissected contained mature larvae. The inheritance of susceptibility was studied by crossing this strain with uniformly refractory strains and testing the susceptibility of the F_1, F_2, and backcross offspring (Macdonald 1962b). The results of these tests were not entirely clear-cut but they suggested that susceptibility was controlled by a recessive sex-linked factor. Alternatively it could have been said that refractoriness was controlled by a dominant sex-linked factor. Macdonald designated the recessive susceptibility gene f^m and its dominant allele F.

Since only 85% of his selected strain supported maturation of *malayi* larvae Macdonald (1963af) tested seemingly refractory individuals and found that they were indeed homozygous for *f^m^*. This shows that *f^m^* is incompletely penetrant and suggests that there are modifying genes present. This is another way of saying that although susceptibility can be explained largely in terms of one gene it is still polyfactorial. This also follows from the observation that the refractory strain was not completely refractory and the F_1 hybrids showed somewhat greater susceptibility than the refractory parents.

Macdonald (1963b) tested the rate of recombination of f^m with sex and found that it was about 8%. This was repeated in a study with Sheppard (1965) in which red-eye (*re*) was used as a marker. The order of the three factors was found to be red-eye--sex--f^m; there was 2.1 to 5.7% recombination between *re* and sex (*M*) and 15.9 to 33.7% recombination between *M* and f^m, a value about twice as large as that previously found. These authors remarked on the large amount of variation in recombination in different crosses and suggested that the variation be accounted for by assuming the presence of two inversions of the "normal" chromosome. The recognition of heterogeneous data here is important since it has been found when sought in most genetic work with mosquitoes; it seems unlikely to be due principally to inversions. That susceptibility is under polygenic control is suggested by the decrease in penetrance of f^m in backcrosses.

Macdonald and Ramachandran (1965) showed that *aegypti* homozygous for f^m were susceptible not only to subperiodic *Brugia malayi* but to all strains of microfilariae tested which developed in the flight muscles (periodic *B. malayi*, *B. pahangi*, periodic and subperiodic *Wuchereria bancrofti*) but not to forms which developed in the Malpighian tubules (*Dirofilaria immitis*, *D. repens*). McGreevy and McClelland (1973) found that f^m homozygotes were also susceptible to *Dirofilaria corynoides* which develops in the fat body. The finding that f^m homozygotes are susceptible to all muscle-parasitic filariae tested is surprising since natural vectors of filariasis do not show this property; they are usually susceptible to only a single kind of filarial worm. The catholic susceptibility of f^m homozygotes suggests that selection has altered a basic defense mechanism of the mosquito so that it can no longer defend itself against these muscle-inhabiting worms.

Rodriguez and Craig (1973) studied the geographical distribution of the f^m gene and found that it occurred commonly in East African strains of *Aedes aegypti* but only rarely in those from other parts of the world.

Beckett and Macdonald (1971) studied the development of microfilariae in their selected strain of *aegypti*. They found that most *Brugia pahangi* which entered the thoracic muscles developed to mature larvae but only 20-25% of *Brugia malayi* survived. Most of the latter larvae died 2 to 3 days after being ingested by the mosquito even though they had succeeded in reaching the thoracic muscles.

McGreevy and McClelland (1973) used Macdonald's selected strain to select a strain susceptible to *Dirofilaria immitis*. They found that susceptibility to this worm was also controlled by a sex-linked recessive gene but a different one than f^m. Zielke (1973) reported the same results and noted that the two susceptibility genes were at different loci.

Similar studies have not been done with other vectors of filariae. Thomas and Ramachandran (1970) reported the selection of strains of *Culex pipiens quinquefasciatus* for susceptibility to infection with an anopheline transmitted strain of *Wuchereria bancrofti*. Initially some 7 to 20% of the mosquitoes were susceptible; after six generations of selection some 50 to 100% were susceptible. Inheritance studies were not done on these strains. Partono and Oemijati (1970) reported the failure of an Indonesian strain of *Culex pipiens quinquefasciatus* to respond to selection for either susceptibility or refractoriness to *Wuchereria bancrofti*. Macdonald (1973) in an informal meeting reported that susceptibility of *Culex pipiens* to infection with *Brugia pahangi* was controlled by a sex-linked recessive gene.

Inheritance of susceptibility to malarial infection in mosquitoes. Huff (1929) was the first to show that the susceptibility of mosquitoes to malarial parasites responded to selection. He selected strains of *Culex pipiens* for susceptibility and for resistance to *Plasmodium cathemerium*. By crossing susceptible and refractory strains and examining the F_1 and F_2 he (1931) showed that susceptibility disappeared in the F_1 but reappeared in the F_2 in a ratio of 1 susceptible to 3 refractory individuals. Although the data were not entirely consistent, the inheritance of susceptibility could be accounted for principally by a single major recessive gene.

Ward (1963) studied the susceptibility of *Aedes aegypti* to infection with *Plasmodium gallinaceum*. The Malayan strain of *aegypti* used appeared to be essentially homozygous for susceptibility; selection for susceptibility had no effect. Selection for refractoriness, on the other hand, was successful and after 26 generations a line was established that seemed to be essentially homozygous for refractoriness. Hybrids of the susceptible and refractory lines showed intermediate susceptibility. F_2 and backcross analyses also indicated that susceptibility and refractoriness were governed principally by a single pair of co-dominant alleles. Ward used mean oocyst for his interpretation. Kilama and Craig (1969) repeated these experiments using strains of *aegypti* bearing genetic markers. They used a discriminating infection to enable classification of mosquitoes as susceptible (more than 10 oocysts on gut) or refractory (10 or fewer oocysts on gut). In their interpretation susceptibility is controlled by a dominant allele. Backcrosses with marked strains indicated that this allele is autosomal and some 7-8 crossover units from *silver mesonotum* in one direction and 17 crossover units from dieldrin-resistance in the other.

Inheritance of susceptibility to viral infection in arthropods. Storey (1932) studied the inheritance of susceptibility to the virus which causes streak disease of corn in the leafhopper *Cicadulina mbila*. He found that susceptibility was dominant to refractoriness and was linked with sex.

TABLE 1

Inheritance of susceptibility to parasites in various arthropods.

Parasite	Vector	Susceptibility	Reference
Brugia malayi	*Aedes aegypti*	recessive	Macdonald 1962a, b
Brugia pahangi	*Culex pipiens*	recessive	Macdonald 1973
Dirofilaria immitis	*Aedes aegypti*	recessive	McGreevy & McClelland 1971, Zielke 1973
Plasmodium cathemerium	*Culex pipiens*	recessive	Huff 1931
Plasmodium gallinaceum	*Aedes aegypti*	co-dominant	Ward 1963
Plasmodium gallinaceum	*Aedes aegypti*	dominant	Kilama & Craig 1969
bluetongue virus	*Culicoides variipennis*	recessive	Jones 1973
streak virus	*Cicadulina mbila*	dominant	Storey 1932

Jones (1973) by selection developed lines of *Culicoides variipennis* susceptible and resistant to infection with bluetongue virus of sheep. In preliminary studies he found that resistance was dominant to susceptibility.

Field aspects. In the studies cited there is no general pattern of inheritance of susceptibility (Table 1). In most cases the susceptible condition is recessive but in some cases it is dominant. It seems likely that the reported cases include mostly those that yielded more or less easily to analysis. There is no well studied instance of polyfactorial inheritance included, perhaps because of the difficulty of doing such a study. Even in the reported studies, however, there is always evidence of polyfactorial inheritance; strains presumably homozygous for susceptibility always show some failure, and strains presumably homozygous for refractoriness always permit development of some parasites. Penetrance is frequently higher in selected strains than in backcrosses, which suggests the loss of modifying genes. It must be borne in mind, therefore, that reference is always to major genes rather than to complete genetic systems.

If a useful genetic factor is found, now can it be propagated in the field? If, for example, *Aedes aegypti* were the vector of *Wuchereria bancrofti* in East Africa, could a vector population be modified by introducing dominant resistance genes into that population? According to the Hardy-Weinberg principle gene frequencies tend to remain constant unless acted on by some selective factor. Introduced genes would therefore tend to remain at the frequency at which they were introduced unless they possessed some sort of selective advantage or disadvantage. A gene for refractoriness to filarial infection should protect the mosquito from the damaging effects of filariae and should therefore be selectively advantageous in the presence of filarial infection, providing that the gene does not have disadvantageous effects as well. On the other hand if the parasite has no obvious pathological effect it is unlikely that refractoriness will be selectively advantageous. Mechanisms must therefore be found for making desireable genetic factors increase in frequency in populations, perhaps by linking them with insecticide resistance and using the insecticide as a selective agency, or by inducing translocations in the desireable stock and then using translocation homozygotes to replace the field population.

REFERENCES

Beckett, E. B., and Macdonald W. W. (1971). The survival and development of subperiodic *Brugia malayi* and *B. pahangi* larvae in a selected strain of *Aedes aegypti*. Trans. Roy. Soc. Trop. Med. *Hyg.*, *65*:339-346.

Huff, C. G. (1929). The effects of selection upon susceptibility to bird malaria in *Culex pipiens* Linn. Ann. Trop. Med. *Parasitol.*, *23*:427-442+p1.

Huff, C. G. (1931). The inheritance of natural immunity to *Plasmodium cathemerium* in two species of *Culex*. J. Prev. *Med.*, *5*:249-259.

Jones, R. H. (1973). Selection of genetic variants from *Culicoides* populations with reference to vector competence. Abstracts, U. S.-Japan Program Workshop, Univ. Calif. Berkeley, p. 30.

Kilama, W. L., and Craig, G. B. Jr. (1969). Monofactorial inheritance of susceptibility to *Plasmodium gallinaceum* in *Aedes aegypti*. Ann. Trop. Med. *Parasitol.*, *63*:419-432.

Macdonald, W. W. (1962a). The selection of a strain of *Aëdes aegypti* susceptible to infection with semi-periodic *Brugia malayi*. Ann. Trop. Med. *Parasitol.*, *56*:368-372.

Macdonald, W. W. (1962b). The genetic basis of susceptibility to infection with semi-periodic *Brugia malayi* in *Aëdes aegypti*. Ann. Trop. Med. *Parasitol.*, *56*:373-382.

Macdonald, W. W. (1963a). Further studies on a strain of *Aedes aegypti* susceptible to infection with sub-periodic *Brugia malayi*. Ann. Trop. Med. *Parasitol.*, *57*:452-460.

Macdonald, W. W. (1963b). A preliminary cross-over value between the gene f^m (filarial susceptibility, *Brugia malayi*) and the sex locus in *Aedes aegypti*. Ann. Trop. Med. *Parasitol.*, *57*:461-465.

Macdonald, W. W. (1973). Genetics of mosquito susceptibility to filariae. Abstracts, U. S.-Japan Program Workshop, Univ. Calif. Berkeley, pp. 26-29.

Macdonald, W. W., and Ramachandran, C. P. (1965). The influence of the gene f^m (filarial susceptibility, *Brugia malayi*) on the susceptibility of *Aedes aegypti* to seven strains of *Brugia*, *Wuchereria* and *Dirofilaria*. Ann. Trop. Med. *Parasitol.*, *59*:64-73.

Macdonald, W. W., and Sheppard, P. M. (1965). Cross-over values in the sex chromosomes of the mosquito *Aedes aegypti* and evidence of the presence of inversions. Ann. Trop. Med. *Parasitol.*, *59*:74-87.

McGreevy, P. B., and McClelland, G. A. H. (1973). The inheritance of susceptibility to *Dirofilaria immitis* in *Aedes aegypti* with observations on the behaviour of filariae in refractory mosquitos. WHO/FIL/73.110, 2 pp.

Partono, F., and Oemijati, S. (1970). Susceptibility of *Culex pipiens fatigans* to *Wuchereria bancrofti* in Djakarta, Indonesia. S. E. Asian J. Trop. Med. Publ. *Hlth.*, *1*:516-518.

Rodriguez, P. H., and Craig, G. B. Jr. (1973). Susceptibility to *Brugia pahangi* in geographic strains of *Aedes aegypti*. Amer. J. Trop. Med. *Hyg.*, *22*:53-61.

Storey, H. H. (1932). The inheritance by an insect vector of the ability to transmit a plant virus. Proc. Roy. *Soc.*, *B*, *112*:46-60.

Thomas, V., and Ramachandra, C. P. (1970). Selection of *Culex pipiens fatigans* for vector ability to the rural strain of *Wuchereria bancrofti* - a preliminary report. Med. *J. Malaya*, *24*:196-199.

Ward, R. A. (1963). Genetic aspects of the susceptibility of mosquitoes to malarial infection. Exper. *Parasitol.*, *13*:328-341.

Zielke, E. (1973). Untersuchungen zur Verebung der Empfänglichkeit gegenüber der Hundefilarie *Dirofilaria immitis* bei *Culex pipiens fatigans* und *Aedes aegypti*. Ztschr. Tropenmed. *Parasitol.*, *24*:36-44.

PHAGOCYTOSIS IN INVERTEBRATES

Frederik B. Bang

Department of Pathobiology
The Johns Hopkins University
School of Hygiene and Public Health
Baltimore, Maryland

Phagocytosis as a defense mechanism in invertebrates necessarily covers a sequence of events. These are basically (i) recognition, (ii) ingestion, and (iii) digestion or disposal. It is necessary to put them in sequence to avoid misunderstanding in nomenclature and confusion in analysis (Figure 1). Ingestion may be followed by a number of events, including digestion, which will be discussed by Dr. Anderson.

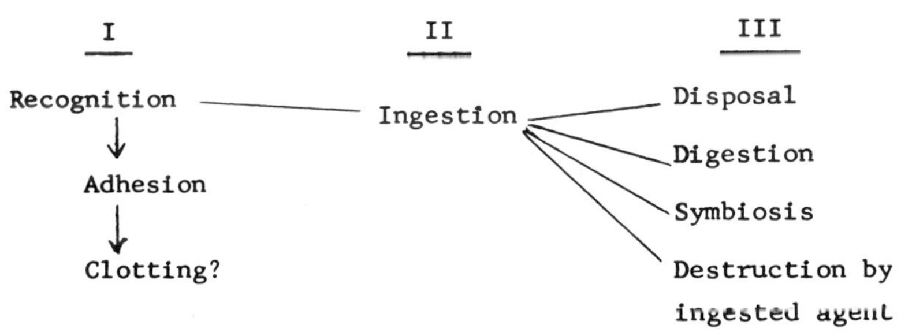

Figure 1. Recognition, ingestion and disposal in invertebrates.

In addition, the subject of autophagocytosis, whereby the cell encloses a portion of its own cytoplasm within a membrane and then digests this, might be considered.

There is extensive literature on this, particularly as related to insects but, since it is not clear that it is a defense mechanism against a foreign invader, it will not be discussed here.

There is an important *caveat* which needs continual emphasis. Invertebrates are not one group of animals, and only a few of them are on any sort of direct evolutionary sequence leading to the vertebrates. "The invertebrata have long since ceased to constitute one of the primary divisions in the scientific classification of the Animal Kingdom. Their name is now no more than a convenience for designating a group of phyla with which it is often necessary to deal as a whole .... Two main stocks stand out fairly clearly: The annelid superphylum which contains annelids, molluscs, and arthropods, and the echinoderm superphylum which contains the echinoderm, protochordates, and chordates." (Kerkut, 1958). Thus the origin of immune processes in man is not to be found in insects or earthworms, though there may be most interesting analogies. Homologous (having an origin in a direct evolutionary sequence) reactions include phagocytosis, but again we must be careful in this. Although most, if not all, of those wandering specialized cells that we call phagocytes are, in Willmer's terms, 'amoebocytes,' this does not mean that all amoebocytes are good phagocytes. Most famous among these are the amoebocytes of *Limulus*, which have not yet been shown to be good phagocytes.

In this connection, amoebocytes are distinguished from most other cells, mechanocytes and epitheliocytes, by the fact that they normally circulate as free cells that do not stick to each other or to a foreign substance. The ability to stick to foreign and invading cells is often used as a criterion or recognition. For instance, when sea urchin cells are injected into a starfish, the foreign cells are immediately recognized by the starfish, inducing massive clumping and subsequent phagocytosis of the sea urchin cells. Failure to realize that cells other than the circulating amoebocytes may also play a role in this sticking has delayed the determination of specific areas in the host in which phagocytic cells predominate. Thus, the analogue of the reticuloendothelial system of the vertebrate has recently been reemphasized, especially in molluscs (Stuart, 1968). However, 'phagocytic organs' had been reported on the basis of uptake of carmine by Cantacuzène (1897) in a number of marine worms as early as 1897.

It is preferable to discuss phagocytosis not simply as a general phenomenon existent among all invertebrates, but to examine it within the different broad groups or orders of animals. When we consider it in evolutionary terms, we may avoid some large, unconscious biases. This is not particularly revolutionary. After all, Metchnikov (1892) did this in his famous "Lectures in Comparative Pathology" but, because of the lack of knowledge of evolution on the part of many of today's medical scientists, it is not always followed.

Protozoa. Phagocytosis of foreign material by protozoa is, after all, a way of life. Two examples of experimental work are worth discussion in the context of this Workshop. The first concerns the ability of a free living soil amoeba, *Dictyostelium discoideum*, to ingest 'pathogenic' bacteria. Gerisch (1973) and Gerisch et al. (1967) have shown that these amoeba fail to ingest smooth (virulent) strains of *Salmonella*, but do ingest the rough forms in suspension cultures. If specific antiserum is added to the cultures, the smooth forms are also readily phagocytosed. This differential effect is apparently due to differential adhesion, for both smooth and rough forms are ingested when the amoeba are placed on agar plates on which the bacteria are growing. This situation then is an exact analogue of the action of an opsonin during infection of the vertebrate. A previous study of ingestion of *Salmonella* by *Amoeba proteus* (Savanat and Pavillard, 1964a, b) showed only modest opsonic effect of antiserum, an effect which might have been due to agglutination. Incidentally, Beug et al. (1973) have shown that the change from a stage of free living amoeba to colony formation and a crude organization is blocked by specific antibody, thus preventing cell to cell adhesion.

Turning to the third phase of phagocytosis, that is, disposal of the ingested material, here again there is a specificity of interaction in protozoa as well as metazoa. Karakashian and Karakashian (1965) have shown that the symbiotic association between *Paramecium* and *Chlorella* is directly dependent upon the particular strain of *Paramecium* which ingests the algae *Chlorella*. Two different strains of *Paramecium* ingest the *Chlorella* readily, but in one strain, the algae disappear in five hours, whereas in the second, after a similar early drop in the number of surviving *Chlorella*, there is multiplication of the algae, and a subsequent development of symbiosis. *Chlorella* strains also differ in these properties, and the authors suggest that they are pre-adapted for this situation, that is, development of a symbiotic relationship.

This kind of relationship has been described in a number of ciliates (summarized in Blackbourn et al., 1973) and it would be of great interest to know whether the chloroplast derived from the algae develops resistance to digestion by the host, or whether the host ciliate fails to form specific enzymes for this digestion.

Hydra. This animal which has been studied so intensively in terms of regeneration does have specific predator-disease organisms. An amoeba which is peculiarly parasitic on these animals (Reynolds and Looper, 1928; Stiven, 1964) eats the surface cells and gradually destroys the animal. Since there are no mobile cells within the hydra, no organized defense has been seen.

The capacity of hydra to recognize those algae which later become symbionts has been studied by injecting these into the gut and determining which of them were recognized, phagocytosed and developed, thus yielding green hydras (Pardy and Muscatine, 1973).

Sponges. Cheng and his associates (1968a, b, c) have reviewed the defense mechanisms of sponges and injected particulate material, such as India ink and carmine, into the body wall of these animals. The foreign particles formed extracellular clumps in the first hour, but were subsequently phagocytosed by archaeocytes. The ink- and carmine-laden cells moved through the mesoglea and migrated out through the excurrent canal. In this way the carmine particles were completely eliminated by the 24th hour and ink by the 96th hour. There apparently was no change in differential numbers of the five types of parenchymal cells during this cleaning of the body wall. The implantation of foreign biological material was accomplished by placing erythrocytes, molluscan muscle and other tissues in the body wall. An incomplete collection of the cells around the heterografts occurred, forming the beginning of a capsule. In addition, there was a concentration of parenchymal cells around the cut surfaces, but no fusion of these surfaces. There are no studies on the reaction of sponges to natural disease processes, even though several epidemic diseases in sponges in the Caribbean have been reported (Galtsoff, 1942; Smith, 1941).

Molluscs. Although the molluscs have a tremendous variety of forms, and may vary from relatively simple and even parasitic forms to highly developed ones such as the squid and the octopus, in this chapter we will not attempt any differential consideration of the different families or taxons.

Perhaps because of its economic importance, attention has been focused chiefly on the oyster. There are a number of important natural infectious diseases of oysters, but as yet it has not been possible to reproduce any of them experimentally, so that the sequence of changes may be followed.

Molluscs have been a favorite subject for study of phagocytosis, and the rapid ingestion of all sorts of injected particulate bodies, including bacteria, has been recorded by a variety of authors. Preeminent has been the group at Rutgers, originally led by Stauber (1950), and now continued by Feng (1968) (Figures 2 and 3) and Tripp (1970). Rates of clearance of bacteria, bacteriophage, India ink, etc., from the blood have been determined. There is no increased rate of disposal with repeated injections (Bayne and Kime, 1970) (Figure 4). The role of temperature in influencing heart rate and the number of circulating cells has been determined. Inflammation, both artificially produced and in response to invading parasites, has been recorded, and a number of different diseases, including several virus diseases, have been described (Sparks, 1972). But again, I am unaware of any definitive studies as to how the amoebocytes combat infection in any natural diseases of a mollusc. Tripp (1970) has summarized much of this material in the Symposium of the Reticuloendothelial Society held in 1969, and in this chapter I will cover only selected topics and some papers that have been published since then.

Phagocytosis in vitro by a variety of molluscan blood cells may be readily observed, but it is important to emphasize that it does not always occur when observations are made in vitro. In this connection, it is important to remember that cells studied on glass have already been stimulated to convert into the active stage (Fauré-Fremiet, 1927) and, therefore, their surfaces are changed at least on one side. This may decrease the ability of such an active cell to phagocytose. In addition, there is the specific problem of opsonins which will be covered by Dr. Tripp in his review on humoral factors. Suffice it to say that no complement-like substances which function to increase phagocytosis have been clearly documented in any mollusc.

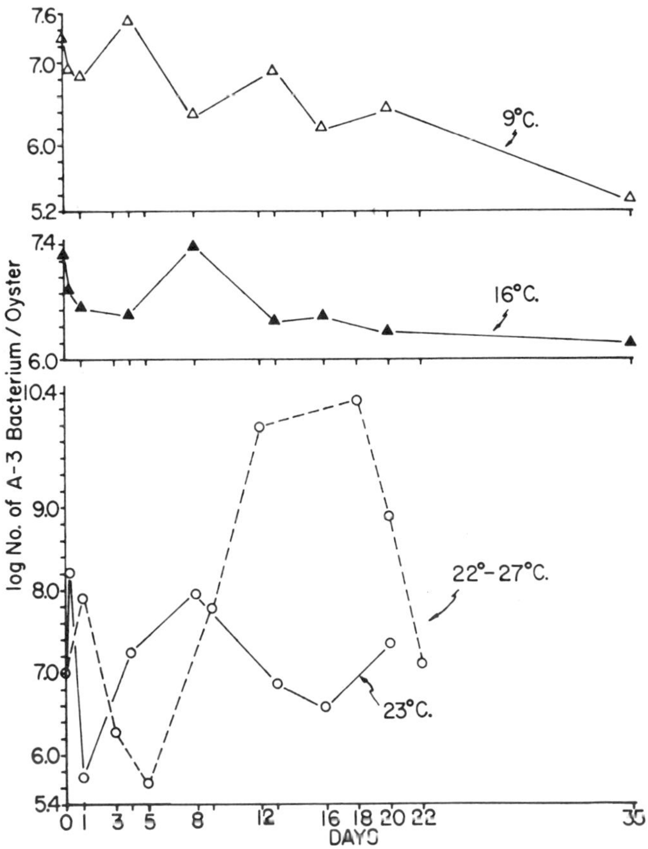

Figure 2. The effect of temperature on the course of experimental infection (Pseudomonas-like A-3) in oysters. Each point on the graph represents the average of six replicate plate counts from two oysters. (from Feng, 1966)

A number of papers have now emphasized the localization of particulate matter which occurs after injection either directly into the heart or into any sinus. These organs of localization have been designated as part of a reticuloendothelial apparatus in the octopus (Stuart, 1968) and have been shown by Reade (1968) to be present in the liver-like digestive diverticula in snails as well as in other species (Bayne, 1973) (Figure 5) (Brown and Brown, 1965) (Figure 6). The exact relationship of this localization to a system of fixed phagocytic cells is not established in every case, and it is not clear that all particulate material is definitely phagocytosed. It may well stick within the open vascular system and remain there for months. However, other work (Reade and Reade, 1972) emphasizes an old finding that phagocytic cells may migrate out of the animal and that this constant loss of particle-laden phagocytes is eventually able to eliminate most, if not all, of the foreign particles.

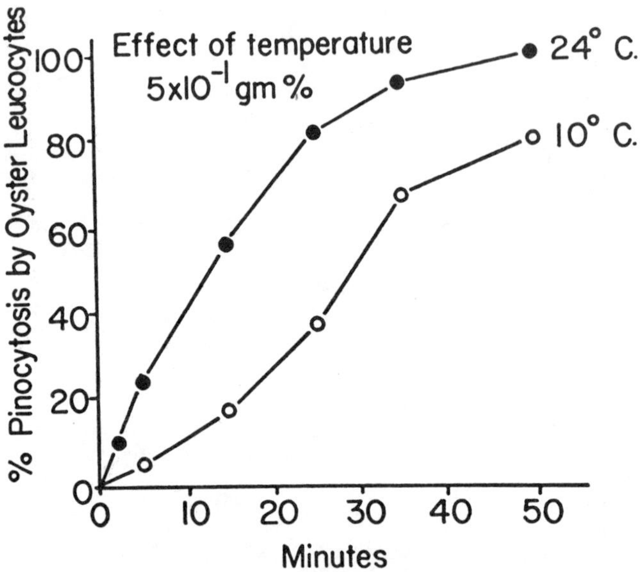

Figure 3. The effect of temperature and time of exposure on the pinocytosis of rhodamine-labelled human gamma globulin by oyster leukocytes. Each point on the graph represents the median of three samples. (from Feng, 1968)

Attempts to produce immunization of the mollusc by repeated injections have failed so far. One careful study on the land snail *Helix* by Bayne and Kime (1970) shows that the elimination of bacteria after injection was not changed by a series of three injections. Following the injection of the bacteria, the number of circulating amoebocytes decreased within the first hour, remaining low while the bacteria were at their peak. The amoebocytes returned to the pre-injection level by 6 hours after inoculation. This response was not altered by repeated injections (Figure 4).

Certain molluscs are well known to carry symbiotic algae but, unlike paramecia and hydra, no experimental studies on the recognition of these symbionts have been published.

Experimental bacterial infections of the oyster have been particularly followed by the Rutgers school (Feng, 1966), and in their studies attention has been paid to bacteria which cause natural infections. Bacteria are not all phagocytosed in vitro (Bang, 1961). Some bacteria may be caught by the amoebocytes by the adherence of their

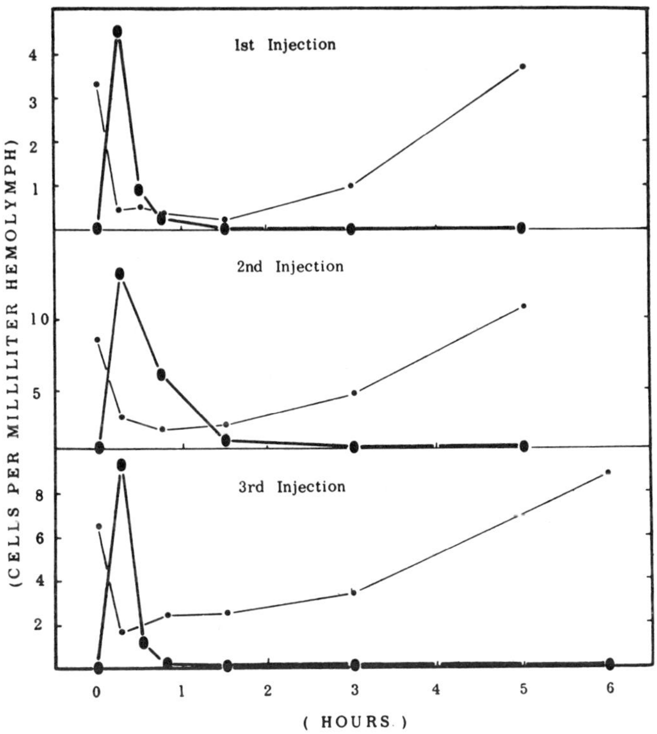

Figure 4. Three injections in 1 snail. The 2nd injection was received 2 weeks after the 1st, and the 3rd was received 1 week later. 0 0 = bacteria. o o = hemocytes. Note the similarity of the reaction in each case. 1st: bacteria and hemocytes X 10^5. 2nd: bacteria X 10^4, hemocytes X 10^5. 3rd: bacteria 0.4 X 10^5, hemocytes X 10^5. (from Bayne and Kime, 1970)

flagella to the amoebocytes. Some bacteria are able to destroy the larvae of clams (Guillard, 1959), and clearance of these bacteria has been studied following injection of normal animals (Tripp, 1960). No studies of the mechanism of pathogenicity, however, have been made.

Sipunculus, and other sipunculoids, make up a remarkable group of unsegmented marine worms with a large fluid cavity, which is usually free of bacteria, and which contains a large variety of cells, some of which are highly phagocytic. These animals have been extensively studied from the standpoint of mechanisms which keep the blood free of bacteria, which recognize foreign substances, and cause lysis of marine ciliates such as *Anophrys* (Bang, 1966). The lytic substance appears in the body cavity fluid of *Sipunculus* after the injection of a number of substances including the ciliates themselves, but apparently there are reservoirs of the material stored in the cells, possibly

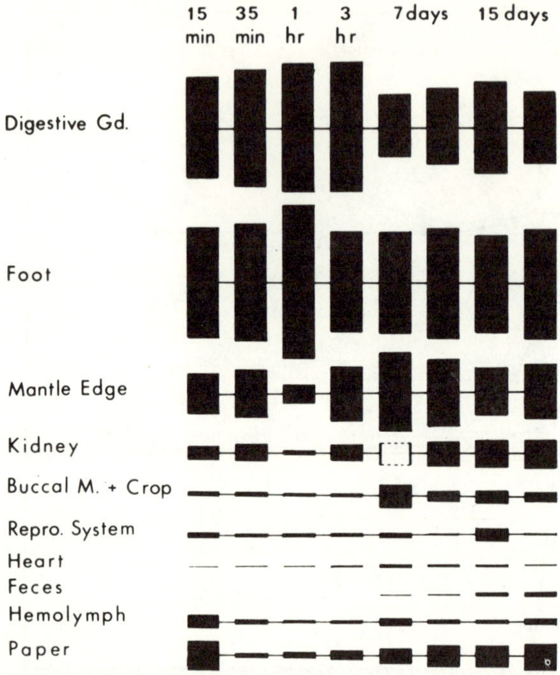

Figure 5. The percent of total DPM present in each of the tissues. To obtain the values plotted, DPM values were summed for all tissues in a snail; then the value for each tissue was divided by the total value and multiplied by 100. The kidney value of the first snail sacrificed at 7 days was in error, and the value shown was estimated by comparison with other data including other experiments. (from Bayne, 1973)

the innominate cells, for the lysin may be evoked in fresh normal blood in vitro by adding a large number of ciliates (*Anophrys*) to the blood (Bang, 1966).

Despite the variety of mechanisms for clearance of foreign particles by the blood, including the urn cells and the dominant role of phagocytosis by the amoebocytes in this, there has been little work on the phagocytic phenomenon itself. Cantacuzène (1928) emphasized the agglutination of the amoebocytes themselves so that spheroidal masses are formed with short spiny pseudopods at the periphery. Hyalin and granular amoebocytes are the only cells within these masses and phagocytosis takes place within the mass.

Earthworms. The earthworms have been studied particularly by Cooper (1973), whose extensive elegant experiments on transplantation have stimulated many people to make further attempts to see if anamnestic responses do occur among the invertebrates. The early work on phagocytosis in earthworms was done by Cameron

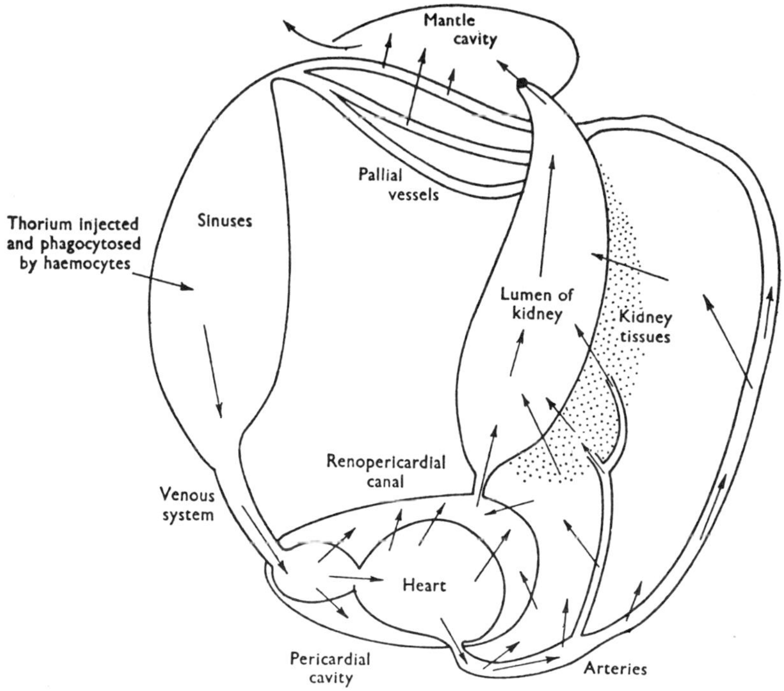

Figure 6. Diagram of main routes taken by migrating haemocytes. The chief pathway is through the heart wall into the pericardial cavity, via the renopericardial canal into the lumen of the kidney and out through the nephropore. Other cells join this route by migrating through the pericardial wall, while yet others enter the kidney lumen through its tissues. A distinct pathway, shorter yet apparently of secondary importance, is into the mantle cavity from the pappial system of vessels. (from Brown and Brown, 1965)

(1932), a noted British pathologist, who undertook to analyze inflammation both in the caterpillar and the earthworm. In the latter, he found that phagocytosis of bacteria, foreign particles and carmine, readily took place, starting within 30 minutes and gaining its maximum at 24 hours. The coelomocytes participated in the process, but the chloragon cells did not. Red cells, as well as tubercle bacilli, were readily phagocytosed. The bovine and avian strains of tubercle bacilli persisted in the tissues for as long as 21 days, but human tubercle bacilli disappeared in 5 days. He emphasized the independence of the different segments of the animal, the continuing role of phagocytes within the segments, and the exit of many bacteria by way of phagocytosis. He found that amoebocytes laden with foreign material could be easily detected by making a light press-preparation on the surface of the earthworm, thus catching the coelomocytes as they left the animal.

Arthropods. The classic work remains that of Metchnikov (1892), and the lack of any great advances since his time is probably due to the fact that no one else has studied natural infections in a similar small transparent, readily available animal. Metchnikov not only observed the process of phagocytosis minutely, but pointed out that there was a direct correlation between the success of phagocytosis and the survival of the water flea (Metchnikov, 1884). More recent studies on the way in which *Daphnia* are stimulated to form hemoglobin in the circulating fluid in response to the lack of oxygen (Fox, 1948; Fox and Phear, 1953) has not provoked a study of the relation of this to the activity of phagocytes. There is a great need to return to this classical animal and reinvestigate the total way in which it overcomes infection.

The extensive work on phagocytosis and defense mechanisms in insects will not be reviewed here because of the existence of the very complete and classic review by Salt (1970), but it is particularly important to take note of his additional assay on defense mechanisms of parasites against their hosts (Salt, 1968), in which it is recognized that defense against phagocytosis by the host is one of the most effective means of protecting the parasite.

Also noteworthy is the work of Wigglesworth (1970), who indicates that the pericardial cells of insects act as an analogue of the reticuloendothelial system of vertebrates.

Phagocytosis is now being actively studied in several freshwater arthropods, particularly in the Australian crayfish, by McKay and his associates (1970). They have worked with several gram-negative bacteria which are pathogenic for the crayfish, and believe that they have shown that a series of injections with non-virulent bacteria will immunize the animal against lethal infections. Since this immunization was produced by gram-negative bacteria, the effect of endotoxin was studied and it was shown that multiple doses of this material did decrease the mortality of experimentally inoculated animals (McKay et al., 1973) (Table I).[1] Since endotoxin causes extensive changes in the amoebocytes of a variety of marine arthropods (Bang, 1956), this effect must be taken into account in relation to immunization.

Echinoderms. To return to the other branch of the evolutionary tree, we will discuss first the echinoderms and then the ascidians. The echinoderm larva was studied first by Metchnikov (1892), and later one of his students, Metalnikov (1927), showed that as soon as the larve had developed to the stage where the mesoderm was active, phagocytosis was readily demonstrable. Recognition of foreign particles precedes phagocytosis. Ghiradella (1965) showed that tissues from other species of sea star were recognized as foreign when they were placed in the coelom of *Asterias*. This recognition was manifest by the accumulation of amoebocytes around the foreign tissue and its subsequent death. The injection of sea urchin cells, which carry a marker in terms of

[1] (from McKay et al., 1973)

their own pigment, into the sea star allows one to observe directly the reaction of the host (Reinisch and Bang, 1971). Immediately after injection, the foreign amoebocytes are surrounded by phagocytes; clumps of cells are formed and these eventually become stuck in various places on the wall of the coelom. Many of these clumps occur at the tip of the papulae and this is followed by necrosis of these tips so that the foreign material migrates out of the injected star. This process of elimination of foreign material by phagocytes was clearly described by Durham (1888) in earlier work.

Sea stars may also be readily infected with bacteria and when the physiological conditions are unfavorable, great numbers of the bacteria are found within the coelomic cavity, which is normally kept sterile by the continuing process of phagocytosis (Bang and Lemma, 1962). Gram-negative bacteria were poorly phagocytosed in vitro by the coelomocytes of sea urchins, whereas gram-positive (Gaffkya) were readily phagocytosed (Johnson, 1969).

TABLE I

EFFECT OF IMMUNIZATION WITH VACCINES PREPARED FROM VARIOUS BACTERIA ON THE RESISTANCE OF THE CRAYFISH (PARACHAERAPS BICARINATUS) TO INFECTION WITH PSEUDOMONAS CP

Vaccine*	Percentage mortality	P
Staphylococcus aureus	85	NS†
Staphylococcus albus	89	NS
Micrococcus lysodiekticus	96	NS
Bacillus subtilis	95	NS
Pseudomonas CP	45	$0.001 < P < 0.005$
Salmonella typhimurium	35	< 0.001
Escherichia coli	42	$0.001 < P < 0.005$
Pseudomonas fluorescens	37	< 0.001
None	92	--------

*All vaccines given at a dose equivalent to 2×10^7 bacteria.
†NS = not significant.

The most direct link between work on the echinoderms and the vertebrate is that of Prendergast and associates (1970, 1974; Willenborg and Prendergast, 1974). Starting with the phenomenon of clumping of cells produced by extracts of amoebocytes, they found that two components are present in these extracts. One is responsible for the clumping or cellular clotting of the amoebocytes, and the other produces a remarkable effect on vertebrate immune cell mechanisms. This fraction is able to prevent migration of vertebrate macrophages and apparently prevents lymphokin-release by lymphocytes in the vertebrate. This work is described by Dr. Prendergast later in this monograph.

Ascidians. Ascidians are the most important group of animals to be studied in connection with the evaluation of immune mechanisms, but probably have been the most neglected because they are of no economic significance. Cantacuzène (1919) did study active infection of one of these ascidians. In the presence of active bacterial infection, he showed that the amoebocytes developed the capacity to form a substance which reacted with the bacteria and formed a jelly-like covering around them. These cells increased in activity during the course of the infection and were apparently responsible for eliminating the bacteria, even though phagocytosis was apparently incomplete.

Recently Anderson (1971) has injected particulate material into another ascidian and shown that it was rapidly phagocytosed.

Freeman (1970), in a classic study on the analogue of the reticuloendothelial system in the ascidian, has demonstrated that there are localized areas of activity and that certain cells, which appear morphologically like lymphocytes, are radiosensitive. Furthermore, these cells play a unique role in the regeneration of the tunicate and thus have the capacity to multiply following a particular stimulus. The importance of work on this group of animals is now recognized by a number of scientists, and we may look forward to rapid accumulation of information in the near future.

Finally, I would like to return to the question of transparent animals and the value of work on these. There are a number of these animals available. One of these is the transparent oyster, *Placenta placenta*, which is found off the coast of India and other oriental areas. This animal is almost completely transparent so that when we injected particulate material directly into the heart, the distribution of this throughout the blood vessels of the mantle and the gills was clearly observed and highly delineated. Particles of carbon ingested by the young oysters could be followed along ciliary pathways, and passage out of the anus was clearly observed. The reaction of this animal to active infections could easily be followed directly.

In addition, there are the numerous sand fleas at the edge of any beach. They have been studied for circulation of amoebocytes, and indeed a luminescent bacterial infection of these has been reported (Giard and Billet, 1889; Inman, 1927). Recently, at a marine station in South India, we discovered a number of such naturally infected luminescent sand fleas and could observe directly the mass of bacteria accumulating within the open vascular system (Das et al., 1973).

There are many other transparent animals to be studied and, no doubt, each will present its own peculiar mechanism of response.

REFERENCES

Anderson, R. S. (1971). *Biol. Bull., 141*:91.

Bang, F. B. (1956). *Bull. Johns Hopkins Hosp., 98*:325.

Bang, F. B. (1961). *Biol. Bull., 121*:57.

Bang, F. B. (1966). *J. Immunol., 96*:960.

Bang, F. B., and Lemma, A. (1962). *J. Insect Pathol., 4*:401.

Bayne, C. J. (1973). *J. Comp. Physiol., 86*:17.

Bayne, C. J., and Kime, J. B. (1970). *Malacol. Rev., 3*:103.

Beug, H., Katz, F. E., and Gerisch, G. (1973). *J. Cell Biol., 56*:647.

Blackbourn, D. J., Taylor, F. J. R., and Blackbourn, J. (1973). *J. Protozool., 20*:286.

Brown, A. C., and Brown, R. C. (1965). *J. Exp. Biol., 42*:509.

Cameron, G. R. (1932). *J. Pathol. Bacteriol., 35*:933.

Cantacuzène, J. (1897). *C. R. Acad. Sci., 125*:326.

Cantacuzène, J. (1919). *C. R. Soc. Biol., 82*:1019.

Cantacuzène, J. (1928). *Arch. Roum. Pathol. Exp., 1*:7.

Cheng, T. C., Yee, H. W. F., and Rifkin, E. (1968a). *Pacific Science, 22*:395.

Cheng, T. C., Rifkin, E., and Yee, H. W. F. (1968b). *J. Invert. Pathol., 11*:302.

Cheng, T. C., Yee, H. W. F., and Rifkin, E., and Kramer, M. D. (1968c). *J. Invert. Pathol., 12*:29.

Cooper, E. L. (1973). In "Non-Specific Factors Influencing Host Resistance" (W. Braun and J. Ungar, eds.), pp. 11-23. S. Karger, Basel, Switzerland.

Das, P. K., Bang, F. B., and Bang, B. G. (1973). Personal observations.

Durham, H. E. (1888). *Proc. Roy. Soc. Med., 43*:327.

Faure-Fremiet, E. (1927). *Arch. Anat. Microbiol., 23*:99.

Feng, S. Y. (1966). *J. Invert. Pathol.*, *8*:505.

Feng, S. Y., and Stauber, L. A. (1968). *J. Invert. Pathol.*, *10*:94.

Fox, H. M. (1948). *Proc. Roy. Soc. London (B)*, *135*:195.

Fox, H. M., and Phear, E. A. *Proc. Roy. Soc. London (B)*, *141*:179.

Freeman, G. (1970). *J. Reticuloendoth. Soc.*, 7:183.

Galtsoff, P. S. (1942). *8th Am. Sci. Congr.*, *3*:411.

Gerisch, G. (1973). In: "Non-Specific Factors Influencing Host Resistance" (W. Braun and J. Ungar, eds.), pp. 33-34. S. Karger, Basel, Switzerland.

Gerisch, G., Luderitz, O., and Ruschmann, E. (1967). *Ztschr. Naturforsch.*, *22b*:109.

Prendergast, R. A., and Suzuki, M. (1970). *Nature*, *227*:277.

Prendergast, R. A., Cole, G. A., and Henney, C. S. (1974). *Ann. New York Acad. Sci.*, (in press).

Reade, P. C. (1968). *Aust. J. Exp. Biol. Med. Sci.*, *46*:219.

Reade, P., and Reade, E. (1972). *J. Reticuloendoth. Soc.*, *12*:349.

Reinisch, C. L., and Bang, F. B. (1971). *Cell. Immunol.*, *2*:496.

Reynolds, B. D., and Looper, J. D. (1928). *J. Parasitol.*, *15*:23.

Salt, G. (1968). *Biol. Rev.*, *43*:200.

Salt, G. (1970)."The Cellular Defence Reactions of Insects." Cambridge Monogr. Exp. Biol. No. 16, Cambridge University Press, Cambridge, England.

Savanat, T., and Pavillard, E. R. J. (1964a). *Aust. J. Exp. Biol. Med. Sci.*, *42*:615.

Savanat, T., and Pavillard, E. R. J. (1964b). *Aust. J. Exp. Biol. Med. Sci.*, *42*:625.

Smith, F. G. W. (1941). *Ecology*, *22*:415.

Sparks, A. K. (1972). "Invertebrate Pathology: Noncommunicable Diseases." Academic Press, New York.

Stauber, L. A. (1950). *Biol. Bull.*, *98*:227.

Stiven, A. E. (1964). *Ecol. Monogr., 34*:119.

Stuart, A. E. (1968). *J. Pathol. Bacteriol., 96*:401.

Tripp, M. R. (1960). *Biol Bull., 119*:210.

Tripp, M. R. (1970). *J. Reticuloendoth. Soc.,* 7:173.

Wigglesworth, V. B. (1970). *J. Reticuloendoth. Soc.,* 7:208.

Willenborg, D. O., and Prendergast, R. A. (1974). *J. Exp. Med., 139*:820.

PHAGOCYTOSIS BY INVERTEBRATE CELLS IN VITRO: BIOCHEMICAL EVENTS AND OTHER CHARACTERISTICS COMPARED WITH VERTEBRATE PHAGOCYTIC SYSTEMS

Robert S. Anderson

Sloan-Kettering Institute for Cancer Research
Donald S. Walker Laboratory
145 Boston Post Road
Rye, New York

Introduction. Phagocytosis is an important mechanism for clearance of foreign materials from the blood, and for protection against infection, in both vertebrate and invertebrate animals. This paper reviews much of the research, to date, concerning the *in vitro* uptake of foreign particulates by phagocytes of non-vertebrate animals.

Data will be presented in roughly phylogenetic order from acellular organisms through arthropods, and through the invertebrate series leading to the vertebrates, including echinoderms and urochordates. The evidence for *in vitro* phagocytosis by invertebrates, including insects, will be discussed in detail. The role of hemolymph factors as opsonins in certain experimental systems will also be considered. Data from studies of metabolic events associated with phagocytosis by insect hemocytes will be compared to data from similar studies of mammalian phagocytes. Evidence of complement components in invertebrate hemolymph, as well as speculation concerning their possible opsonic role, will be presented. The paper concludes by drawing attention to certain similarities between invertebrate phagocytes and mammalian mononuclear cells.

In vitro phagocytosis by non-vertebrate animals. Before considering the results of studies of *in vitro* phagocytosis in more advanced invertebrate species, some mention of similar experiments using acellular organisms should be made. Much of the literature on phagocytosis by acellular organisms concerns food gathering but some provides information concerning the evolution of phagocytosis as a defense mechanism, as shown by the recognition of non-food particulates.

Phagocytosis of polystyrene beads (PS) by the cellular slime mold *Polysphondylium pallidum* has been reported by Githens and Karnovsky (1973) The slime molds resemble plants only in their reproductive stages; in their vegetative existence the body is a naked protoplasmic mass composed of ameboid cells. The phagocytosis of PS was quantified by ultraviolet spectrophotometry of dioxane extracts of washed cell pellets. In typical experiments over 95% of the cells phagocytized at least one bead during a 15 minute incubation. The uptake of beads was a linear function of time for at least 30 minutes over a fairly broad range of PS concentrations. If PS concentrations were held constant and the number of cells varied, the uptake of PS was proportional to the number of cells. However, the rate of uptake per cell was unchanged. Indeed, it was shown that uptake was a function of PS concentration, within the concentration limits examined. Similar results have been obtained with *Acanthamoeba* and polymorphonuclear leukocytes (PMN). The effect of PS bead size on the rate of uptake was determined by assaying the weight ingested/unit time by amoebae incubated with equal concentrations, by weight, of PS ranging from 0.109 to 1.857 μm in diameter. Beads approximately 1 μm in diameter seemed to be the optimal size for ingestion.

The kinetics of phagocytosis by slime mold amebae were reported in the same paper, and estimates of the PS concentration required for half-maximal phagocytic velocity (K_p) and maximal velocity (V_p^{max}) were made. Such data could be compared with similar data produced by studies of *Acanthamoeba* and PMN. These comparisons indicated that the slime mold cells were the most avid phagocytes (in terms of affinity for PS) while the leukocytes are the most effective phagocytes (in terms of maximal uptake velocity per unit surface area).

Weisman and Korn carried out kinetic studies of PS phagocytosis by *Acanthamoeba* sp. No requirements for specific organic molecules for uptake could be demonstrated; however, appropriate osmotic conditions were required. The rate of uptake was a linear function of time of incubation and cell concentration. The rates of uptake of various bead sizes were proportional to the concentration of beads. The values for K_p and V_p^{max} were similar for all sizes of beads examined. Therefore, the kinetics of uptake are a function of bead mass not number, surface area or diameter. There was little ingestion of incubation medium accompanying particle uptake (as indicated by a lack of uptake of C^{14} glucose or I^{131} albumin from the medium during PS phagocytosis). The data indicate that PS is bound to the cell surface and accumulates until an optimum volume is attained; this volume is approximately constant regardless of bead size. When the appropriate mass accumulates, it is ingested by a mechanism that is highly selective. Electron micrographs show that beads of diameter 1.3 μm and greater are ingested individually; those of diameter 0.557 μm and smaller are accumulated at the cell surface and ingested in groups. Often more than 1,000 PS beads could be seen tightly packed within one membrane-bound vessicle.

Recently the details of PS bead ingestion by *Acanthamoeba castellanii* have been graphically revealed by the use of the scanning electron microscope by Goodall and Thompson (1971). The changes in surface morphology associated with ingestion of the beads such as food cup formation and membrane invagination are shown.

Rabinovitch and DeStefano (1971a) showed that *Acanthamoeba* sp. displays phagocytic discrimination, unrelated to feeding, by a series of experiments involving the uptake of mammalian erythrocytes (E). The amebae showed clear differences in the uptakes of red cells from different species: for example, the uptake of horse E was about 100-fold greater than the uptake of sheep or human E. Furthermore, chemical modification of the E frequently resulted in an augmented rate of phagocytosis. Glutaraldehyde-treated horse E became attached to the amebae in about 1 min and ingestion was begun within 3-5 min. The process of phagocytosis was shown to be temperature-sensitive; at 0-1°C attachment did not occur, at 5-9°C attachment was observed but with minimal ingestion, at 10-25°C attachment and ingestion increased with increasing temperatures. Treatments other than glutaraldehyde which increased erythrocyte phagocytosis included: formaldehyde, tannic acid, carbodiimide, poly-l-lysine and salt-free sucrose; other treatments (such as high salt, chymotrypsin, trypsin, etc.) were without effect. The amebae were found to ingest materials also ingested by mammalian macrophages in the absence of serum such as: titanium dioxide, iron, carbon, glass, carmine, crystals of uric acid or tyrosine, heat denatured albumin or immunoglobin and starch or yeast cell walls.

Phagocytosis serves a major protective role in metazoa. The majority of observations of phagocytosis by invertebrate hemocytes have been made *in vivo* or *in vitro* using cells from animals which have previously received injections of foreign particulates; this literature will be reviewed elsewhere in this volume. Only those papers clearly concerned with *in vitro* phagocytosis by invertebrate cells will be considered here.

One of the earliest reports of *in vitro* phagocytosis was by Cameron (1932) in his classic description of inflammation in earthworms. The annelid coelomocytes were shown to take up various bacteria, and the process was specific in that the phagocytes could distinguish between quite similar biological materials.

In vitro phagocytosis has been reported in all of the major molluscan groups. Bang (1961) recorded *in vitro* phagocytosis of bacteria by the oyster (*Crassostrea virginica*). Bacterial uptake was variable, but in certain preparations massive adherence of bacteria to the ameboyctes was observed, followed by ingestion. Extracellular clotting was sometimes observed, a phenomenon which might serve to immobilize the bacteria prior to phagocytosis.

Gastropod molluscs also possess blood cells which are actively phagocytic *in vitro*. Ameboyctes from the hemolymph of *Helix aspersa* will readily take up formalized yeast cells and formalized sheep E in normal serum (Prowse and Tait, 1969). Evidence was presented for a serum factor which is required for phagocytosis in this system. In the presence of serum, phagocytosis was rapid, with a mean of 3-5 particles ingested per amebocytes by 1 hour. Preliminary data from work currently in progress in our laboratory indicate that hemocytes from another snail (*Otala lactea*) can take up various foreign particles in the absence of hemolymph; however, the presence of hemolymph may stimulate phagocytosis in certain cases. Also, *O. lactea* phagocytes can discriminate between various species of bacteria and between untreated and chemically modified erythrocytes.

Cells from the hemolymph of the California sea hare have been used to study the *in vitro* uptake of chicken red blood cells by Pauley, et al. (1971). Washed hemocytes phagocytized chicken E incubated in serum more rapidly than untreated E or E exposed to heated serum.

Stuart (1968) has studied the phagocytic hemocytes of a cephalopod mollusc, the lesser octopus. He notes that this animal, like other invertebartes, lacks lymphatics and has no discrete foci of lymphoid tissue. Treatment of the hemocytes with phytohemagglutinin was without effect; therefore, it was concluded that the cells were unrelated to the lymphoid series. These blood cells were actively phagocytic in culture, ingesting human E and concentrating nile blue sulfate. The hemocytes are thought to be identical to those in phagocytic depots in the salivary glands, gills and white body of the octopus. As in the case of other molluscs, a humoral agent potentiating phagocytosis is reported in this species.

One of the earliest observations of the phenomenon of phagocytosis was that of Metchnikoff (1884) in the crustacean, Daphnia. Since that time many reports of phagocytosis by various arthropods have appeared. The crustaceans and insects have been most extensively studied, and the results of *in vitro* experiments with representatives of these classes will be discussed in the following paragraphs. The evidence for *in vitro* phagocytosis by insect hemocytes in particular will be discussed in detail.

McKay and Jenkin (1970a) used hemocyte monolayer preparations to study phagocytosis of sheep E by crayfish (*Parachaeraps bicarinatus*) cells. They found that a serum factor was required for adherence and phagocytosis. This monolayer system was used to examine the correlation of phagocytic activity with resistance to infection (McKay and Jenkin; 1970b). It was shown that a degree of protection against infection could be produced in the crayfish by injection of bacterial vaccines. Hemocytes from "immunized" crayfish showed increased phagocytic activity when compared to cells from normal crayfish.

There is considerable evidence that insect cell cultures are capable of phagocytizing a wide range of foreign particles. Grace (1962) demonstrated the development of a cytoplasmic polyhedrosis in cultured *Antheraea eucalypti* cells following *in vitro* phagocytosis of the polyhedra. Starch grains and India ink suspensions were added to other *A. eucalypti* cell cultures; 1 1/2 hr later starch grains were seen in the cytoplasm, and most of the cells were packed with carbon. Both the adhesion and ingestion stages of phagocytosis were observed directly and photographed. Using time-lapse photography, Grace and Day (1963) filmed the elimination of dead cells by phagocytosis in lepidopteran cell cultures. Grace (1971) concludes that there is little doubt that these cultured phagocytes are capable of ingesting dead or foreign material, virus particles, and bacteria. Vago (1964) also has filmed various aspects of the morphology and physiology of cultured insect phagocytes. In a discussion of this film (Vago 1972), he refers to the fact that different stages of phagocytosis of bacteria may be followed in the motion picture.

In vitro reactions of phagocytic plasmatocytes from the blood of the cockroach (*Periplaneta Americana*) were studied by Scott (1971a). Monolayers of cockroach hemocytes consisted of essentially homogeneous cell populations which readily adhered to glass and rapidly accumulated neutral red, characteristics common to mammalian macrophages as well. The cells remained in good condition during the course of the experiments on the basis of morphology, uptake of vital dye, exclusion of trypan blue and extent of spreading. Cells in these monolayers formed rosettes with either sheep or chicken red cells. Ingestion of the erythrocytes was only occasionally observed; this was attributed to a combination of sub-optimal culture conditions and the large size of the target particle. Adherence was maximal at about 20°C; however, it was reduced at both lower and higher temperatures. The active uptake of vertebrate E onto the surface of the hemocytes was interpreted as the first stage of phagocytosis.

Rabinovitch and DeStefano (1970) carried out comparative studies of phagocytosis of erythrocytes by wax moth (*Galleria mellonella*) hemocytes and mouse macrophages, *in vivo* and *in vitro*. *In vivo* there was no, or minimal, uptake of untreated erythrocytes; however, formalized, glutaraldehyde-treated effete, and antibody-sensitized E were phagocytized. Untreated sheep or horse E were minimally taken up *in vitro*. Effete and sensitized E were also not extensively phagocytized *in vitro*. Certain treatments that caused marked changes in E surface conformation caused enhanced phagocytosis *in vitro*; exposure to other chemicals was without effect. These phagocytosis-stimulating treatments included: formaldehyde, tannic acid, saponin, colloidal silica, ferrous sulfate and carbodiimide. In most cases red cells treated so as to be ingested by insect hemocytes in the absence of hemolymph were also attached to or phagocytized by mouse macrophages in the absence of serum. These reactions carried out in the absence of humoral recognition factors constitute "non-immunological" phagocytosis. Possibly common features of the cell membranes of these phylogenetically distant phagocytes may underlie this form of recognition. *Galleria* cells *in vivo* took up a wider range of modified E than were ingested *in vitro* (10 types *in vivo* vs. 7 *in vitro*). It remains to be shown whether or not this difference can be ascribed to hemolymph recognition factors.

The *in vitro* bactericidal capacity of the cockroach *Blaberus craniifer* was assayed by Anderson, et al. (1973a). Hemocytes, maintained in short-term culture, were capable of phagocytizing and destroying 5 species of bacteria. Considerable ingestion and killing of these bacteria occurred during the first 10 min of incubation. Maximal bactericidal activity against *Staphylococcus aureus* 502A was attained at a hemocyte: bacteria ratio of 8:1, at pH 7.0 and 26°C. Under these conditions, the hemocytes destroyed about 25% of the initial *S. aureus* innoculum in 10 min, less than 50% survived after 60 min. The measured decrease in viable bacteria was shown to be caused entirely by intracellular killing by the hemocytes. Bacteria were seen in the hemocyte cytoplasm surrounded by vacuoles, and gradually lost morphological definition and normal staining characteristics as a result of intracellular digestion. Minimal cell lysis during incubation might release potentially bactericidal agents into the medium. However, aqueous extracts of experimentally lysed hemocytes were shown not to be bactericidal. Incubation of the bacteria in the suspending medium in the absence of cells did not decrease their numbers. To assay for possible production of bactericidal substances by the hemocytes, bacteria were incubated in cell-free medium in which unstimulated hemocytes had been maintained. This medium was not bactericidal. Hemocytes in suspension were phagocytically stimulated with latex beads or bacteria; after 1 hr the cells were removed. This medium was also not bactericidal. Hemolymph, at the concentration found in the hemocyte-bacteria mixtures, was not bactericidal. Undiluted, cell-free, hemolymph (and hemolymph concentrated by ultrafiltration) was not bactericidal for 4 of the 5 bacteria killed by the hemocytes and was very weakly bactericidal for *S. aureus*.

Four other species of bacteria were phagocytosed but not killed by *Blaberus* hemocytes. One of these, *Pseudomonas aeruginosa* is an established insect pathogen. However, deficient bactericidal capacity probably does not entirely correlate with pathogenicity since certain bacteria pathogenic to insects were destroyed in this *in vitro* system.

Anderson, et al. (1973b) used the *in vitro* system just described to study the metabolism of phagocytizing insect cells. The results of this work will be discussed in detail in section IV of this paper; however, some data from this study give additional information on the process of *in vitro* phagocytosis in general. Biochemical assays of glucose utilization, glycogen breakdown and lactate production by *Blaberus* hemocytes during phagocytosis of latex beads indicate that the glycolytic pathway supplies energy for particle engulfment. This is also the case for mammalian phagocytes. By using metabolic inhibitors, at concentrations which had no effect on the bacteria in question, it was possible to block phagocytosis and bactericidal activity by the hemocytes. This is additional evidence that the source of antimicrobial activity in the hemocyte suspensions resides in the cells themselves and is a result of *in vitro* phagocytosis.

On the limb of the phylogenetic tree which leads toward the vertebrates are several interesting groups including the echinoderms and the urochordates. *In vitro* phagocytosis by sea urchin (*Strongylocentrotus*) amebocytes has been the subject of several recent papers (Johnson, 1969 and Johnson, et al., 1970). In hanging drop preparations coelomocytes, called "bladder amoebocytes" could readily take up Gram-positive bacteria. The Gram-negative marine bacterium "pi" was phagocytosed by 4-day old hanging drop cultures. This process, as well as the uptake of a Gram-positive form (*Bacillus imomarinus*), followed by electron microscopy. Certain Gram-negative bacteria were usually not phagocytized but were walled off and encapsulated.

Reports of *in vitro* phagocytosis by tunicate hemocytes are rare. There is an early study by Fulton (1920) that reports *in vitro* uptake of carmine particles by the cells of *Ascidia atra*. Phagocytosis did not occur in sea water alone but required the presence of tunicate plasma. Results of work in progress in the author's laboratory suggest that studies of *in vitro* phagocytosis can be performed using tunicate (*Halocynthia pyriformis*) blood cells. These cells can recognize and phagocytize a wide range of foreign particles.

The role of opsonins in in vitro phagocytosis by invertebrate hemocytes. Phagocytosis by mammalian blood cells frequently involves the participation of antibody and components of complement. This process has sometimes been referred to as "immune" phagocytosis. "Non-immune" phagocytosis occurring in the absence of serum also may be observed in mammals. It is the goal of this section of the review to indicate that both "immune" and "non-immune" phagocytosis can be studied *in vitro* using invertebrate cells. The possible role of complement will be discussed in a subsequent section (V).

It is generally agreed that invertebrates lack mammalian-type immunoglobulins. However, most invertebrates studied have been shown to possess an array of humoral factors (antisomes) which functionally mimic antibodies. There are numerous reports on serum requirements for phagocytosis among the invertebrates which suggest the presence of a type of humoral recognition factor or opsonin.

Opsonizing activity has been most frequently associated with agglutinins, particularly hemagglutinins, although other factors have also been suggested. Hemagglutinins (lectins) with varying degrees of specificity have been found in plants and a huge number of invertebrates. A complete review of the distribution and physical-chemical properties of hemagglutinins in invertebrate animals cannot be given here. Only their role as *in vitro* opsonins will be discussed in detail.

Phagocytosis by protozoans is usually thought to require no specific organic molecules in the medium as long as the appropriate osmotic conditions are maintained. Weisman and Korn (1967) never observed any requirement for added protein for the uptake of latex beads by *Acanthamoeba*. Latex beads can also be phagocytized by mammalian leukocytes in the absence of serum proteins.

In vitro uptake of bacteria by mammalian blood cells often requires the presence of serum proteins in the medium or prior opsonization of the bacteria. There is, of course, no absolute requirement for mammalian antibodies for uptake of bacteria by slime mold amebae or protozoans. However, Gerisch, et al. (1967) report that antibodies promote phagocytosis of *Salmonella* sp. by *Dictyostelium discoideum*. The rough forms of *Salmonella* were quickly and completely phagocytized by the slime mold amebae without prior opsonization. On the other hand, untreated smooth forms were phagocytized slowly, if at all. If the smooth forms were opsonized with rabbit anti-*Salmonella* serum (1:100, 30 min, 4°C), they were readily phagocytized. Smooth forms of pneumococcus require opsonization to be phagocytized by mammalian cells in cell suspension; the smooth form can be ingested without serum if they are immobilized on agar. This same phenomenon was shown for phagocytosis of unopsonized smooth forms of *Salmonella* on agar by *Dictyostelium*.

Results similar to those reported for slime mold were also reported in *Acanthamoeba* by Rabinovitch and DeStefano (1971b) who showed that antibody and plant agglutinins stimulate phagocytosis of sheep erythrocytes. The addition of rabbit anti-sheep E serum consistently stimulated red cell uptake at serum dilutions that did not cause hemagglutination. Phagocytosis was enhanced when the E were treated with immune serum and washed prior to exposure to the amebae. The effect was caused by antibody because normal rabbit serum was ineffective and antiserum absorbed with sheep E had no opsonic activity. The plant agglutinins, phytohemagglutinin M or concanavalin A, also increased phagocytosis of red cells.

Opsonization has not been studied among the *Porifera*, although hemagglutinins may be found in certain species (Khalap, et al., 1970). R. Brown, et al. (1968) tested 54 invertebrate species, including coelenterates and annelids for blood group specific agglutinins and concluded that the occurrence of strong, specific hemagglutinins, such as those found in certain molluscs, is uncommon. Nonspecific agglutination and weak anti-H reactions were more usually encountered. Red cell agglutinins have been found in numerous invertebrates, including polychetes, by Tyler (1946). The physiological role of many of these factors has yet to be determined. In contrast to hemagglutinins in other phyla, those of molluscs and arthropods are often quite strong and specific. *In vitro* phagocytosis by mollusc and arthropod cells has been more extensively studied than other groups, so it is not surprising that most of the information on invertebrate opsonins comes from studies of these animals.

Hemagglutinins have been described in a number of pelecypod molluscs. An agglutinin specific for human blood group A was found in the Butter Clam (*Saxidonus giganteus*) by Johnson(1964). Weak agglutinins for human A, A_2, A, B and A_2 B red cells was found in the swan mussel (*Anodonta cygnea*) by Pemberton (1969). Jenkin and Rowley (1970) purified a hemagglutinin to rat and rabbit E by density gradient centrifugation from the hemolymph of the Murray mussel (*Velesunio ambiguus*). It was found that this mussel has several hemagglutinins which behave as a homogeneous group of proteins. These proteins lack tyrosine and sulfur-containing amino acids and are rich in aspartic acid; phytohemagglutinin has similar amino acid composition.

A hemagglutinin against several vertebrate E and all human A B O (H) antigens was reported in the blood of the oyster, *Crassostrea virginica* by Tripp (1966). This factor was a nondialyzable, heat-sensitive protein that migrated electrophoretically between α and β human globulins. Phagocytosis of rabbit E by oyster cells could proceed in hemagglutinin-free medium, but when hemagglutinin-containing shell liquor was added, phagocytosis was stimulated. In the presence of hemagglutinin, both the numbers of cells containing E and the numbers of E/phagocyte increased. Tripp and Kent (1967) observed that about 90% of the phagocytes contained 1 or more E by 40 min in the serum-free medium. The presence of 1:2 oyster hemolymph depressed phagocytosis by extensive agglutination of red cells. The rate and extent of phagocytosis was stimulated by 1:16 homologous hemolymph, particularly during the first 20 min of the reaction. Phagocytosis was inhibited by 10% calf serum.

Oyster hemagglutin has been further characterized by McDade and Tripp (1967) who found that its reactive site on human cells is the ABO blood group hapten. The hemagglutinin required Ca^{++} to clump human E, but no such requirement was needed for other vertebrate E. None of 13 different saccharides was universally effective in inhibiting oyster hemagglutinin. Immunoelectrophoresis gives 3 bands in the region of human β_1-a_2 (Acton and Evans, 1967), which emphasizes the non-immunoglobulin nature of this humoral factor. The hemagglutinin was further shown to be associated with a 33.4 S molecule, which is a polymer of similar units noncovalently bonded to each other. Native hemagglutinin can be dissociated by dialysis vs 5M guanidine-HC1, pH 7.5, to yield these subunits, which are about 1S and have a molecular weight of about 20,000 (Acton, et al., 1969).

Gastropod molluscs possess hemagglutinins of a wide range of strength and specificity. Hemagglutinins are present in limpets (Pemberton, 1970a), slugs (Pemberton, 1970b), and many snails (Cheng and Sanders, 1962; Boyd and Brown, 1965; Boyd, et al., 1966; Prokop, et al., 1967; Gilbertson and Etges, 1967; Lee-Potter, 1969; Schnitzler and Kilias, 1970; Bizot, 1971). The snail *Otala lactea* has a strong, specific lectin which agglutinates blood group A. When hydrolyzed by papain, three smaller fragments are obtained, all of which possess anti-A activity. This result suggests that the native lectin has at least three combining sites (Matsubara and Boyd, (1974). Another snail (*Helix pomatia*) has six specific combining groups in its anti-A lectin (Hammarström and Kabat, 1971). Most antibodies have two combining sites; therefore, the larger number in some lectins probably represents a significant difference between vertebrate antibodies and invertebrate lectins.

The best evidence to date for the existence of an opsonic factor in the hemolymph of a gastropod is that of Prowse and Tait (1969) in *Helix aspersa*. In contrast to oyster hemagglutinin which influences the rate of *in vitro* phagocytosis, the snail opsonic factor is absolutely required for particle uptake. Prowse and Tait (1969) never speculate on the nature of the opsonin. Whether or not it is a hemagglutinin remains to be shown; hemagglutinins have been described in many species of *Helix*.

Large numbers of hemocytes were obtained from the perivisceral hemocoel of *Helix aspersa*. The cells were allowed to spread on glass for 1 hr at room temperature and washed free of serum and nonadhering cells prior to experimental manipulations. Formalized yeast cells were readily phagocytized in the presence of normal serum, but no phagocytosis was observed in serum which had been absorbed with formalized yeast. Yeast that had been opsonized in normal serum (1 hr room temperature) were phagocytized in absorbed serum. There was some degree of specificity of the opsonin for the two foreign particles. There was no phagocytosis of formalized yeast or formalized sheep E in serum absorbed with the homologous particle; however, phagocytosis occurred in serum absorbed with the heterologous particle and in normal serum.

A substance in the serum of the sea hare (*Aplysia californica*) which agglutinates marine bacteria and vertebrate red blood cells was characterized by Pauley, et al. (1971a). The agglutinating activity was caused by a heterogeneous group of high molecular weight molecules. The activity was associated with two substances having sedimentation coefficients of about 18.5 and 31 S. The active moiety is at least partially protein and is sensitive to heat, pH extremes, and extraction with 2-mercaptoethanol, phenol, chloroform and trichloracetic acid. The agglutinin is not specific in that any of the marine bacteria tested are capable of completely absorbing out the agglutinating activity of normal serum for itself and the other marine bacteria and the vertebrate red cells. However, *Serratia marcescens* is not agglutinated by normal sea hare serum and is incapable of absorbing out any agglutinating activity against the marine bacteria or the erythrocytes. Study of the physical and chemical properties of this lectin shows that it differs from other invertebrate agglutinins and from vertebrate antibodies.

Opsonic factors in the hemolymph of *Aplysia californica* were described by Pauley, et al. (1971b). Sea hare hemocytes were washed and allowed to spread on glass for 1 hr in moist chambers. About 4% phagocytosis of chicken E was observed at the end of 2 hr in the absence of homologous serum; about 30% phagocytosis took place in the same period of time in the presence of normal serum. Heating (100°C) the serum abolished most opsonic activity.

If the assumption is made that the opsonin and the natural agglutinin in sea hare serum are the same substance, certain observations can be easily explained. *Serratia marcescens* persists at almost an unchanged level in sea hare hemolymph for at least 16 days after injection, whereas the other bacteria tested were cleared by 1 day. Of the bacteria tested, only *S. marcescens* was not agglutinated by sea hare serum. *S. marcescens* can cause a fatal septicemia in *A. californica*. The titer of agglutinins drops dramatically after injection of all bacteria tested except *S. marcescens*. Therefore, it is suggested that soon after entering the hemolymph marine bacteria interact with an agglutinin, causing a concomitant decrease in the circulating agglutinin. The bacteria are opsonized by the agglutinin and are actively phagocytized; a reduction in the number of circulating hemocytes follows the period of intense phagocytosis. The bacteria may be destroyed intracellularly or are eliminated after migration of the phagocytes across epithelial borders.

During his study of the reticuloendothelial system of the lesser octopus (*Eledone cirrosa*) Stuart (1968) found that phagocytosis of human red cells required a serum factor. No phagocytosis occurred in serum-free suspending medium, but there was marked uptake in normal serum. Human E sensitized with *Eledone* serum were phagocytized in serum-free medium. Red cells sensitized with rabbit antiserum and effete red cells were not phagocytized in serum-free medium. It is probable that *Eledone* phagocytes lack receptors for mammalian antibodies.

The fact that some component of *Eledone* serum interacts with the surface of the red cells could be demonstrated by opsonizing the E in octopus serum, washing them, and treating them with rabbit anti-*Eledone* serum. Erythrocytes treated in this fashion agglutinate; the same effect is seen with yeast and *E. coli*, suggesting the presence of a single polyvalent opsonin with broad-spectrum affinity for a wide range of particles.

The octopus lacks natural agglutinins. Indeed, gel diffusion and immunoelectrophoresis show that the only major macromolecule that is present in the blood is hemocyanin. Stuart (1968) suggests that hemocyanin might act both as a respiratory pigment and as an opsonin for foreign material.

Hemagglutinins have been described in a number of crustaceans such as the spiny lobster (Tyler and Metz, 1945; Tyler and Scheer, 1945), and other species (Cornick and Stewart, 1968; Miller, et al., 1972). A hemagglutinin described by Cohen, et al. (1965) has been characterized in detail by Marchalonis and Edelman (1968) in the primitive arachnid *Limulus*. This natural agglutinin is a protein of 400,000 MW and sedimentation coefficient of about 13.5 S. This unit can be dissociated to give eighteen 1.5 S subunits of about 22,500 MW. This agglutinin is clearly dissimilar to classical immunoglobulin in that it is composed of a single class of homogeneous subunits, which are not joined covalently in the intact molecule. The configuration of this molecule was shown by electron microscopy by Fernández-Morán, et al. (1968).

Pauley (1973) has studied the natural agglutinins for mammalian red cells found in the Blue Crab (*Callinectes sapidus*). An attempt was made to increase the hemagglutinin titer by injection of E suspensions; however, the slight increases observed were not specific and of short duration, and might only be a result of trauma. A natural mammalian red cell hemagglutinin associated with the hemocytes of the lobster was found by Cornick and Stewart (1973).

Extensive work with the Yabbie, or crayfish (*Parachaeraps bicarinatus*), has shown that a blood factor which interacts with mammalian red cells is involved in opsonization *in vitro*. McKay, et al. (1969) showed that the percentage of hemocytes with adhering mouse E was dependent on the opsonization of the E by crayfish serum. The second stage of phagocytosis (ingestion) was not observed in the preparations probably because EDTA was present in the culture fluid; divalent cations are frequently required for ingestion. About five times more mouse E sensitized with Yabbie serum were taken up on the hemocytes than untreated E. Yabbie serum absorbed with mouse red cells lost its opsonic activity for mouse cells, whereas, serum absorbed with sheep E could still opsonize the uptake of mouse erythrocytes. *In vivo* experiments showed that crayfish hemocytes could extensively ingest mammalian red cells and that a drop in hemagglutinin titer was concomitant with red cell clearance.

McKay and Jenkin (1970a) showed that adherence of sheep erythrocytes to crayfish hemocytes was dependent on a serum factor and that no phagocytosis was observed unless the E were first treated with serum. The process of opsonization evidently is rather specific since only crayfish or crab sera could opsonize red cells in the crayfish hemocyte system.

No opsonins for bacteria could be found in crayfish hemolymph (McKay and Jenkin, 1970c). Injection of bacterial vaccines produced increased protection against infection by the same microorganisms. However, no bactericidal factors were present in normal or immune sera. The clearance of bacteria treated by prior exposure to immune serum was no more rapid than that of untreated bacteria. McKay and Jenkin (1970b) showed that the phagocytic activity of hemocytes from immunized crayfish was increased over non-immunized animals and suggest that this is the basis for the protective action of their vaccines. Hence, it was possible to correlate the extent of phagocytosis and degradation of mammalian red cells by the hemocytes *in vitro* with induced resistance to infection. It is interesting to note that an increased titer of red cell opsonin did not accompany immunization.

Rabinovitch and DeStefano (1970) observed that a slightly greater variety of chemically-modified erythrocytes were phagocytized by *Galleria in vivo* than *in vitro* (in the absence of hemolymph). Whether this difference is due to less than optimal conditions in the *in vitro* medium or to a lack of recognition factors is as yet unknown. Red cells sensitized by antibodies were not ingested *in vitro* by *Galleria* hemocytes, but were extensively phagocytized by mouse macrophages. This result is not surprising and indicates that insect cells lack receptors for mammalian antibody.

Naturally occurring hemagglutinins are present in the hemolymph of a number of insects, some of the early reports on these substances were by Bernheimer (1952), Bernheimer, et al. (1952), and Feir and Walz (1964). More recently, Scott (1971b) described a hemagglutinin in the American cockroach (*Periplaneta Americana*) which is a heat-labile, non-dialysable, euglobulin protein. The agglutinin had a degree of specificity for vertebrate red cells from certain species, although marked cross reactivities were evident. No specificity was seen for human ABO blood group antigens. The protein is not a γ globulin and is not inducible by injection of red cells; it does not fix complement. Bacterial agglutinins were rarely encountered in cockroach hemolymph (only 1 of 10 tested was agglutinated) and no agglutinating activity for yeast could be demonstrated. Anderson, et al. (1972) reported a heat-labile agglutinin with specificity for certain mammalian red cells in the cockroach *Blaberus craniifer*. The lectin was present in both late instar nymphs and adults. It is active in the absence of divalent cations and the activity is stable for several days at 0° and for prolonged periods at -70°C. Naturally occurring hemolysins were not present in *Balberus craniifer*.

Despite the presence of hemagglutinins in insect hemolymph, there is, at this time, no indication of the presence of opsonic factors. Scott (1971a) showed that active uptake of mammalian E onto the surfaces of cockroach phagocytes could occur *in vitro* in the absence of hemolymph. Sheep E, incubated in the highest hemolymph concentration which did not grossly clump the cells, were added to the washed hemocyte monolayer preparations. These sensitized cells were not as extensively taken up as were untreated controls. Therefore, the hemolymph was shown to be without opsonic activity. Red cells sensitized with rabbit anti-sheep E serum did not adhere to the cockroach hemocytes, but were avidly taken up by mouse peritoneal macrophages. These studies indicate that *Periplaneta* hemocytes do not possess a receptor site for γ-globulin on their surfaces.

Anderson, et al. (1973a) studied the *in vitro* phagocytosis and killing of a number of bacterial species by hemocytes of a cockroach (*Blaberus craniifer*). Various attempts to demonstrate an agent in the hemolymph with opsonic activity for *Staph. aureus* 502A (a microorganism that is readily phagocytosed by the hemocytes) were unsuccessful. Bacteria "opsonized" in hemolymph, or hemolymph concentrated by ultrafiltration, were not killed by hemocyte suspensions any more efficiently than were untreated bacteria. The hemocyte suspensions contained hemolymph diluted 50 to 100 fold; attempts to increase bactericidal activity of the cells by the addition of concentrated hemolymph was without result. The hemocytes actively phagocytized bacteria in the absence of hemolymph, in thoroughly washed monolayer preparations. Certain bacteria were not killed by the hemocytes; this was apparently not a result of impaired phagocytic recognition because these species are ingested. Therefore, it is suggested that intracellular killing is deficient for these bacteria, implying possible pathogenicity and a possible connection between the lack of hemocytic bactericidal capacity and the role of roaches as potential disease vectors.

Naturally-occurring hemagglutinins have been reported in several echinoderms (Tyler, 1946); their possible role in opsonization has not been determined.

Serum factors may be required for carmine phagocytosis by hyaline amoebocytes of the tunicate *Ascidia atra* (Fulton, 1920). Tunicates possess natural hemagglutinins (Tyler, 1946; Fuke and Sugai, 1972). *Ciona intestinalis* also has a mammalian red cell lectin (Wright, 1973). Duck or human A red cells injected beneath the tunic of *C. intestinalis* are agglutinated and phagocytosized rapidly.

Metabolic events associated with phagocytosis by mammalian phagocytes and insect hemocytes. Recently many of the details of the biochemistry of phagocytosis and post-phagocytic events in mammalian leukocytes have been defined; similar data for invertebrate species is meager. Anderson, et al. (1973b) and Anderson (1974) have begun a series of investigations on the metabolism of phagocytizing invertebrate blood cells, using the cockroach *Blaberus craniifer*. The data on insect cells presented in this section of the review derives from these papers, unless otherwise indicated.

Two major hypotheses to explain the source of energy required for phagocytosis have been formulated. The first of these emphasizes the role of surface forces. Fenn (1921-1922) and Ponder (1928) suggested that the energy required for ingestion could be provided by the surface-free energy associated with the phagocytized particle. Most present-day investigation centers around the second hypothesis emphasizing alterations in metabolism in phagocytizing cells. Attention was drawn to this idea by the work of Stahelin, et al. (1956, 1957) who reported that the addition of tubercle bacilli to guinea pig polymorphonuclear leukocytes (PMN) caused a marked increase in respiration and an increased flow of glucose through the direct oxidative pathway or hexose monophosphate shunt (HMP). Sbarra and Karnovsky (1959) went on to show that glycolytic inhibitors inhibited phagocytosis; observations that suggested that the energy for phagocytosis may be supplied by glycolytic pathways. Subsequent work indicated that chemical inhibition of cytochrome-linked respiration, Krebs cycle activity, and/or oxidative phosphorylation did not alter particle uptake of the accompanying metabolic activity (Iyer, et al., 1961). The demonstration that impaired phagocytic function could be correlated with specific metabolic deficiencies in certain human disease conditions (Holmes et al., 1967; Baehner and Nathan, 1967; Good et al., 1968) served to further stimulate research in the area of phagocyte biochemistry.

The data so far collected indicate that mammalian PMN, whether circulating or peritoneal, from various species behave quite similarly, in a metabolic sense. This is not the case for mononuclear phagocytes of mammalian origin, which differ considerably from one another in terms of their metabolic response to phagocytosis (Karnovsky et al., 1970). The observed differences depend in part on species and mode of harvesting the monocytes. The details of these metabolic alterations will be included in the following discussion.

Information on cockroach hemocyte metabolism was obtained using hemocyte suspensions which were shown to readily phagocytize bacteria *in vitro* (Anderson et al., 1973a). In some of these studies, changes associated with the uptake of latex particles rather than bacteria was measured to obviate any biochemical contribution by bacteria to the chemical assays. The cells avidly phagocytized latex as shown by spectrophotometric quantitation of latex after dioxane extration of the hemocytes by a modification of the technique of Roberts and Quastel (1963).

The oxygen consumption by non-phagocytizing cockroach cells (1-2.5 x 10^6 cells/m1) was determined (at 26°C with constant mechanical mixing) using an oxygen electrode. Particles (heat-killed *Staph. aureus* 502A or latex beads) were introduced into the incubation, and the rate of respiration during phagocytosis was recorded for at least 15 min. The calculated oxygen consumption (μ1 $0_2/10^7$ cells/m) for stimulated cells was slightly augmented during phagocytosis. This increment was not comparable to the marked increase in oxygen uptake seen in mammalian PMN (Sbarra and Karnovsky, 1959; Cohn and Morse, 1960; Holmes, et al., 1967) and casein-elicited peritoneal macrophages (Karnovsky, et al., 1970).

Casein-elicited mouse and guinea pig peritonial monocytes (MC) show a PMN-like respiratory burst during phagocytosis, but this increase is absent in phagocytizing MC obtained by lavage with saline and in cultured MC. Alveolar macrophages (AM) also show a minimal, or no, increase in respiration during phagocytosis (Karnovsky, et al., 1970).

To measure glycogen, glucose and lactate metabolism, insect hemocyte-latex suspensions were incubated on a reciprocating water-bath shaker at 26°C. Hemocytes not stimulated by particulates were also analyzed. Aliquots were deproteinized, neutralized and analyzed. Particle ingestion stimulated glycogen breakdown, glucose consumption and lactate production. Hence, the chemical data suggest that glycolysis accompanies phagocytosis in insect hemocytes, as well as in mammalian PMN and macrophages.

To further test metabolic involvement in phagocytosis, the effects of various inhibitors on killing of bacteria by the hemocytes were determined. This experimental approach was used by Cohn and Morse (1960) to study the effect of inhibitors on the bactericidal capacity and biochemistry of rabbit PMN. The inhibitors were shown to have no effect on the viability of the bacteria, at the concentrations used in the hemocyte suspensions.

The respiratory chain inhibitor KCN had no effect on the uptake or killing of *Staph. aureus* 502A by the hemocytes. In PMN, cyanide inhibited the respiration of resting cells but had no effect on intracellular killing or the associated, non-cytochrome linked, increment in oxygen uptake. Similar results are obtained from monocytes in which interference with respiration or oxidative phosphorylation does not affect phagocytosis; however, interference with oxidative metabolism has a definite effect in alveolar macrophages (Karnovsky, et al., 1970).

Inhibition of the glycolytic pathway in *B. craniifer* hemocytes by antimetabolites such as arsenite, iodoacetate, and fluoride resulted in marked decreases in the bactericidal capacity of the cells. Microscopic examination indicated that the decrease in bacterial killing was a result of inhibited engulfment of bacteria. Similar results were obtained (Sbarra and Karnovsky, 1959; Cohn and Morse, 1960) with mammalian PMN. MC and AM are also very sensitive to glycolytic inhibitors. Therefore, it would appear that glycolysis can supply a substantial proportion of the energy required for phagocytosis by both insect and mammalian cells.

Oxidative glucose metabolism by hemocytes was measured in terms of $^{14}CO_2$ production from ^{14}C-glucose incubated with resting and phagocytizing cells. The $^{14}CO_2$ production from uniformly labeled ^{14}C-glucose was stimulated about 18% during phagocytosis. This represents the sum of the activities of the direct oxidative pathway and the tricarboxylic acid cycle, and is low when compared to mammalian phagocytes. Trehalose (a glucose disaccharide) is the usual blood sugar of insects; therefore, similar experiments were performed using uniformly labeled ^{14}C-trehalose. However, $^{14}CO_2$ production from ^{14}C-trehalose was comparable to that from ^{14}C-glucose.

Many investigators have confirmed the observations of Stahelin, et al. (1957) that phagocytosis by PMN induces a great increase in CO_2 production from carbon 1 of glucose relative to carbon 6. This is thought to reflect stimulated utilization of glucose via the direct oxidative pathway relative to the pathway via pyruvic acid and the Krebs cycle. When cockroach hemocytes were incubated with glucose-1-^{14}C, no increased $^{14}CO_2$ production accompanied phagocytosis. Phagocytosis caused a slight increment in $^{14}CO_2$ production from glucose-6-^{14}C. C-1:C-6 ratios for mammalian phagocytes were calculated; representative values were given by Karnovsky, et al. (1970). The C-1:C-6 ratios for PMN, MC and AM are about 2.7, 0.9 and 1.1, respectively. Clearly the increased hexose monophosphate pathway activity seen in phagocytizing PMN is absent in roach hemocytes, which in this respect would appear to more closely resemble macrophages.

The exact biochemical pathway leading to the observed alterations in PMN biochemistry during phagocytosis is not completely understood; however, the following scheme is one which is currently favored (Reed, 1969; Holmes, et al., 1970). The increased respiration is mediated by flavoprotein oxidase resulting in the generation of H_2O_2 and oxidized nicotinamide adenine dinucleotide phosphate ($NADP^+$). The H_2O_2 is reduced to water after reacting with reduced glutathione mediated by glutathione peroxidase. The oxidized glutathione produced above reacts with NADPH + H^+, in the presence of glutathione reductase, to generate reduced glutathion and $NADP^+$. The increase in available $NADP^+$ produced in the above reactions stimulates the hexose monophosphate pathway.

Flavoprotein oxidase utilizing NADH will reduce nitroblue tetrazolium (NBT) as shown by Cagan and Karnovsky (1964). NBT reduction is associated with the respiratory burst and H_2O_2 production by PMN. Normal human leukocytes markedly reduce NBT, but cells from patients with a genetic defect in phagocytic function do not reduce NBT (Nathan, et al., 1969). These patients have an increased frequency of infection with certain bacteria and their cells fail to show the usual respiratory increment, direct oxidation of glucose, and H_2O_2 production during phagocytosis (Holmes, et al., 1967). Reduced NBT in PMN results in large blue-black formazan deposits in the cytoplasm. Monocytes will also reduce NBT; however, the result is speckled distribution of small formazan granules throughout the cytoplasm (Freeman and King, 1972). *Blaberus* hemocytes failed to reduce NBT to any great extent; occasionally a faint blue color could be seen in some phagocytic vacuoles, but discrete formazan deposits were not seen. The stimulated NBT test, in which the reaction is intensified by the ingestion of particulates, was also essentially negative. In this connection, it is interesting to note that subsequent, as yet unpublished, work in this laboratory indicates that both tunicate and gastropod mollusc hemocytes reduce NBT in a manner comparable to mammalian monocytes. To date attempts to show a stimulated NBT test have been unsuccessful.

The increased oxidative metabolic activities of phagocytizing PMN may be related to bactericidal events. Anaerobic conditions and/or inhibitors of HMP such as phenylbutazone inhibit bactericidal activity (Selvaraj and Sbarra, 1966; Strauss, et al., 1968). Hydrogen peroxide (a by-product of HMP) is antimicrobial; it is found in various phagocytes and is generated during phagocytosis (Iyer, et al., 1961; Paul, et al., 1970). A potential antimicrobial system has been described in PMN which involves H_2O_2, myeloperoxidase (MPO), and a halide (Kelbanoff, 1968). Iodide can function as the halide *in vitro*, but chloride is the probable substrate for MPO activity *in vivo*. A MPO-H_2O_2-$C1^-$ complex is formed which reacts with bacterial amino acids yielding unstable amino acid chloramines which are spontaneously decarboxylated and deaminated to yield antimicrobial aldehydes (Sbarra, et al., 1972).

Efforts to detect components of the MPO-halide-H_2O_2 antimicrobial system in insect phagocytes were largely unsuccessful. Hemocyte smears and human PMN smears were stained for MPO by the method of Kaplow (1965). The human PMN contained numerous MPO positive granules in the cytoplasm of every cell. The roach cells contained many granules but less than 1% of the hemocytes contained some MPO-positive material. MPO acts in synergism with a halide and H_2O_2. Human cells in a medium containing $Na^{125}I$ will iodinate zymosan particles by the action of MPO and H_2O_2. Since some MPO activity was present in *Blaberus* phagocytes, quantitative iodination tests, by the method of Pincus and Klebanoff (1971), were performed. Iodination was essentially lacking in the cockroach hemocytes, whereas it was marked in human PMN run at the same time. Obviously, insect cells utilize other antimicrobial mechanisms, possible mechanisms include lysozyme and granule-associated antibacterial cationic proteins, such as those of guinea pig peritoneal exudate PMN described by Zeya and Spitznagel (1966a, 1966b). So far, we have been unable to isolate electrophoretically cationic proteins similar to those of mammalian PMN.

The lack of PMN antimicrobial systems in hemocytes, which have been shown to be quite capable of killing bacteria, may not be surprising. Mammalian monocytes and macrophages are efficient killers of many species of bacteria but contain few, if any, peroxidase granules. Also, monocytes possess little or none of the antibacterial cationic proteins found in PMN. Hirsch and Fedorko (1970) summarize the situation by stating that the mechanism of bactericidal action within macrophages remains completely unknown; the same could be said for insect hemocytes at this time.

Complement and invertebrate phagocytosis. At the present time, there is no direct evidence that complement participates in phagocytosis among invertebrate species. However, its role in phagocytosis and chemotaxis by vertebrate phagocytes is well defined. In addition, there is evidence for the presence of complement components and molecules having functional properties of the complement component 3 proactivator in the hemolymph of certain invertebrates. Therefore, it is proper to include a few words about invertebrate complement and the heat-labile opsonins of vertebrates.

For many years it has been known that antibody and classical complement enhances the phagocytosis of bacteria by mammalian leukocytes. However, in the heat-labile opsonin (HLO) system antibody and the initial complement components are not required for opsonization of Gram-negative bacteria. Jasin (1972) showed that opsonization took place normally in human serum lacking complement component Iq and immunoglobulins. HLO capacity for *E. coli* was markedly reduced in serum heated for 30 min at 50°C, a treatment which inactivates a component of the alternative pathway of complement activation. If purified complement component 3 proactivator is added to heated serum, the opsonic activity is restored.

The participation of the alternate pathway of complement activation (the properdin pathway) in opsonization has been implied by many recent observations. A patient was found who had marked susceptibility to infection and who spontaneously inactivated C3 *in vivo*. His serum had normal levels of immunoglobulins, C1, C4, and C2 (the initial complement components), but did not opsonize bacteria even in the presence of purified C3 (Alper, et al., 1969). His serum was shown to have a defect in the properdin (C3-C9) pathway (Alper, et al., 1971; Abramson, et al., 1971). Other studies have shown that guinea pigs deficient in C4 (Ellman, et al., 1970) and human beings deficient in C2 (Klemperer, et al., 1967) are in good health.

Unlike the classical complement pathway, the alternate pathway can be activated by polysaccharides in the absence of antibody (Pillemer, et al., 1955). Stossel, et al. (1973) described quantitative studies of phagocytosis by guinea pig peritoneal exudate PMN of paraffin oil droplets coated with bacterial lipopolysaccharide. These particles can be opsonized in the absence of C4 and C2, but requires C3 and the properdin pathway. Lipopolysaccharide constitutes the outer wall of gram-negative bacteria; the paraffin oil provides a stability and configuration to the lipopolysacchardie that may produce a particle quite similar to the parent bacteria (*E. coli*). These particles are rapidly ingested by both guinea pig and human PMN and human monocytes after opsonization by fresh homologous serum. This activity was resistant to dialysis but sensitive to heat, hydrazine, and zymosan. Opsonization was associated with the fixation of ^{125}I labeled C3 to the particles by means of the properdin system; hence the ingestion of these particles constitutes a quantitative functional assay of this pathway.

Morgun (1950) reported that invertebrate hemolymph contained a factor similar to C3, suggesting the possible presence of the terminal complement components.

Day, et al. (1970) described aspects of complement-like activity in lower vertebrates and invertebrates. She used a factor isolated and purified from cobra venom (CVF) which in concert with a serum factor called C3 proactivator (C3PA) activated the terminal complement components to lyse erythrocyte target cells. Total hemolytic complement (as indicated by incubating sera with antibody-sensitized erythrocytes) was not present in the invertebrate species tested. Lysis-inducing activity of purified CVF was found in the sera of horseshoe crabs, sipunculid worms and starfish. This data suggest that the terminal components of complement may antedate the appearance of the immune (antibody) system. Possibly this system contributes to bodily defense systems through its involvement with phagocytosis and inflammation.

A humoral factor similar to C3PA of mammals is also present in the cockroach *Blaberus craniifer* (Anderson, et al., 1972). The formation of the roach hemolymph factor-CVF complex requires divalent cations and is heat sensitive. This complex could mediate lysis of unsensitized sheep erythrocytes via activation of the terminal components of frog serum. CVF alone cannot lyse target cells; hemolymph alone also was without effect. Erythrocyte lysis by the complex was shown to be complement-dependent; if the complement in frog serum was inactivated by heating or by CVF, no lysin could be induced by cockroach proactivator.

A starfish (*Asterias forbesi*) hemolymph factor which activates vertebrate complement in the presence of CVF has been reported by Day, et al. (1972a). As with other invertebrates, a CVF-hemolymph factor complex could be shown to activate the terminal components of frog complement. The hemolymph factor (HLF) was functionally similar to C3PA; however, its molecular weight was estimated to be about 2000, whereas C3PA is about 80,000 MW. Such a small size almost precludes the presence of enzymatically activity on HLF similar to that on C3PA; therefore, more work must be done to elucidate the mode of action of CVF-invertebrate serum factor complexes on complement. This work was extended by Day, et al. (1972b) who reported that the interaction of starfish hemolymph, CVF and purified C3 results in the generation of anaphylatoxic activity in guinea pig ileum preparations.

The importance of using pure CVF in complement studies has recently been emphasized by Hall, et al. (1972). Both hemagglutinating and CVF-activated activities were found in lobster (*Homarus americanus*) hemolymph. These activities were independent and functionally and chemically separate. The hemolytic activity was not complement dependent but was mediated by phospholipase contamination of the CVF used in this study. Lecithin in lobster lipoproteins was converted by phospholipase to lysolecithin, which is a potent hemolysin.

The preceding work indicates the possible presence of a complement system akin to the properdin system in invertebrates, which have been shown to lack classical antibodies. Evidence was also presented for the presence in vertebrate species of a system of heat-labile opsonins which operate in the absence of immunoglobulins and require the properdin system. Many invertebrate opsonins are heat sensitive. The possible role of complement components in phagocytosis by invertebrate hemocytes has yet to be defined.

Similarities between invertebrate phagocytes and mammalian mononuclear cells. Our knowledge of phagocytosis and immune mechanisms, in general, in higher vertebrates greatly exceeds our understanding of similar mechanisms in invertebrate animals. This is to be expected if one considers the relative amounts of effort so far expended in the two fields of research, and the great diversity of form and function displayed among the invertebrates. However, it may not be premature to make some correlations between vertebrate and invertebrate immunological effector systems. Such comparisons are rather difficult in the area of humoral defense mechanisms because of the lack of mammalian-type immunoglobulins in invertebrates, as well as the frequent lack of specificity and "memory" seen in these reactions. However, one may draw a number of possibly meaningful comparisons between invertebrate and vertebrate phagocytes. The predominant mammalian circulating phagocyte is the polymorphonuclear neutrophil. Circulating invertebrate cells are also phagocytically active, but apparently differ greatly in morphology, in metabolism, in bactericidal mechanisms, and in other ways, from PMN. Mononuclear cells, present in mammalian blood, are also phagocytic and give rise to highly phagocytic tissue macrophages. Data available at this time suggest that invertebrate hemocytes resemble mononuclear cells more closely than PMN.

Recently, the mononuclear phagocytic system has been carefully defined and differentiated from the reticulo-endothelial system; the advantages of this change in nomenclature are discussed by Langevoort, et al. (1970). The morphology of these cells may vary dependent on their anatomical location; their most specific morphological characteristic is a ruffling of the plasma membrane. Invertebrate hemocytes with multilobed nuclei like PMN have not been described; only mononuclear cells have been observed to date. The characteristic ruffled plasma membrane has been observed on the periphery of the hemocytes of several invertebrate species by the author.

The functional criteria for inclusion in the mononuclear phagocyte system are avid phagocytosis and firm attachment to glass surfaces. There is ample evidence for active *in vivo* phagocytosis by invertebrate blood cells; a number of examples of *in vitro* phagocytosis have been given in this review, in which many instances of the cells attaching to glass are reported.

In mammals, kinetic studies have demonstrated that mononuclear phagocytes originate from precursor cells in bone marrow, circulate in the peripheral blood, and finally become tissue macrophages. There is no evidence that mononuclear macrophages are derived from lymphocytes. The exact anatomic origin of invertebrate phagocytes is generally not well known; however, the fixed phagocytes of certain invertebrates often resemble hemocytes and are probably identical to them (Lange, 1932; Nutting, 1951; Wigglesworth, 1959; Scharrer, 1966; Stuart, 1968). There are no true lymphocytes in invertebrates (the "lymphocytes" reported in many invertebrates are so named solely because of a superficial morphological resemblance to mammalian lymphocytes); there is no evidence for a lymphoid system in invertebrate species.

The phylogeny of mononuclear phagocytes is reviewed by Stuart (1970), with emphasis on molluscs, annelids, and arthropods. The uptake by cells and eventual fate of foreign particulates in representatives of these major phyla are described. It is interesting that colloidal particles cleared from the blood of many invertebrates are eventually excreted by a number of unique mechanisms, rather than sequestered in various organs as is the case in mammals.

Very similar patterns of particle recognition by insect phagocytes and mouse macrophages in the absence of hemolymph or serum were recorded (Rabinovitch and DeStefano, 1970; Rabinovitch, 1970). In addition, *in vitro* uptake of glutaraldehyde-treated red cells by *Galleria* hemocytes was stimulated when the red cells were pretreated with chymotrypsin and was inhibited by ovalbumin treatment; similar results were obtained with macrophages.

A number of metabolic characteristics of phagocytizing insect hemocytes resemble those of mammalian macrophages (Anderson, 1973b). Some of these shared characteristics include: (1) lack of a respiratory increment; seen in non-elicited monocytes and alveolar macrophages; (2) stimulation of glycolysis during particle ingestion; (3) little hexose monophosphate shunt activity accompanying phagocytosis;

(4) similar patterns of nitroblue tetrazolium reduction; (5) little intracellular peroxidase, (6) lack of bactericidal cationic proteins.

CONCLUSION

The studies reviewed here indicate that invertebrate cells, under the proper conditions, are capable of recognizing and ingesting a wide variety of foreign particles. In many instances recognition is rather non-specific and can occur in the absence of any humoral factors. However, in other cases hemolymph recognition factors can facilitate, or are required for, phagocytosis.

It is clear that phagocytosis is not an automatic response which always may be observed after hemocyte-particle contact *in vitro*. Indeed, usually the hemocytes respond to a more extensive assortment of foreign substances *in vivo* than *in vitro*. This is not surprising and probably reflects: organic deficiencies in the media; improper conditions of pH, temperature, osmolarity, ionic composition, etc.; deficient opsonization; mechanical trauma to the cells experienced during collection; improper particle: cell ratios; presence of inhibitors of phagocytosis, such as some anticoagulents, in the medium; and many other problems commonly encountered in *in vitro* systems. Nevertheless, it has been quite possible to study *in vitro* uptake of particles such as yeast, bacteria, viruses, erythrocytes, latex bead, etc. by many types of invertebrate cells. Many of these studies have been performed in recent years and have already contributed considerable information which may be useful in viewing the ubiquitous phenomenon of phagocytosis in phylogenetic perspective. It is helpful, and possibly imperative, to compare data from invertebrate work to the comparatively vast amount of information already known about mammalian phagocytes. A number of similarities between phagocytizing insect hemocytes and mammalian macrophages have been discussed in this paper, however, much more experimental evidence is required before any generalized statements about phagocytosis in invertebrates, as a group, can be made. Perhaps it will prove to be impossible to generalize on the details (especially metabolic details) of this mechanism in group as large and diverse as the invertebrates.

To summarize what is now known about *in vitro* phagocytosis among invertebrate species, one may confidently state that the surface characteristics of the particle to be taken up are very important in many experimental systems. Chemical alteration and/or opsonization may substantially effect the rate of ingestion of a given particle. Invertebrate (and in some cases plant) hemagglutinins are frequently implicated as opsonins. There is no evidence as yet to indicate that the hemocytes of phylogenetically advanced invertebrates (such as molluscs and arthropods) possess receptors for mammalian immunoglobulin, although antibody does stimulate phagocytosis by slime mold ameba and protozoans. As in vertebrates, phagocytosis consists of an adherence stage and an ingestion stage. Also, in vertebrates and in insects, phagocytosis is an energy requiring event which depends on glycolysis. However, insect phagocytes lack the post-phagocytic respiratory increment and increased hexose monophosphate pathway activity seen in vertebrate PMN. The phagocytosis-related metabolism of insect cells seems to resemble that of mononuclear cells more than that of polymorphonuclear cells. Insect hemocytes also do not utilize the myeloperoxidase-hydrogen peroxide-halide antibacterial system described in mammalian PMN, and lack bactericidal cationic proteins.

The utilization of *in vitro* systems for the study of phagocytosis by invertebrate hemocytes will make available data which is virtually impossible to obtain in any other way. It is quite possible that this information will contribute substantially not only toward a more complete understanding of phagocytosis, in general, but toward advances in other areas such as host-parasite interactions, invertebrate pathology, the role of invertebrates as vectors of disease, biological control of undesirable species and the phylogeny of immunity. It is important to utilize, where appropriate, the techniques and to profit from experience and knowledge already gained from the study of vertebrate mechanisms. It will also be necessary to develop new techniques, in the course of elucidating the details of invertebrate phagocytosis, which may find application in the study of vertebrate phagocytes.

REFERENCES

Abramson, N., Alper, C.A., Lachman, P. J., Rosen, F. S., and Jandl, J. H. (1971). *J. Immunol., 107*:19.

Acton, R. T., Bennett, J. C., Evans, E. E., and Schrohenloher, R. (1969). *J. Biol. Chem., 244*:4128.

Acton, R. T., and Evans, E. E. (1967). *In Vitro, 3*:139.

Alper, C. A., Abramson, N., Johnston, B., Jr., Jandl, H., and Rosen, F. S. (1969). *New Engl. J. Med., 282*:349.

Alper, C. A., Abramson, N., Johnston, B., Jr., Jandl, H., and Rosen, F. S. (1970). *J. Clin. Invest., 49*:1975.

Anderson, R. S. (1974). *Contemp. Topics Immunol., 4*:47.

Anderson, R. S., Day, N. K. B., and Good, R. A. (1972). *Infect. Immun., 5*:55.

Anderson, R. S., Holmes, B., and Good, R. A. (1973a). *J. Invertebr. Pathol., 22*:127.

Anderson, R. S., Holmes, B., and Good, R. A. (1973b). *Comp. Biochem. Physiol., 46B*:595.

Baehner, R. L., and Nathan, D. G. (1967). *Science, 155*:835.

Bang, F. B. (1961). *Biol. Bull., 121*:57.

Bernheimer, A. W. (1952). *Science, 115*:150.

Bernheimer, A. W., Caspari, E., and Kaiser, A. P. (1952). *J. Exp. Zool., 119*:23.

Bizot, M. (1971). *Vox Sang., 21*:465.

Boyd, W. C., and Brown, R. (1965). *Nature (Lond.), 208*:593.

Boyd, W. C., Brown, R., and Boyd, L. G. (1966). *J. Immunol., 96*:301.

Brown, R., Almodovar, L. R., Bhatia, H. M., and Boyd, W. C. (1968). *J. Immunol., 100*:214.

Cagan, R. H., and Karnovsky, M. C. (1964). *Nature (Lond.), 204*:255.

Cameron, G. R. (1932). *J. Pathol. Bacteriol., 35*:933.

Cheng, T. C., and Sanders, B. G. (1962). *Proc. Pa. Acad. Sci., 36*:72.

Cohen, E., Rowe, A. W., Wissler, F. C. (1965). *Life Sci., 4*:2009.

Cohn, Z. A., and Morse, S. I. (1960). *J. Exp. Med., 111*:667.

Cornick, J. W., and Stewart, J. E. (1968). *J. Fish. Res. Bd. Can., 25*:695.

Cornick, J. W., and Stewart, J. E. (1973). *J. Invertebr. Pathol., 21*:255.

Day, N., Geiger, H., Finstad, J., and Good, R. A. (1972a). *J. Immunol., 109*:164.

Day, N. K. B., Gewurz, H., Johannsen, R., Finstad, J., and Good, R. A. (1970). *J. Exp. Med., 132*:941.

Day, N. K. B., Good, R. A., and Müller Eberhard, H. J. (1972b). *Fed. Proc., 31*:788

Ellman, L., Green, I., and Frank, M. (1970). *Science, 170*:74.

Feir, D., and Walz, M. (1964). *Ann. Entomol. Soc. Am., 57*:388.

Fenn, W. O. (1921-1922). *J. Gen. Physiol., 4*:373.

Fernández-Morán, H., Marchalonis, J. J., and Edelman, G. M. (1968). *J. Mol. Biol., 32*:467.

Freeman, R., and King, B. (1972). *J. Clin. Pathol.*, *25*:912.

Fuke, M. T., and Sugai, T. (1972). *Biol. Bull.*, *143*:140.

Fulton, J. F. (1920). *Acta Zoolog.*, *1*:381.

Gerisch, G., Lüderitz, O., and Ruschmann, E. (1967). *Z. Naturforsch., (B)* *22*:109.

Gilbertson, D. E., and Etges, F. J. (1967). *Ann. Trop. Med. Parasitol.*, *61*:144.

Githens, S., III, and Karnovsky, M. L. (1973). *J. Cell Biol.*, *58*:536.

Good, R. A., Quie, P. G., Windhorst, D. B., Page, A. R., Rodey, G. E., White, J. E., and Holmes, B. (1968). *Semin. Hematol.*, *5*:215.

Goodall, R. J., and Thompson, J. E. (1971). *Exp. Cell Res.*, *64*:1.

Grace, T. D. C. (1962). *Virology*, *18*:33.

Grace, T. D. C. (1971) In "Invertebrate Tissue Culture" (C. Vago, ed.), Vol. I, pp. 171-209. Academic Press, New York.

Grace, T. D. C., and Day, M. F. (1963). Film C.S.I.R.O.

Hammarström, S., and Kabat, E. A. (1971). *Biochemistry*, *10*:1684.

Hirsch, J. G., and Fedorko, M. E. (1970). In "Mononuclear Phagocytes" (R. van Furth, ed.), pp. 7-28. F. A. Davis, Philadelphia.

Holmes, B., Page, A. R., and Good, R. A. (1967). *J. Clin. Invest.*, *46*:1422.

Holmes, B., Park, B. H., Malawista, S. E., Quie, P. G., Nelson, D. L., and Good, R. A. (1970). *New Engl. J. Med.*, *283*:217.

Iyer, G. Y. N., Islam, M. F., and Quastel, J. M. (1961). *Nature (Lond.)*, *192*:535.

Jasin, H. E. (1972). *J. Immunol.*, *109*:26.

Jenkin, C. R., and Rowley, D. (1970). *Aust. J. Exp. Biol. Med. Sci.*, *48*:129.

Johnson, H. M. (1964). *Science*, *146*:548.

Johnson, P. T. (1969). *J. Invertebr. Pathol.*, *13*:42.

Johnson, P. T., Chien, P. K., and Chapman, F. A. (1970). *J. Invertebr. Pathol.*, *16*:466.

Kaplow, L. S. (1965). *Blood, 26*:214.

Karnovsky, M. L., Simmons, S., Glass, E. A., Shafer, A. W., and D'arcy Hart, P. (1970). In "Mononuclear Phagocytes" (R. van Furth, ed.), pp. 103-117. F. A. Davis, Philadelphia.

Khalap, S., Thompson, T. E., and Gold, E. R. (1970). *Vox Sang., 18*:501.

Klebanoff, S. J. (1968). *J. Bacteriol., 95*:2131.

Klemperer, M. R., Austen, K. F., and Rosen, F. S. (1967). *J. Immunol., 98*:72.

Lange, H. H. (1932). *Z. Zellforsch. Mikrosk. Anat., 16*:753.

Langevoort, H. L., Cohn, Z. A., Hirsch, J. G., Humphrey, J. H., Spector, W. G., and van Furth, R. (1970). In "Mononuclear Phagocytes" (R. van Furth, ed.), pp. 1-6. F. A. Davis, Philadelphia.

Lee-Potter, J. P. (1969). *Vox Sang., 16*:500.

Marchalonis, J. J., and Edelman, G. M. (1968). *J. Mol. Biol., 32*:453.

Matsubara, S., and Boyd, W. C. (1974). *Science, 183*:339.

McDade, J. E., and Tripp, M. R. (1967). *J. Invertebr. Pathol., 9*:523.

McKay, D., and Jenkin, C. R. (1970a). *Aust. J. Exp. Biol. Med. Sci., 48*:139.

McKay, D., and Jenkin, C. R. (1970b). *Aust. J. Exp. Biol. Med. Sci., 48*:609.

McKay, D., and Jenkin, C. R. (1970c). *Aust. J. Exp. Biol. Med. Sci., 48*:599.

McKay, D., Jenkin, C. R., and Rowley, D. (1969). *Aust. J. Exp. Biol. Med. Sci., 47*:125.

Metchnikoff, E. (1884). *Virchows Arch. Pathol. Anat. Physiol., 46*:177.

Miller, V. H., Ballback, R. S., Pauley, G. B., and Krassner, S. M. (1972). *J. Invertebr. Pathol., 19*:83.

Morgun, G. I. (1950). *Mikrobial. Zh. (Kiev), 11*:43.

Nathan, D. G., Baehner, R. L., and Weaver, D. K. (1969). *J. Clin. Invest., 48*:1895.

Nutting, W. L. (1951). *J. Morphol., 89*:501.

Paul, B. B., Strauss, R. R., Jacobs, A. A., and Sbarra, A. J. (1970). *Infect. Immun.*, *1*:338.

Pauley, G. B. (1973). *Experientia*, *29*:210.

Pauley, G. B., Granger, G. A., and Krassner, S. M. (1971a). *J. Invertebr. Pathol.*, *18*:207.

Pauley, G. B., Krassner, S. M., and Chapman, F. A. (1971b). *J. Invertebr. Pathol.*, *18*:227.

Pemberton, R. T. (1969). *Vox Sang.*, *16*:457.

Pemberton, R. T. (1970a). *Vox Sang.*, *18*:71.

Pemberton, R. T. (1970b). *Vox Sang.*, *18*:74.

Pillemer, L., Schoenberg, M. D., Blum, L., and Wurz, L. (1955). *Science*, *122*:545.

Pincus, S. H., and Klebanoff, S. J. (1971). *New Engl. J. Med.*, *284*:744.

Ponder, E. (1928). *J. Gen. Physiol.*, *11*:757.

Prokop, O., Uhlenbruck, G., and Schnitzler, St. (1967). *Bibl. Haematol.*, *27*:108.

Prowse, R. H., and Tait, N. N. (1969). *Immunology*, *17*:437.

Rabinovitch, M. (1970). In "Mononuclear Phagocytes" (R. van Furth, ed.) pp. 299-313. F. A. Davis, Philadelphia.

Rabinovitch, M., and DeStefano, M. (1970). *Exp. Cell Res.*, *59*:272.

Rabinovitch, M., and DeStefano, M. (1971a). *Exp. Cell Res.*, *64*:275.

Rabinovitch, M., and DeStefano, M. (1971b). *Nature (Lond.)*, *234*:414.

Reed, P. W. (1969). *J. Biol. Chem.*, *244*:2459.

Roberts, J., and Quastel, J. M. (1963). *Biochem J.*, *89*:150.

Sbarra, A. J., and Karnovsky, M. L. (1959). *J. Biol. Chem.*, *234*:1355.

Sbarra, A. J., Paul, B. B., Jacobs, A. A., Strauss, R. R., and Mitchell, G. W. (1972). *J. Reticuloendothel. Soc.*, *12*:109.

Scharrer, B. (1966). *Z. Zellforsch Mikrosk. Anat.*, *69*:1.

Schnitzler, St., and Kilias, R. (1970). *Blut., 20*:221.

Scott, M. T. (1971a). *Immunology, 21*:817.

Scott, M. T. (1971b). *Arch. Zool. Exp. Gen., 112*:73.

Selvaraj, R. J., and Sbarra, A. J. (1966). *Nature (Lond.), 211*:1272.

Stahelin, H., Karnovsky, M. L., Farnham, A. E., and Suter, E. (1957). *J. Exp. Med., 105*:265.

Stahelin, H., Suter, E., and Karnovsky, M. L. (1956). *J. Exp. Med., 104*:121.

Stossel, T. P., Alper, C. A., Rosen, F. S. (1973). *J. Exp. Med., 137*:690.

Strauss, R. R., Paul, B. B., and Sbarra, A. J. (1968). *J. Bacteriol., 96*:1982.

Stuart, A. E. (1968). *J. Pathol. Bacteriol., 96*:401.

Stuart, A. E. (1970). In "Mononuclear Phagocytes" (R. van Furth, ed.) pp. 316-334. F. A. Davis. Philadelphia.

Tripp, M. R. (1966). *J. Invertebr. Pathol., 8*:478.

Tripp, M. R., Bisignani, L. A., and Kenny, M. T. (1966). *J. Invertebr. Pathol., 8*:137.

Tripp, M. R., and Kent, V. E. (1967). *In Vitro, 3*:129.

Tylor, A. (1946). *Biol. Bull., 90*:213.

Tyler, A., and Metz, C. B. (1945). *J. Exp. Zool., 100*:387.

Tyler, A., and Scheer, B. (1945). *Biol. Bull., 89*:193.

Vago, C. (1964). "Culture de tissus d'invertébrés." Service du Film de Recherche Scientifique, Paris.

Vago, C. (1972). In "Invertebrate Tissue Culture" (C. Vago, ed.), Vol. II, pp. 245-278. Academic Press, New York.

Weisman, R. A., and Korn, E. D. (1967). *Biochemistry, 6*:485.

Wigglesworth, V. B. (1959). *Ann. Rev. Entomol., 4*:1.

Wright, R. K. (1973). Ph.D. dissertation, University of California at Santa Barbara.

Zeya, H. I., and Spitznagel, J. K. (1966a). *J. Bacteriol., 91*:750.

Zeya, H. I., and Spitznagel, J. K. (1966b). *J. Bacteriol., 91*:755.

CELLULAR DEFENSE MECHANISMS AND TISSUE ALTERATIONS IN GASTROPOD MOLLUSCS

Edward H. Michelson
Department of Tropical Public Health
Harvard School of Public Health
Boston, Massachusetts

The Phylum Mollusca constitutes one of the major divisions of the Animal Kingdom and is of unusual interest both in regards to diversity of organization and in the multitude of living species, conservatively estimated to be 47,000 (Boss, 1971). All of the existing higher taxa have been recognized since the Pre-Cambrian and this stability reflects, in part, the ability of the molluscs to cope successfully with the repeated insults imposed by pathogens and parasites. Members of the Class Gastropoda (snails & slugs), at least numerically, are the most successful of all molluscs and are of special concern in that they serve as intermediate and paratenic hosts for helminth parasites causing disease in man and his domestic animals.

Metchnikoff (1892) first recognized the existence of cellular defense mechanisms (phagocytosis) in molluscs and other invertebrates; however, interest in these phenomena in molluscs waned and was not rekindled until the 1950's when Stauber and Mackin and their co-workers initiated a series of studies on pathological and immunological responses in the oyster. During the subsequent two decades there has been a continuing interest in the defense mechanisms and histopathologic responses of the Mollusca. The literature which has evolved has been ably collated and reviewed by Stauber (1961), Cheng and Sanders (1962), Tripp (1963, 1969, 1970), Wright (1966), Feng (1967), Brooks (1969), Bang (1973), Cheng and Rifkin (1970), and Hildemann and Reddy (1973). In addition, Pauley (1969) reviewed the occurrence of neoplasia in molluscs, Johnson (1968) prepared a comprehensive bibliography on invertebrate pathology, and Sparks (1972) published an excellent text on invertebrate pathology. One might logically assume, from the number of reviews published, that considerable progress has been made in our understanding of the cellular defense mechanisms and histopathologic responses of molluscs. Unfortunately, this is not the case and one is disappointed to find that essentially the same studies are continually cited without having stimulated logical extensions of work that might be expected. In this respect, I must stand with the guilty. An attempt will be made, therefore, to review not only our present state of knowledge of the cellular defense mechanisms of gastropod molluscs, but also to document problems which have impeded progress in our understanding of these mechanisms.

CELLULAR RESPONSES

The histopathologic responses observed in gastropods, as well as in other molluscs, may be classified into six general categories: 1) phagocytosis by amoebocytes and other tissue cells, (Fig. 1) 2) fibrotic encapsulation (Fig. 2), 3) diffuse fibrosis and metaplasia, 4) isolation of foreign bodies by pearl formation of nacreous deposition, 5) atrophy, necrosis and tissue liquification, (Fig. 3) and 6) biochemical alterations. There is no evidence to suggest that these responses are specific and all may be elicited by "inert" substances as well as by dead and living organisms. In some cases, there appears to be little specific response to the presence of invading or established organisms.

Phagocytosis: Phagocytic cells are found in the hemolymph and tissues of gastropods and have been considered to be an important component of the cellular defense system (Tripp, 1963; Cheng et al., 1969). The activity of the phagocytic cells of vertebrates is initiated and accelerated by antibodies; however, invertebrates appear to lack an organized lymphoid system (Good and Papermaster, 1964) and presumably the ability to synthesize immunoglobulins and antibodies. Efforts to demonstrate either tissue or humoral antibodies in molluscs have been unsuccessful. Tripp (1966), however, reported the presence of a natural hemagglutinin in oyster hemolymph which acted as an opsonin and accelerated phagocytosis by amoebocytes. Recently, Prowse and Tait (1969) described the presence of an opsonic factor(s) in the hemolymph of *Helix aspersa* which controls the phagocytic activity of amoebocytes and is necessary for the phenomenon to occur. Preliminary experiments (Michelson and DuBois, unpublished) suggest that a hemolymph factor occurs in *Biomphalaria glabrata* and that heating of hemolymph destroys this factor and diminishes the phagocytic activity of amoebocytes. The nature and origin of these factors remain unknown and warrant further investigation.

The principle phagocytic cell of gastropods and of other molluscs is the amoebocyte (= leucocyte and hemocyte of authors) which wanders freely both in the hemolymph and through the loose connective tissues. These cells are considered to be multifunctional and, in addition to phagocytosis, are believed to participate in such diverse functions as digestion, excretion, wound-healing, shell repair, calcium and other ion transport, glycogen storage and transport, and in the initiation of encapsulation (Wagge, 1955; Michelson, 1961; Tripp, 1970; Stang-Voss, 1970; Sminia, 1972; Sparks, 1972). The relative importance of their role, *in vivo*, as phagocytes with respect to their other functions cannot be assessed.

Confusion still exists as to the types of cells found in the hemolymph of gastropod molluscs. Although Muller (1956) reported the presence of 4 cell-types in *Lymnaea stagnalis* hemolymph, recent studies suggest but a single cell-type, the amoebocyte, in this species and in all other pulmonates (Pan, 1958; Stang-Voss, 1970; Sminia, 1972; Fernandez, 1971). On the other hand, several investigators (George and Ferguson, 1950; Brown and Brown, 1965; Cheng et al., 1969) have described more than one cell-type from the hemolymph of prosobranch snails. These studies, by and large, were performed only with light microscopy and further studies with electron microscopy are indicated.

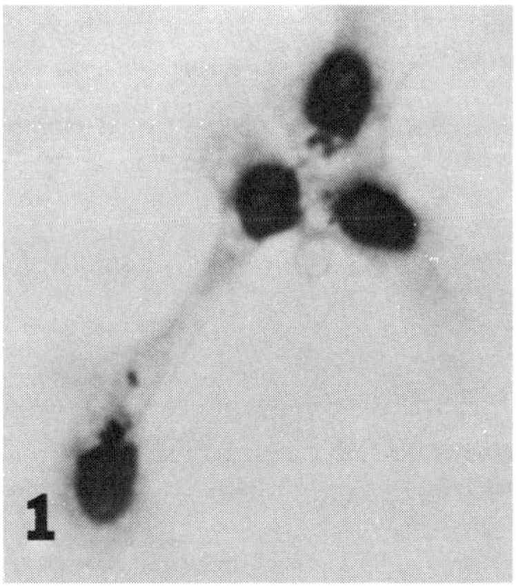

Fig. 1. Amoebocytes of Biomphalaria glabrata. Cells exposed 30 minutes in vitro to a suspension of carbon particles. Note that only a few particles have been ingested. Giemsa (1125x).

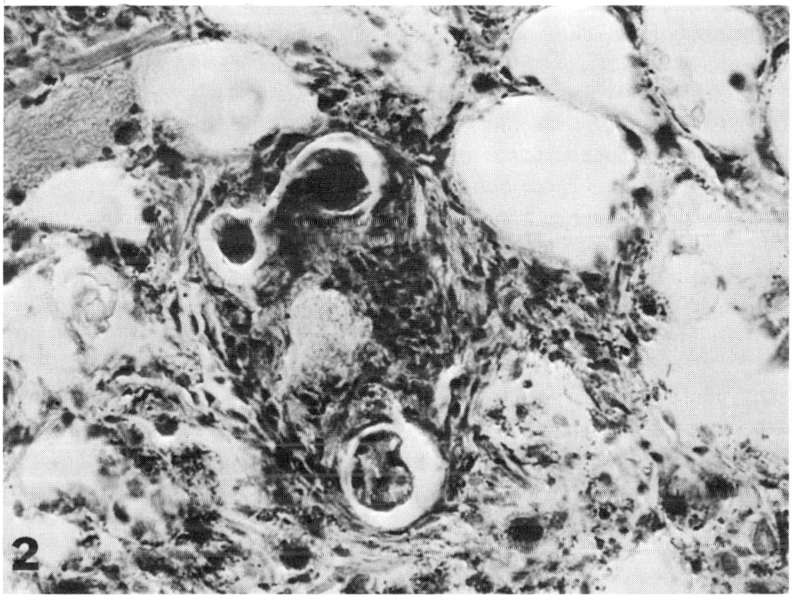

Fig. 2. Early formation of a fibrotic capsule around the nematode, Angiostrongylus cantonensis. Section of worm in mantle tissue of Biomphalaria glabrata 3 days post-infection. Hemalum, azure II-eosin (300x).

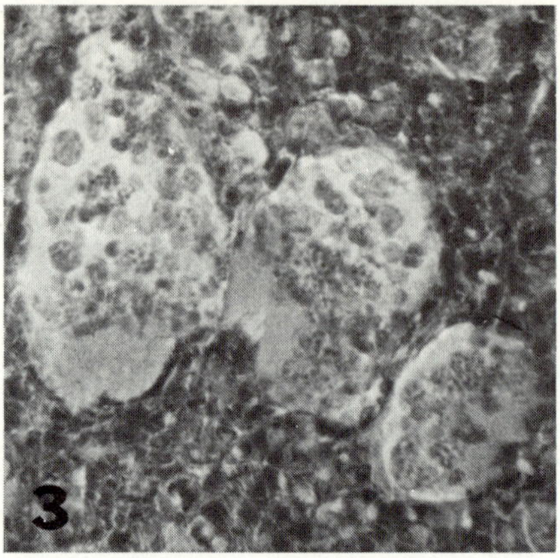

Fig. 3. Necrotic lesions and tissue liquification in the albumin gland of Physa heterostropha. Snail infected with Plistophora husseyi. Note that lesions are filled with masses of spores. Hemalum, azure II-eosin (325x).

Factors initiating phagocytic activity in gastropod amoebocytes remain vague and poorly defined. Other than the possible effects of hemolymph opsonins, only particle size and fixation have been observed to influence phagocytosis (Tripp, 1961). Particles or organisms exceeding 15 μ are not subject to phagocytosis, but may initiate aggregation and pre-encapsulation. Gastropod amoebocytes are capable also of distinguishing between allografts and xenografts and attack the latter more rapidly and intensely than homologous tissues (Tripp, 1961; Cheng and Galloway, 1970). The fate of ingested material and its degradation or disposal also requires further study. Tripp (1961) noted that particle-laden amoebocytes of *B. glabrata* migrated through epithelial surfaces to the exterior. A similar phenomenon was noted in *Helix aspersa* (Brown, 1967), *Bullia* (Brown and Brown, 1965) and in *Littorina scabra* (Cheng et al., 1969), but appeared to be of minor importance in *Lymnaea stagnalis* (Sminia, 1972). Degradation of biotic material has been observed in gastropod amoebocytes (Tripp, 1961), but the mechanisms of intracellular digestion have received relatively little attention. Amoebocytes of bivalves are known to contain intracellular amylases, proteases and lipases; however, these enzymes have yet to be demonstrated in phagocytes of gastropods (Owen, 1966). Recently, Sminia (1972) reported the occurrence of acid phosphatase and peroxidases in amoebocytes from *L. stagnalis*. An increase in acid phosphatase occurred subsequent to phagocytosis. It was further postulated that the peroxidases of the amoebocytes may function in a manner similar to those in mammalian leucocytes--i.e., as bactericidal, virucidal, and fungicidal agents.

Leucocytosis (an increase in either tissue or hemolymph amoebocytes) in response to invading organisms or injected substances has been reported to occur in bivalve molluscs (Cheng et al., 1969; Farley, 1968), but appears to be a rare phenomenon in gastropods. Two exceptions are an increase in eosinophilic cells in the digestive gland and hemolymph of *Viviparus japonicus* infected with *Aspidogaster conchicola*

(Michelson, 1970), and a recent observation (Michelson, unpublished) on *Schistosoma mansoni* infected *B. glabrata*, in which a two-fold increase in circulating amoebocytes occurred 18 days post-infection. Brown and Brown (1965) also noted an increase in *Bullia* haemocytes following an injection of thorium dioxide.

Speculation has arisen with respect to the function of connective tissue cells as "fixed" phagocytes and their role in phagocytosis and cellular defense mechanisms (Reade, 1968; Bayne, 1973). The lack of a standardized nomenclature for the cells of the connective tissues of molluscs and a paucity of information as to their functions have impeded clarification of the issue. Recent studies, employing electron microscopy, (Plummer, 1966; Wondrak, 1969; Stang-Voss, 1970; Buchholz et al., 1971; Fernandez, 1971; Sminia, 1972; Sminia et al., 1972) offer a new insight on the function of fixed phagocytes. Sminia's (1972) classification of the connective tissue cells of *Lymnaea stagnalis* appears applicable to most gastropods and should provide a firm basis for future work. Aside from tissue residing amoebocytes, Sminia found that only one other cell-type, the Pore Cell, functioned as a phagocyte; this cell type, however, operates only in a restricted sense--ingesting trypan blue and ferritin, but not zymosan particles. Earlier, Buchholz et al., (1971) noted that the Blasenzellen or Globular Cells (= Pore Cells) of *Cepaea nemoralis* and *Helix pomatia* ingested both ferritin and trypan blue, and the cells were considered to be phagocytic (Wolburg-Buchholz, 1972). In addition to phagocytic activities, the Pore Cell of gastropods is thought to store glycogen (Fernandez, 1966; Wondrak, 1969), produce collagen and give rise to fibroblasts (Plummer, 1966), and synthesize hemocyanin (Sminia, 1972) and hemoglobin (Sminia et al., 1972). The connective tissue sheaths surrounding ganglia in *Helix pomatia* were found to contain both mobile and "groups" of amoebocytes (Fernandez, 1971); however, it could not be determined if these "groups" were fixed cells and an integral part of the sheath or had originated therein. Studies by Reade (1968) and by Bayne (1973) suggest that various tissues and organs of gastropods, particularly the digestive gland, play an important role in the clearing of injected particles from the hemolymph, but the responsible cells were not identified.

The origin of the amoebocyte remains unknown. Szabo and Szabo (1934) believed that in *Limax flavus*, Leydig Cells gave rise to the amoebocytes. Pan (1958), on the other hand, postulated their origin, in *Biomphalaria glabrata*, from fibroblasts of a "hemopoietic" region situated near the saccular portion of the kidney.

Encapsulation: Encapsulation may be defined as the isolation of invading organisms and "foreign bodies" from host tissue by the formation of a fibrous envelope or capsule. Although the phenomenon is recognized as a primary means of cellular defense in gastropods, it is not fully understood. The analysis of the sequential process in insects, as summarized by Salt (1970), has no real counterpart in molluscs. Nevertheless, Cheng and his associates (see review by Cheng and Rifkin, 1970) have made important contributions to our understanding of this phenomenon in molluscs as has Pan (1958, 1963, 1965), Meuleman (1972), and others. Most observers agree that encapsulation is initiated by the migration of amoebocytes through the tissues of the site of the "foreign body." Subsequently, the amoebocytes appear to surround the object and may even adhere to it. Fibroblast-like cells appear soon after. The origin of the fibroblasts has been the subject of considerable speculation and the argument is far from resolved (Haughton, 1934; Wagge, 1951, 1955, Pan, 1958). Recent evidence suggests that the ultrastructure of the amoebocyte and fibroblast are very similar and it is likely that both originate from a common mesenchymal precursor (Sminia, 1972). Myofibers, collagenous fibers, and other tissue elements may appear even later and contribute to capsule formation (Brooks, 1953; Cheng and Rifkin, 1970).

The literature on encapsulation in gastropods strongly suggests that, as in insects, more than one development process may occur, with the mechanism being dependent on the host, the nature of the initiator, and possibly environmental factors. Reviews by both Brooks (1969) and Cheng and Rifkin (1970) cite numerous initiators of encapsulation in gastropods. It suffices here to note, that abiotic and biotic particles and organisms too large to be phagocytized tend to initiate encapsulation. In general, living organisms rarely elicit this reaction or if they do, it is delayed. Thus, in schistosome-infected snails viable sporocysts are seldom encapsulated and cercariae only when they begin to degenerate or clump and form emboli (Pan, 1965). Similarly, eggs, larvare, and adults of *Daubaylia potomaca* are not encapsulated in the normal snail host as long as they remain viable (Chernin et al., 1960. The nematode *Angiostrongylus cantonensis* reacts quite differently and living larvare are encapsulated shortly after invasion and remain viable for months or years (Richards and Merritt, 1967). In our experience, encapsulation of *A. cantonensis* in *B. glabrata* begins as early as 3 days post-infection. Other species of lungworms apparently elicit a similar type of response (Hobmaier and Hobmaier, 1934). If amoebocytes are incapable of containing an infection by phagocytosis, as in the case of the acid-fast bacteria described by Michelson (1961), encapsulation may occur.

Information is scarce concerning the histopathologic response of gastropods to metacercarial infections and subsequent encapsulation. Lal and Baugh (1955) noted that a fibrous outer cyst surrounded plagiorchid metacercaria in the tissues of *Viviparus bengalensis*, but their histopathologic description is too superficial to evaluate. Whether or not true encapsulation occurs in such cases is not known. Likewise, the reaction of snails to species of *Hartmanella* (Richards, 1968) may represent a different phenomenon as it is likely that both host and parasite contribute to the capsule. Dying and degenerating *Hartmanella* do, however, initiate the formation of a true capsule.

Diffuse fibrosis and metaplasia: In addition to focal encapsulation, Pan (1963, 1965) described a more generalized tissue reaction (Type II) in *Biomphalaria glabrata*. The reaction was associated with long-standing *S. mansoni* infections and was characterized by "extensive proliferation of connective tissue elements"; i.e., the appearance of numerous hypertrophic amoebocytes and fibroblasts. The response also was characterized by the appearance of large granulomata (encapsulation) around trapped cercariae and cercarial emboli occluding arteries and veins. Meuleman (1972) could not confirm Pan's observation in schistome-infected *Biomphalaria pfeifferi*; however, her infections were less intense, she limited her observations to the digestive gland, and she was dealing with a different host. Diffuse reactions and a similar type of metaplasia have been noted in viviparids infected with *Aspidogaster conchicola* (Michelson, 1970) and in species of slugs infected with *Tetrahymena limacis* (Brooks, 1968, Michelson, 1971).

Pearl formation and nacreous deposition: In contrast to bivalves, the deposition of nacre and the formation of pearls in response to irritants and larval helminths rarely occurs in gastropods. Wright (1966) and later Cheng and Rifkin (1970) reviewed the role of larval helminths in initiating pearl formation, but did not cite an instance of this reponse in a gastropod host. To my knowledge, *Biomphalaria glabrata* is the only gastropod reported to form pearls (Richards, 1970, 1972). Pearl formation in *B. glabrata* is thought to be a genetic trait controlled by a recessive gene, and it is not known if some "foreign" object serves as a seed or if they are spontaneously secreted. In any case, the role of this mechanim in the overall cellular defense of gastropods appears minimal.

Atrophy, necrosis and tissue liquification: Atrophy, necrosis and tissue liquification occur in concert with other tissue reactions and are difficult to delimit from the overall pathologic response. On occasion, these manifestations may occur so quickly that they appear to predominate in the apparent absence of other reactions. As noted by Sparks (1972), our concept of the necrotic process in invertebrates is based primarily on knowledge derived from vertebrates and little attention has been directed towards studying the phenomenon in lower animals. Necrosis in gastropods occurs in association with a variety of etiologic agents -- trematodes (Agersborg, 1924; Pan, 1965; Wright, 1966), bacteria (Michelson, 1961), and protozoa (Michelson, 1963). The specific cause of cell death, however, be it trauma, cytolytic enzymes, elaborated toxins, or other mechanisms, is not known.

Atrophy of the tentacles is frequently observed in snails subjected to environmental stress (Michelson, 1958; Harry and Aldrich, 1963) and may occur also as a consequence of infection (Mead, 1956; Michelson, 1961). Again, it is difficult to separate atrophy *per se* from necrosis or other changes.

Biochemical alterations: Our knowledge of histochemical alterations which occur in gastropod tissues, as a consequence of infection, is sketchy. Quantitative data are almost non-existent and information is based primarily upon histochemical and electrophoretic studies. In addition, few species of snails have been studied and these have been, for the most part, aquatic pulmonates. Cells and tissues of the digestive gland have been the target of most investigations and essentially nothing is known with respect to biochemical alterations occurring in other organs and tissues.

Cells of the digestive gland of *Helisoma trivolvis*, infected with either *Glypthelmins pennsylvaniensis* or *Echinoparyphium* sp., exhibit a loss of glycogen and an increase in acid and alkaline phosphatases (Cheng and Snyder, 1962a, b; Cheng, 1963, 1964).

Similar alterations in the digestive gland concentrations of glycogen and phosphatses were noted in *Schistosoma mansoni*-infected *Biomphalaria glabrata* (Von Brand and Files, 1947; Michelson and DuBois, 1973, Christie et al., 1974). Coles (1966), on the other hand, noted a decrease in alkaline phosphatase in *Biomphalaria sudanica* infected with an unidentified plagiorchid. *Physa occidentalis* infected with *Echinostoma revolutum* are reported to show a decrease in glycogen and in increase in lipids in the foot musculature (Hurst, 1927). Arginine was observed to be reduced in foot-tissue extracts from *Schistosoma haematobium*-infected *Bulinus nasutus productus*, but not in extracts from similarly infected *Physopsis globosus* or in two strains of *Schistosoma mansoni*-infected *Biomphalaria glabrata* (Senft, 1967). These results are open to question, since the snails were apparently infected in the field and neither the age nor intensity of infection was controlled. Moreover, the author reported studies on *Schistosoma haematobium*-infected *Biomphalaria sudanica tanganyikiensis* and, to my knowledge, no species of *Biomphalaria* serves as a host for *S. haematobium*.

Prosobranch snails infected with various trematodes show biochemial alterations similar to those found in pulmonates, that is, depletion of glycogen and free amino acids in foot tissues, digestive gland, and gonads (Richards, 1969; Robson and Williams, 1971; Reader, 1971). In contrast to observations on the pulmonate *Helisoma trivolvis* (Cheng and Snyder, 1962b), a decrease in the lipids of the digestive gland of *Bithynia tentaculata* was noted after infection with *Cercaria helvetica* XII (Reader, 1971).

COMMENTS

I have elected to conclude this review by citing some areas in which there has been a paucity of effort or in which further study is indicated. To be sure, the selection is arbitrary and undoubtedly biased. Nonetheless, it may serve to stimulate interest in acquiring an understanding of cellular defense mechanisms in gastropods and other molluscs, one of the goals of this workshop.

It is axiomatic that progress in an understanding of the mechanisms of cellular defense in molluscs must rest on a sound basis of knowledge of cell morphology and function. Studies in cytology, histology, and histochemistry, both at the light and ultra-microscopic level, should be encouraged. It is surprising and regrettable that studies on other molluscs comparable to Pan's (1958) on the histology of *Australorbis glabratus* (=*Biomphalaria glabrata*) have not been forthcoming during the past decade, particularly for species of medical and veterinary importance. To be sure, there has been a continuing series of excellent studies on the ultrastructure of molluscan cell-types, but these have been limited to a few species and have been done largely by our European colleagues. Comparable studies by American cellular morphologists are almost non-existent.

A glance at Spark's book (1972) reveals that most of our information relevant to non-infectious causes of pathology in molluscs is based on observations in bivalves. The situation is equally poor with respect to pathology in gastropods initiated by infectious agents. Although many gastropods serve as hosts for digenetic trematodes, the pathologic responses of the snails to these insults have been ignored or only superficially described.

There is an obvious need to employ naturally occurring pathogens in experimental studies on phagocytosis and other defense mechanisms. The use of esoteric agents and unnatural particles (charcoal, polystyrene, thorium, etc.) has its value, but the responses elicited may be atypical and misleading.

A continued effort to understand the cellular defense mechanisms of gastropods and other molluscs appear warranted and deserves further support. Not only may we gain insight into the more complex defense mechanisms employed by higher animals, but we may find also a means for controlling the intermediate snail-hosts of diseases causing economic hardship and misery to millions of individuals.

REFERENCES

Agersborg, H. P. K. (1924). Studies on the effect of the parasitism upon the tissues. I. With special reference to certain gasteropod molluscs. Quart. J. microscop. *Sci.*, *68*:361-401.

Bang, F. B. (1973). Immune reactions among marine and other invertebrates. *Bio. Science*, *23*:584-589.

Bayne, C. J. (1973). Molluscan internal defense mechanisms: the fate of C^{14}-labelled bacteria in the land snail *Helix pomatia* (L.) J. Comp. *Physiol.*, *86*:17-25.

Boss, K. J. (1971). Critical estimate of the number of recent mollusca. *Occ. Paps. Moll.*, *3*:81-135.

Brooks, CP. (1953). A comparative study of *Schistosoma mansoni* in *Tropicorbis havanensis* and *Australorbis glabratus*. *J. Parasit.*, *39*:159-163.

Brooks, W. M. (1968). Tetrahymenid ciliates as parasites of the gray garden slug. *Hilgardia*, *39*:205-276.

Brooks, W. M. (1969). Molluscan immunity to metazoan parasites. In "Immunity to Parasitic Animals", v. 1, Appelton-Century Crofts, New York, pp. 149-171.

Brown, A. C. (1967). Elimination of foreign particles by the snail, *Helix aspersa*. *Nature*, *213*:1154-1155.

Brown, A. C. and Brown, R. J. (1965). The fate of thorium dioxide injected into the pedal sinus of *Bullia* (Gastropoda: Prosobranchiata). *J. Exp. Biol.*, *42*:509-514.

Buchholz, K., Kuhlmann, D., and Nolte, A. (1971). Aufnahme von Trypanblau und Ferritin in die Blasenzellen des Bindegewebes von *Helix pomatia* und *Cepaea nemoralis* (Stylommatophora, Pulmonata). *Z. Zellforsch., 113*:203-215.

Cheng, T. C. (1963). The effects of *Echinoparyphium* larvae on the structure of and glycogen deposition in the hepato-pancreas of *Helisoma trivolvis* and glycogenesis in the parasite larvae. *Malacologia, 1*:291-303.

Cheng, T. C. (1964). Studies on phosphatase systems in hepatopancreatic cells of the molluscan host of Echinoparyphium sp. and in the rediae and cercariae of this trematode. *Parasitology, 54*:73-79.

Cheng, T. C. and Galloway, P. C. (1970). Transplantation immunity in mollusks: the histoincompatibility of *Helisoma duryi normale* with allografts and xenografts. *J. Invert. Path., 15*:177-192.

Cheng, T. C. and Rifkin, E. (1970). Cellular reactions in marine molluscs in response to helminth parasitism. In "A Symposium on Diseases of Fishes and Shellfishes", Amer. Fish. Soc., Spec. Publ. No. 5, pp. 443-496.

Cheng, T. C. and Snyder, R. W., Jr., (1962a). Studies on host-parasite relationships between larval trematodes and their hosts. I. A review. II. The utilization of the host's glycogen by the intramolluscan larvae of *Glypthelmins pennsylvaniensis* Cheng, and associated phenomena. Trans. Am. Microscop. *Soc., 81*:209-228.

Cheng, T. C. and Snyder, R. W., Jr., (1962b). Studies on Host-parasite relationships between larval trematodes and their hosts. III. Certain aspects of lipid metabolism in *Helisoma trivolvis* (Say) infected with the larvae of *Glypthelmins pennsylvaniensis* Cheng and related phenomena. Trans. Am. Microscop. *Soc., 81*:327-331.

Cheng, T. C., Thakur, A. S., and Rifkin, E. (1969). Phagocytosis as an internal defense mechanism in the Mollusca: with an experimental study of the role of Leucocytes in the removal of ink particles in *Littorina scabra* Linn. In "Proceedings of the Symposium on Mollusca", Marine Biol. Assoc. India, Pt. 2, pp. 546-563.

Chernin, E., Michelson, E. H., and Augustine, D. L. (1960). *Daubaylia potomaca*, a nematode parasite of *Helisoma trivolvis*, transmissible to *Australorbis glabratus*. *J. Parasit., 46*:599-607.

Christie, J., Foster, W. B., and Stauber, L. A. (1974). The effect of parasitism and starvation on carbohydrate reserves of *Biomphalaria glabrata*. *J. Invert. Path.,* (in press).

Coles, G. C. (1969). Isoenzymes of snail livers - I. Hydrolysing enzymes and peroxidase.

Comp. Biochem. *Physiol., 29*:403-411.

Farley, C. A. (1968). *Minchinia nelsoni* (Haplosporidia) disease syndrome in the American oyster *Crassostrea virginica*. *J. Protozool., 15*:585-599.

Feng, S. Y. (1967). Responses of molluscs to foreign bodies, with special reference to the oyster. *Fed. Proc., 26*:1685-1692.

Fernandez, J. (1966). Nervous system of the snail *Helix aspersa*. I. Structure and histochemistry of ganglionic sheath and neuroglia. *J. Comp. Neurol., 127*:157-182.

Fernandez, J. (1971). Nervous system of the snail *Helix aspersa*. II. Fine structure of vascular channels and amebocytes associated with the ganglionic sheath. *Z. Zellforsch., 118*:512-524.

George, W. C. and Ferguson, J. H. (1950). The blood of gastropod molluscs. *J. Morph., 86*:315-327.

Good, R. A. and Papermaster, B. W. (1964). Ontogeny and phylogeny of adaptive immunity. Advanc. *Immunol., 4*:1-115.

Harry, H. W. and Aldrich, D. V. (1963). The distress syndrome in *Taphius glabratus* (Say) as a reaction to toxic concentrations of organic ions. *Malacologia, 1*:283-289.

Haughton, I. (1934). Amoebocytes and allied cells in Invertebrata. J. Roy. Microscop. *Soc., 54*:246-262.

Hildemann, W. H., and Reddy, A. L. (1973). Phylogeny of immune responsiveness: marine invertebrates. *Fed. Proc., 32*:2188-2194.

Hobmaier, A., and Hobmaier, M. (1934). The route of infestation and the site of localization of lungworms in mollusks. *Science, 80*:229.

Hurst, C. T. (1927). Structural and functional changes produced in the gastropod mollusk, *Physa occidentalis* in the case of parasitism by the larvae of *Echinosto revolutum*. Univ. Calif. Publ. *Zool., 27*:321-404.

Johnson, P. T. (1968). "An Annotated Bibliography of Pathology in Invertebrates other than Insects". Burgess Publ. Co., Minneapolis, 322 pp.

Lal, M. B., and Baugh, S. C. (1955). Studies in histopathology-effects of the presence of a plagiorchid metacercaria on the tissues of the snail *Vivipara bengalensis* (Lamarck). Proc. Indian Acad. *Sci., 42B*:114-122.

Mead, A. R. (1956). Disease in the Giant African Snail *Achatina fulica*. Science,

123:1130-1131.

Metchnikoff, E. (1892). Lecons sur la pathologie compareé de l'inflammation. Paris. (Reprinted by Dover Publications Inc., New York, 1968 as Lectures on the Comparative Pathology of Inflammation).

Meuleman, E. A. (1972). Host-parasite interrelationships between the freshwater pulmonate *Biomphalaria pfeifferi* and the trematode *Schistosoma mansoni*. Netherlands *J. Zool., 22*:355-427.

Michelson, E. H. (1958). A method for the relaxation and immobilization of pulmonate snails. Trans. Am. Microscop. *Soc., 77*:316-319.

Michelson, E. H. (1961). An acid-fast pathogen of fresh-water snails. Am. J. Trop. Med. *Hyg., 10*:423-433.

Michelson, E. H. (1963). *Plistophora husseyi* sp. n., a microsporidian parasite of aquatic pulmonate snails. *J. Insect Path., 5*:28-38.

Michelson, E. H. (1970). *Aspidogaster conchicola* from freshwater gastropods in the United States. *J. Parasit., 56*:709-712.

Michelson, E. H. (1971). Distribution and pathogenicity of *Tetrahymena limacis* in the slug *Deroceras reticulatum*. *Parasitology, 62*:125-131.

Michelson, E. H., and DuBois, L. (1973). Increased alkaline phosphatase in the tissues and hemolymph of the snail *Biomphalaria glabrata* infected with *Schistosoma mansoni*. Comp. Biochem. *Physiol., 44B*:763-767.

Müller, G. (1956). Morphologia, Lebenslauf und Bildungsort der Blutzellen von *Lymnaea stagnalis* L. *Z. Zellforsch., 44*:519-556.

Owen, G. (1966). Digestion. In "Physiology of Mollusca", v. 2, Academic Press, New York and London, pp. 53-96.

Pan, C-T. (1958). The general histology and topographic microanatomy of *Australorbis glabratus*. Bull. Mus. Comp. *Zool.* Harvard College, *119*:237-299.

Pan, C-T. (1963). Generalized and focal tissue responses in the snail, *Australorbis glabratus*, infected with *Schistosoma mansoni*. Ann. N.Y. Acad. *Sci., 113*:475-485.

Pan, C-T. (1965). Studies on the host-parasite relationship between *Schistosoma mansoni* and the snail *Australorbis glabratus*. Am. J. Trop. Med. *Hyg., 14*:931-976.

Pauley, G. B. (1969). A critical review of neoplasia and tumor-like lesions in mollusks.

Nat. Cancer Inst. Monogr. No. 31, pp. 509-539.

Plummer, J. M. (1966). Collagen formation in Achatinidae associated with a specific cell type. Proc. malac. *Soc.* London, *37*:189-198.

Prowse, R. H., and Tait, N. N. (1969). *In vitro* phagocytosis by amoebocytes from the haemolymph of *Helix aspersa* (Müller). I. Evidence for opsonic factor(s) in serum. *Immunology, 17*:437-443.

Reade, P. C. (1968). Phagocytosis in invertebrates. Austr. J. Exp. Biol. med. *Sci., 46*:219-229.

Reader, T. A. J. (1971). Histochemical observations on carbohydrates, lipids, and enzymes in digenean parasites and host tissues of *Bithynia tentaculata. Parasitology, 63*:125-136.

Richards, C. S. (1968). Two new species of *Hartmannella* amebae infecting freshwater molluscs. *J. Protozool., 15*:651-656.

Richards, C. S. (1970). Pearl formation by *Biomphalaria glabrata. J. Invert. Path., 15*:459-460.

Richards, C. S. (1972). *Biomphalaria glabrata* genetics: pearl formation. *J. Invert. Path., 20*:37-40.

Richards, C. S., and Merritt, J. W. (1967). Studies on *Angiostrongylus cantonensis* in molluscan intermediate hosts. *J. Parasit., 53*:382-388.

Richards, R. J. (1969). Qualitative and quantitative estimations of the free amino acids in the healthy and parasitized digestive gland and gonad of *Littorina saxatilis tenebrosa* (Mont.) and in the daughter sporocysts of *Microphallus pygmaeus* (Levinsen, 1881) and *Microphallus similis* (Jägerskïold, 1900) (Trematoda: Microphallidae). Comp. Biochem. *Physiol., 31*:655-665.

Robson, E. M., and Williams, I. C. (1971). Relationships of some species of Digenea with the marine prosobranch *Littorina littorea* (L.) III. The effect of larval digenea on the glycogen content of the digestive gland and foot of the *L. littorea. J. Helminthol., 45*:381-401.

Salt, G. (1970). "The Cellular Defense Reactions of Insects". Cambridge Monogr. Exp. Biol. No. 16, Cambridge Univ. Press, London and New York.

Senft, A. W. (1967). Studies in argenine metabolism by schistosomes-II. Argenine depletion in mammals and snails infected with *S. mansoni* or *S. hematobium*. Comp. Biochem. *Physiol., 21*:299-306.

Sminia, T. (1972). Structure and function of blood and connective tissue cells of the fresh water pulmonate *Lymnaea stagnalis* studied by electron microscopy and enzyme histochemistry. *Z. Zellforsch., 130*:497-526.

Sminia, T., Boer, H. H., and Niemanssverdreit, A. (1972). Haemoglobin producing cells in freshwater snails. *Z. Zellforsch, 135*:563-568.

Sparks, A. K. (1972). "Invertebrate Pathology", Academic Press, New York and London.

Stang-Voss, C. (1970). Zur Ultrastruktur der Blutzellen wirbelloser Tiere. III. Über die Haemocyten der Schnecke *Lymnaea stagnalis* L. (Pulmonata). *Z. Zellforsch, 107*:142-156.

Stauber, L. A. (1961). Immunity in invertebrates, with special reference to the oyster. Proc. Nat. Shellfisheries Assoc., *50*:7-20.

Szabo, I., and Szabo, M. (1934). Epitheliale Geschwulstbildung bei einem wirbellosen Tier, *Limax flavus. Z. Krebsforsch., 40*:540-546.

Tripp, M. R. (1961). The fate of foreign materials experimentally introduced into the snail *Australorbis glabratus. J. Parasit., 47*:745-751.

Tripp, M. R. (1963). Cellular responses of mollusks. Ann. N.Y. Acad. *Sci., 113*:467-474.

Tripp, M. R. (1966). Hemagglutinin in the blood of the oyster *Crassostrea virginica. J. Invert. Path., 8*:478-484.

Tripp, M. R. (1969). General mechanisms and principles of invertebrate immunity. In "Immunity to Parasitic Animals", v. 1, Appleton-Century-Crofts, New York, pp. 111-128.

Tripp, M. R. (1970). Defense mechanisms of mollusks. J. Reticuloendothelial *Soc., 7*:173-182.

Von Brand, T., and Files, V. S. (1947). Chemical and histological observations on the influence of *Schistosoma mansoni* infection on *Australorbis glabratus. J. Parasit., 33*:476-482.

Wagge, L. E. (1951). The activity of amoebocytes and of alkaline phosphatases during the regeneration of the shell in the snail *Helix aspersa*. Quart. J. Microscop. *Sci., 92*:307-321.

Wagge, L. E. (1955). Amoebocytes. Internat. Rev. *Cytol., 4*:31-78.

Wolburg-Buchholz, K. (1972). Blasenzellen im Bindegewebe des Schlundrings von *Cepaea nemoralis* L. (Gastropoda. Stylommatophora). I. Feinstruktur der Zellen. *Z. Zellforsch., 128*:100-114.

Wondrak, G. (1969). Die Ultrastruktur der Zellen aus dem interstitiellen Bindegewebe von *Arion rufus* (L.), Pulmonata, Gastropoda. *Z. Zellforsch, 95*:249-262.

Wright, C. A. (1966). The pathogenesis of helminths in the Mollusca. Helminth. *Abstr., 35*:207-224.

CELLULAR COMPONENTS

Phyllis Johnson and Donald Heyneman*
Department of Interior
Oxford Laboratory
Oxford, Maryland

Morphological studies of insect hemocytes, which are known to function in phagocytosis, encapsulation, etc., have not provided much new information in the past several years. There is continuing controversy about the origin of the various hemocyte types, and speculation as to whether they form a series or arise separately from germinal tissues. The most significant recent advances have been in the biochemical analysis of insect phagocytes. Morphologically and physiologically insect phagocytes share many characteristics with vertebrate monocytes and tissue macrophages. Concerning other arthropods, crayfish hemocyte activity has been studied in vivo, with the finding that opsonic activity may be marked, and that the rate of phagocytosis may be increased by a prior injection of endotoxin.

The invertebrate series most likely to contain elements homologous with vertebrate cellular defense mechanisms includes the echinoderms and protochordates. Several investigators are now searching for homologies between tissue and cell reactions in these groups and those of vertebrates. Some parts of vertebrate mechanisms have already been found. Also encouraging is the more general recognition that experimental approaches to the study of invertebrate cellular mechanisms seldom can be the approaches used in study of vertebrate cellular immunology.

In all invertebrates, we need further investigation of the germinal tissues that produce reactive cells. We must determine whether the cells active in defense form a vertical series or arise separately. Amongst those types which derive from a single stem cell, it is important to know whether, as morphology and function change, the cell is so programmed that it cannot dedifferentiate, or whether it still retains some of the properties of its younger self.

In spite of work on molluscan cellular defenses stemming back to Metchnikoff in 1892, and a substantial body of literature that has accumulated since then and has

---

*The George Williams Hooper Foundation, University of California, San Francisco, California.

been frequently reviewed, the cellular response mechanisms in phagocytosis and immunity in mollusks are still largely unknown. The aspects of cellular response that are generally recognized include: phagocytosis, fibrotic encapsulation, diffuse fibrosis, pearl formation, necrosis, and biochemical alteration. Emphasis was given on the need to study normal invertebrate histology prior to describing histopathology; to extend studies to a greater variety of mollusks to broaden our comparative view; and to employ naturally occurring pathogens rather than arbitrarily selected laboratory pathogens and unnatural particles. The research needs for the study of molluscan defenses parallel the equally pressing ones in other invertebrate groups: 1) a proper definition of basic cell types, 2) studies on the ontological origin of amebocytes, with particular emphasis on the granulocytes, 3) determination of the role of phagocytic cells as fixed phagocytes, 4) the cellular sequence in encapsulation, and 5) biochemical changes in tissues undergoing a defensive reaction.

III. Hemolymph Components in Invertebrate Immunity

HUMORAL FACTORS AND MOLLUSCAN IMMUNITY

M. R. Tripp

Department of Biological Sciences
University of Delaware
Newark, Delaware

I. INTRODUCTION

The last 10 to 15 years have seen considerable research in the general area of immunity of invertebrate animals. Considerable evidence concerning both cellular and humoral responses to infection has been gathered and has been used to make some generalizations. Detailed examination of these data, however, reveals some significant deficiencies. Most notable is that relatively few invertebrate species have been studied. Altogether about half a dozen phyla and a dozen classes have been studied to some degree but less than 10 species have been studied thoroughly. The determining factor seems to be not their phylogenetic position but 1) ease of maintenance in the laboratory, and/or 2) their commercial or public health importance.

This review will consider in detail what is known about biologically active substances in molluscan hemolymph. Related observations about members of other phyla will be cited for comparative purposes. The underlying assumption is that the phylum Mollusca has unique characteristics and that some of these characteristics are shared with other phyla. By considering the spectrum it may be possible to make some useful generalizations.

II. SUBSTANCES WITH DEFENSIVE POTENTIAL FOUND IN THE TISSUES OF NORMAL ANIMALS

A. *Antitumor and Antimicrobial Activity*

The concept of growth regulators in tissues of marine invertebrates is not new. Two such substances are particularly interesting because of their ability to inhibit growth of tumors in vertebrate species. Mercenene, a substance extracted from the tissues of *Mercenaria mercenaria,* the common quahog, is inhibitory for sarcoma 180 (Schmeer and Huala, 1965) and the Krebs-2 carcinoma (Schmeer and Beery, 1965a) in Swiss mice. Similar extracts of *Mercenaria campechiensis* show both *in vivo* (Krebs-2) and *in vitro* (Hela cells) inhibitory action against tumor cells (Schmeer and Beery, 1965b). Mercenene also was effective against virus-induced leukemia in mice, although infected mice all died eventually (Judge, 1966). Chemical characterization of mercenene is incomplete (Schmeer et al, 1966), but it seems to be a small molecule (molecular weight 1 - 5000) and heat stable.

A similar antimicrobial effect was noted by Li, first in abalone (Li, 1960a), and later in several other marine species (both vertebrates and invertebrates) (Li, 1960b). The most definitive studies focused on "paolins" from clams, *Mercenaria mercenaria* (Li et al, 1965) and oysters (probably *Crassostrea virginica;* Prescott et al, 1966). Abalone juice fed to mice infected with poliovirus ameliorated their disease but did not protect them from paralysis or death. The same material inhibited *Staphylococcus aureus* growth as well (Li, 1960a). Fractions obtained through DEAE column chromatography were inhibitory to both Gram positive *(Staphylococcus aureus* and *Streptococcus pyogenes)* and Gram negative *(Salmonella typhi, S. paratyphi* A and *S. paratyphi* B) bacteria (Prescott and Li, 1960). This material was characterized only as a heat stable protein (Prescott and Li, 1960). Subsequent studies suggested that some fractions of abalone juice were effective against bacteria while others exerted antiviral effects (Li et al, 1962a,b). The mechanism of action was not determined for either material. Both abalone and oyster tissue extracts possessed some antiviral activity when fed to infected mice or added to tissue cultures; they were named paolin 1 and paolin 2, to indicate their antibacterial and antiviral activity, respectively (Li et al, 1962a,b). Paolin 2 was found to inhibit tumors in hamsters caused by adenovirus type 12; and to prevent growth of adenovirus 12 and herpes simplex in tissue culture (Li et al, 1962). It consisted of 3 fractions with molecular weights ranging from 700 to 9-10,000, although the chemical nature of these materials was not ascertained (Li et al, 1965). More detailed biochemical studies of oyster paolin 2 failed to clarify the nature of this material although the authors concluded that it was not lysozyme or histone (Prescott et al, 1966).

Despite the range of studies reported by Schmeer and Li with their colleagues, little is clear about the nature of the materials they extracted from marine mollusks or how they act. The action is certainly nonspecific and it is not clear whether the effective extracts are one or more compounds working separately or in unison. Thus it is not easy to decide whether this is a specific effect of possible significance to the mollusk or simply an experimental artifact due to introduction of foreign tissue extracts into vertebrate tissues *in vivo* or *in vitro.* More detailed study should reveal whether these molluscan tissue extracts are similar to fatty acids, sterols, certain porphyrins, basic peptides, histones, globin, protamines, spermine extracted from tissues of normal vertebrates (Hirsch, 1959), and inhibitory against a great variety of microbial agents (Skarnes and Watson, 1952).

B. Lysozyme

Lysozyme is a mucolytic enzyme that splits the bond between N-acetyl glucosamine and N-acetyl muramic acid (Salton, 1957). It is found in many body tissues and secretions; it is of special interest that lysozyme is found in high concentrations in polymorphonuclear leukocytes and alveolar macrophages (Hirsch, 1959). The latter secrete lysozyme, suggesting that reticuloendothelial cells may be the source of secreted lysozyme in serum, urine and tears of vertebrates (Heise and Myrvik, 1967). Although the biochemical action of pure lysozyme is well understood and is apparently limited in scope, under natural conditions its action spectrum may be significantly wider. Lysozyme acting in combination with other agents such as chelators and acids may have lytic, cidal or static effects on a variety of microorganisms (Salton, 1957; Hirsch, 1959).

The role of lysozyme in invertebrates is less thoroughly studied but some points are clear. Lysozyme is widely distributed among invertebrate phyla: it has been found in oysters, *Crassostrea virginica* (McDade and Tripp, 1967b,c); mussels, *Anadonta anatina* (Messner and Mohrig, 1969); various insects (Malke, 1965; Mohrig and Messner, 1968); a marine polychete worm, *Nephthys bombergi* (Jolles and Zuili, 1960) and 9 species of marine, fresh water and terrestrial annelids (Schubert and Messner, 1971).

It is clear that there is no single lysozyme, but that there is a class of basic, low molecular weight proteins that lyse certain bacteria and act as an N-acetylmuramidase (Chipman and Sharon, 1969). Thus, physical differences may be marked from species to species or even within a single species. For example, Jolles and Zuili (1960) found that lysozyme from chicken lung differed from chicken egg white lysozyme in chromatographic behavior. Recently Feng (1974) has found that oyster hemolymph lysozyme activity is associated with acidic proteins, suggesting that a molecule very different from the "traditional" one may be responsible for lysozyme-like activity. Lysozyme is even produced by bacteria *(Escherichia coli)*. This is of special interest because production is controlled by phage DNA and mutant lysozymes may be obtained by treating phage with mutagens; by sequensing amino acids the architecture of the molecule can be determined and correlated with activity. This approach shows that T4 phage lysozyme and egg white lysozyme have no common first-order structure, but the distribution of basic, acidic and hydrophobic amino acids seems to be similar (Inouye and Tsugita, 1966). The result is functional similarity that may well have selective advantage for diverse organisms.

Beyond the question of phylogenetic distribution and molecular configuration of lysozyme, lies the more basic question: What is the function of lysozyme in invertebrates? Lysosomes containing at least some of the usual lysosomal enzymes have been demonstrated in starfish *(Asterias forbesi)*, clam *(Mercenaria mercenaria;* Janoff and Hawrylko, 1964); and oyster *(C. virginica;* Eble and Tripp, 1968) leukocytes. It has not been demonstrated conclusively that these cells are the origin of lysozyme, however. There have been no reports either of what lysozyme may be doing in normal animals. It is known that lysozymes kill various bacteria and some blue-green algea in experimental situations (Crespi et al, 1962), but there is no reason to assume that this necessarily occurs in all cases. In fish lysozyme is found in both leukocytes and epidermal cells suggesting roles as part of both internal and external defense mechanisms (Fletcher and White, 1973). Such evidence exists for the oyster in that lysozyme is found in both hemolymph and on the mantle surface, but direct evidence of *in vivo* function is lacking (Eble and Tripp, 1968; McDade and Tripp, 1967b,c).

A logical hypothesis (if one assumes that leukocytes are the primary source of lysozyme) that could be tested experimentally is that hemolymph lysozyme concentration will change with hemocyte concentration and activity during immune responses. Feng and Canzonier (1970) have presented evidence that such a phenomenon may occur; in oysters *(C. virginica)* infected with *Bucephalus* hemolymph lysozyme increased, but decreased in oysters with *Minchinia* infections; in doubly infected animals the hemolymph concentrations remained relatively unchanged. Further exploration of this reaction would be a worthwhile line of research.

C. *Hemagglutinins*

1. *Distribution among nonmolluscan invertebrates*

Heteroagglutinins in body fluid of invertebrate animals have been known since the early part of the twentieth century. At first the focus was on the ability of these substances to react with gametes (Lillie, 1913), but later detailed studies of erythrocyte agglutination were initiated (Tyler, 1946, Landsteiner, 1945). Studies of lobster *(Panulirus interruptus)* serum showed that hemagglutinating material was protein, had "class specificity" (Tyler and Metz, 1945) and could be isolated by electrophoresis (Tyler and Scheer, 1945). Cushing et al, (1963) found a substance in the serum of the spiny lobster *(P. interruptus)* that agglutinates whale cells carrying the Ju2 antigen. Sindermann and Mairs (1959) used the agglutinin of Atlantic lobster *(Homarus americanus)* serum to identify herring blood groups. Cornick and Stewart (1973) have shown that *H. americanus* agglutinin is a protein destroyed by moderate heat (35-45°C), stabilized by calcium ion and formed in hemocytes; it seems to react with D-glucosamine or molecules of similar configuration on the red cell surface. A protein in the hemolymph of the crayfish *Procambarus clarkii* agglutinates chicken and rabbit red cells (Miller et al, 1972). Another crayfish, *Parachaeraps bicarinatus,* has a substance in its hemolymph that agglutinates and opsonizes sheep red cells (McKay and Jenkin, 1970b).

The most fully characterized of all the invertebrate hemagglutinins is that of the horseshoe crab, *Limulus polyphemus,* first described by Cohen et al, (1965). This material agglutinates erythrocytes of a great variety of species (from axolotl to man) presumably reacting with one (or more) determinants common to cells of those vertebrate species; the active material is a protein with electrophoretic mobility similar to vertebrate alpha and beta globulins (Cohen et al, 1965). Marchalonis and Edelman (1968) found the *Limulus* hemagglutinin to be a protein of molecular weight 400,000 (13.5 S) composed of 6 subunits linked noncovalently, arranged in a ring and with each unit made up of 3 protein molecules (molecular weight (22,500). This concept has been strengthened by electron micrographs showing "doughnuts" of purified agglutinin (Fernandez-Moran et al, 1968). This configuration is unique and very different from the architecture of vertebrate immunoglobulins. Further characterization of primary, secondary, tertiary and quaternary molecular structure by peptide mapping, N-terminal amino acid analysis and circular dichroism analysis (Finstad et al, 1972) confirm the unique form of this agglutinin.

Hemagglutinins have been found in the hemolymph of lepidopteran larvae (Bernheimer, 1952) and other insects (Feir and Walz, 1964). Hemolymph of both larval and adult cockroaches, *Blabarus craniifer,* contains a heat-labile factor that agglutinates sheep, horse and rabbit erythrocytes (Anderson et al, 1972).

The hemolymph of the starfish, *Asterias forbesi,* contains a protein that agglutinates erythrocytes of several vertebrate species, but differs in structure from other invertebrate agglutinins and from vertebrate immunoglobulins (Finstad et al, 1972).

2. *Molluscan Hemagglutinins: Distribution and Structure*

Agglutinins for erythrocytes have been reported in numerous molluscan species. Hemolymph of the giant viviparous freshwater snail, *Viviparus molleatus* selectively agglutinates rabbit red cells (Cheng and Sanders, 1962); the active material was apparently protein but could not be identified specifically. More recently much attention has been focused on materials extracted from the albumin glands of freshwater snails. *Otala lactea* contains a material that specifically agglutinates human A cells and it specifically precipitates soluble A substance (Boyd et al, 1966). Albumin gland extracts of *Helix pomatia* and *H. aspersa* have been shown to react specifically with human A cells (Uhlenbruck et al, 1972). In an extremely thorough investigation of the nature of this material Hammarstrom and Kabat (1969) found on polyacrylamide gel electrophoresis several bands of similar mobility and a molecular weight of 100,000. Activity is best inhibited by Me-α-DGalNAc and less efficiently by D-GalNAc; the size of the combining site may be no larger than an α-linked monosaccharide (Hammarstrom and Kabat, 1969).

A hemagglutinin from the sea hare *Aplysia* has also been described (Pauley et al, 1971). In the native state it seems to be a protein of at least 150,000 molecular weight; ultracentrifugation yields 2 fractions (18.5S and 31S) both of which are active (Pauley et al, 1971).

A protein hemagglutinin from the Murray mussel *Velesunio ambiguus* cross-reacts with erythrocytes of rats and rabbits (as does the *Aplysia* material), has a sedimentation coefficient of 28S and has no sulfur-containing amino acids (Jenkin and Rowley, 1970). Johnson (1964) demonstrated that saline extracts of the butter clam *Saxidomus giganteus* specifically agglutinate human A_1 and A_1B cells. The reactive material is apparently protein and is inhibited by N-acetyl-d-galactosamine and N-acetyl-d-glucosamine.

Undoubtedly the best studied molluscan hemagglutinin is from oyster hemolymph (Tripp, 1966; Li and Flemming, 1967; Acton, 1970). This material reacts with erythrocytes of a great variety of vertebrate species. Maximal inhibition of agglutination of human cells is obtained with n-acetyl-d-glucosamine, n-acetyl-d-galactose-amine, D-galactosamine and only partial inhibition with D-glucosamine (McDade and Tripp, 1967a). Calcium is needed to stabilize the active protein (McDade and Tripp, 1967a). Detailed physiochemical studies show the existence of a large complex molecule (33.4S) composed of subunits of about 20,000 molecular weight not bound covalently (Acton et al, 1969). Amino acid analysis showed relatively high histidine content and absence of lysine; the first 2 amino acids at the NH_2 terminus of the subunit are threonine and alanine; mannose galactose and glucosamine contribute the 13% carbohydrate content (Acton et al, 1969). It seems quite clear that the molecular structure of oyster hemagglutinin at least is fundamentally different from *Limulus* hemagglutinin and vertebrate immunoglobulin, although these 3 substances share biological activity. It may be noted also that lectins, plant extracts with hemagglutinating ability, are unlike any of the animal substances in structure.

III. RESPONSES TO FOREIGN SUBSTANCES

A. Nonmollusks

A great variety of responses of invertebrates to foreign materials has been recorded. Numerous phyla have been sampled, but the total number of species examined in any detail is extremely small. Although diversity of response is obvious there are certain principles that seem to apply generally.

1. Coelenterata

Phillips and Yardley (1960) found that the anemone, *Anthopleura elegantissima,* when injected with bovine serum albumin fraction V, secreted material that interfered with the ability of specific anti-BSA antibody to react with the antigen (Phillips, 1960; Phillips and Yardley, 1960). The nature of this substance, its source and mode of action are not known.

2. Arthropods

Insect hemolymph has been reported to react with foreign substances in various ways, some of the reactive factors being innate, others inducible.

In *Limulus polyphemus* there is a naturally occurring heat-labile bactericidal factor active against Gram-negative (but not Gram-positive) bacteria; this material may originate in hemocytes (Johannsen, et al, 1973). *Limulus* hemolymph also serves as a source of terminal complement components for hemolysis mediated by frog C3PA-cobra venom complex factor (Day et al, 1970). The significance of this as a defense mechanism is not clear.

Studies of insect larvae show that humoral substances protective against bacterial infection can be generated by injection of vaccines prepared in various ways (Stephens, 1963a,b). A few hours after vaccination of 11 species of Lepidoptera, an extremely heat stable, acid and alkali-resistant nonspecific protective factor is detectable (Briggs, 1958). The large milkweed bug *(Onocopeltus fasciatus),* vaccinated with *Pseudomonas,* produces a lytic substance that is protective on passive transfer, and resistant to heat (75°C), 1N HCl and 1N NaOH (1 hr.); it is associated with, but not part of, hemolymph proteins (Gingrich, 1964). Vaccination of wax moth *(Galleria mellonella)* larvae with killed *Pseudomonas aeruginosa* gave apparently paradoxical results; significant protection against challenge with living *P. aeuginosa* was evident at 8 hours and disappeared by 48 hours; antigen in the hemolymph seemed to increase (4X over concentrations in control animals) 1-2 hrs. after vaccination and greatly (10-12 times) increased at 24 hrs. (Stephens, 1959). The explanation for this curious situation is not clear. One possible explanation is that, since the assay involved injected hemolymph along with bacteria into rabbits to raise antibodies, the hemolymph has an adjuvant effect (Stephens, 1959). Undoubtedly some of this result is attributable to change in melanization potential following immunization, although the details of that reaction are not clear either (Stephens, 1962).

Hemolymph of normal larvae of 4 orders contained material with bactericidal properties for 5 different species of test bacteria (Stephens, 1963). The protective substance in the hemolymph of wax moth larvae vaccinated with *Pseudomonas* was dialysable, heat stable, unaffected by trypsin and weakly acidic (Stephens and Marshall, 1962).

Cockroaches *(Periplaneta americana)* injected with *Tetrahymena* produce a factor that immobilizes the ciliates and protects host insects (Seaman and Robert, 1968). The material presumably responsible for this reaction is one or more of 3 proteins found in immune but not in normal hemolymph. However, it is not certain that these fractions could not have originated from lysed protozoa (Seaman and Robert, 1968). The cockroach's *(Blabarus craniifer)* defense against bacteria seems to be mainly cellular with no bactericidal or opsonic effect contributed by hemolymph (Anderson et al, 1973).

Hemolymph of *Chironomus* larvae forms capsules on the surface of nematodes independent of hemocyte encapsulation (Gotz, 1969).

The immunology of decapod crustaceans has been studied extensively, probably because of their large size and commercial value. McKay and Jenkin (1969) found that crayfish *(Parachaeraps bicarinatus)* can be made more resistant to bacterial infection by injection of vaccine or endotoxin. This reaction was nonspecific, temperature dependent and was not associated with an antibody-like material (McKay and Jenkin, 1969). This immunity could be attributed primarily to increased phagocytic cell activity (Reade, 1968; McKay and Jenkin, 1970a) and a hemolymph factor was necessary for increased phagocytic uptake of indicator particles (McKay and Jenkin, 1970b).

More recent work with lobsters *(Homarus americanus)* has demonstrated an inducible bactericidin that is a product of the interaction of plasma components and material within hemocytes (Stewart and Zwicker, 1972). Apparently there is more than one fraction of hemolymph existing in an inactive form until activated by material in the hemocytes (Stewart and Zwicker, 1972). Induced bactericidin production has been reported also in the spiny lobster, *Panularus argus*. It is nonspecific, induced by living or dead bacteria, heat sensitive (65°C for 10 min.) and not complement-dependent (Evans et al, 1968). A quick (maximal titer by 24 hr.) primary response as well as secondary and tertiary responses to killed bacteria were detected, although none of the injections after the first produced significant elevation of bactericidin titer (Evans et al, 1969). The specificity of this material is limited since heterologous antigens could induce it also (Winheimer et al, 1969).

3. *Sipunculida*

The sipunculid worms have also received considerable attention. Early experiments examined the possibility that tissue transplantation rejection, phagocytosis or anitbody production might follow antigenic stimulus; no antibody to cellular or protein antigens was detected in *Dendrostomum zostericolum* (Triplett et al, 1958). Other studies revealed the presence of antibacterial activity in *Golfingia gouldii* coelomic fluid; bacterial inoculation did not enhance its activity (Rabin and Bang, 1964). *Sipunculus nudus* was found to form a lysin active against an infective protozoan (ciliate) *Anophrys magii* (Bang, 1966). This substance could be induced by injecting bacteria and its apparent origin is coelomic cells (Bang, 1966). A bactericidal material has been reported from the coelomic fluid of *D. zostericolum* following injection of killed bacteria (Evans et al, 1969). In a comparative study of *Dendrostomum pyroides* and *G. gouldii,* Krassner and Flory (1970) found a nonspecific bactericidal material apparently protein, but whose mode of action was not elucidated.

B. *Mollusca*

A limited number of molluscan species have been examined for the ability of this hemolymph to neutralize, or otherwise affect, infectious organisms. Cheng et al, (1966a) reported that of 8 species of marine pelecypods tested for their ability to serve as second intermediate hosts for the trematode *Himasthla quissetensis,* only 2 species of oysters *(C. virginica* and *C. gigas)* were refractive. The location of the encysted metacercariae on gill and mantle surfaces suggested inactivation and prevention of penetration by some host-elaborated substance. Further study showed that hemolymph of all species stimulated metacercarial encystment and that the stimulatory factor is heat labile; although it was suggested that the superficial location of metacercaria on oyster surfaces is due to seepage of hemolymph into the mantle cavity, there was no direct evidence that this in fact happens (Cheng et al, 1966b).

The pathology induced in host mollusks by infecting trematodes is extensive (Pan, 1965) and probably induces substantial changes in body fluids. Within one hour after infection with *S. mansoni* the relative concentrations of free amino acids and proteins change significantly (Dusanic and Lewert, 1963). These workers suggested that these changes might affect the rate of movement of the miracidium through the snail tissues but no direct evidence of this was presented. Lee and Cheng (1972) reported that in *S. mansoni*-infected *B. glabrata* the number of hemolymph protein fractions decreased from 10 to 17. More recently Gress and Cheng (1973) found that hemolymph protein of infected snails decreased nonspecifically; by day 70 total hemolymph protein was only about 1/3 that of normal hemolymph. Interestingly, 3 disc gel fractions remained essentially unchanged; the authors speculate that the hemolymph protein may be used by the larval trematode, although direct experimental proof is not available (Gress and Cheng, 1973).

The observation of greatest interest from the immunological point of view was made by Michelson (1963 and 1964). *S. mansoni*-infected *B. glabratus* were dissected and tissues (ovotestis, part of the digestive gland and hemaphroditic duct, and seminal vesicle) were extracted with water. These extracts were then tested for their ability to immobilize *S. mansoni* miracidia. Snails infected with *S. mansoni* nearly always yielded miracida immobilizing substance (MIS), but uninfected or nonspecifically stimulated snails rarely did (Michelson, 1963). It is not clear how much of the tissue extract is hemolymph and how much is the contents of ruptured cells of various sorts. This seems to be an important point to clarify if we are to think of this reaction as an "immune response" in any way. Since Benex and Lamy (1959) have reported that extracts of nonhost snails also immobilize *S. mansoni* miracidia also, it may be that the reaction is nonspecific; whether it is responsible for the refractoriness of these snails to infection is not clear. Characterization of these active materials would be an important contribution.

Studies of the sea hare *(Aplysia californica)* indicate that its hemolymph lacks bactericidal activity against a variety of Gram-positive and Gram-negative species (Johnson and Chapman, 1970). On the other hand, Pauley et al, (1971) found bacterial agglutinins in *A. californica* effective against both Gram-positive and Gram-negative species. The concentration of the agglutinin was not changed significantly by repeated injection of bacteria. There seemed to be accelerated clearance of one bacterial species from the hemolymph following secondary injection of one species of gram negative bacteria, but controls necessary for determining the specificity of the reaction were lacking (Pauley et al, 1971). Sea hare hemolymph contains an opsonin for chicken red blood cells, but whether an opsonin for bacteria is present is not known (Pauley et al, 1971). The clearance of the hemolymph of *Aplysia* is considered mainly a hemocytic reaction wherein there is an initial decrease in numbers of circulating hemocytes followed by a return to normal concentrations within a few hours (Pauley et al, 1971). A similar phenomenon has been reported in the land snail *Helix pomatia* (Reade, 1968; Bayne and Kime, 1970) although these authors did not present evidence for the presence of a humoral factor reactive with bacteria as did Prowse and Tait (1969) in *Helix aspersa.* When *Helisoma duryi normale* is injected with living bacteria no lysin or agglutinin was detected although there was a pronounced decrease in certain fractions of hemolymph protein (Cheng, 1969).

The single outstanding claim for induction of bactericidal substances in mollusks is contained in a brief report by Cushing and others (1971).

In the only published attempt to induce antibody-like materials in a cephalopod *(Octopus doefleini),* Bayne (1973) found that repeated injections of bacteria *(Serratia marcescens)* did not enhance blood clearance and no bactericidal factor was detectable.

One of the mollusks most frequently studied in various immunological aspects is the oyster, particularly *Crassostrea virginica*. The first, and in many ways the most basic, such study was by Stauber (1950). He established that the basic response of the oyster to foreign particles in its tissues is phagocytosis with subsequent migration of the host cells across epithelial to the exterior. This concept has been confirmed repeatedly, and modified to include intracellular digestion by these same host cells (Tripp, 1970).

Specific humoral factors in oysters (aside from lysozyme) that react with bacteria have not been demonstrated directly. Tripp (1960) found that live bacteria, nonpathogenic for the oyster, injected intracardially into *C. virginica* lysed in the hemolymph, but no *in vitro* evidence of bacteriolysis was presented. The possibility exists, however, that tests for acquired immunity in mollusks have not been performed under proper experimental conditions and the proper manifestations sought. In oysters *(C. virginica)* receiving injections of *Hexamitis* some indirect evidence of a lytic factor in hemolymph was found, but only under certain environmental conditions (Feng and Stauber, 1968). A detailed study of the fate of infected bacteria by S. Feng (1966) supports the notion that clearance of the tissues is mediated by cellular action only.

Another technique used in attempts to demonstrate immunological competence of the oyster is enhanced clearance of antigen from the hemolymph. J. Feng (1966) found no specific antiviral factor in hemolymph from "immune" oysters *(C. virginica)* at 25°C in both "immune" and normal hemolymph phage was inactivated over a period of 180 hours. Acton and Evans (1968) found that T2 coliphage was cleared more rapidly following secondary injection into oysters *(C. virginica)*, but they too failed to demonstrate induction of a specific antiviral factor in hemolymph. Similar studies in the giant land snail *Achatina* showed that in animals injected with T1 coliphage, secondary clearance was slower than primary clearance, and no neutralizing antibody was detected in hemolymph (Nelstrop et al, 1968c).

The concept emerging from these studies is that mollusks do not synthesize specific proteins to neurtalize immunogens, although normal hemolymph may contain proteins with biological activity. Motile cells, however, are active and effective in clearing foreign substances from tissues.

IV. WHAT EVIDENCE IS NECESSARY TO DEMONSTRATE AN IMMUNE REACTION IN AN INVERTEBRATE?

A. *What does it mean when an allograft is (or is not) rejected by an invertebrate recipient?*

One of the most sensitive measures of immunological competence among vertebrate animals is their ability to reject transplanted tissue. Among the salient features of these reactions are a characteristic induction period, anamnesis, and specificity. Typically an orthotopic allograft heals into place (as does any accompanying autograft) but after 7 or 8 days the corium is invaded by mononuclear cells, blood flow and lymph drainage stop and the epithelium begins to break up and lose its attachment to the underlying connective tissue. This reaction has been recorded in all orders of vertebrates including cold blooded species such as lizards, amphibians and goldfish. The most primitive vertebrate to exhibit this ability is the hagfish (Hildemann and Theones, 1969). The transplanted tissue is the target of attack by mononuclear immunocytes. Skin transplants are invaded broadly; other organs or tissues are rejected because the focus of attack is the blood vessels serving that tissue. Humoral antibody is not needed, and may inhibit, this reaction.

Perhaps the most significant and extensive transplantation work among invertebrates has been done by Cooper utilizing oligochaete annelids (Cooper, 1969a,b). In an experimental system using *Eisenia foetida* and *Lumbricus terrestris* successful autografts were made routinely; allografts (between individuals of different geographical populations) were variable but usually rejected; xenografts were always rejected. Second set reactions were commonly, but not always, recorded (Cooper, 1968a). The time of rejection varied widely (for example, 6-147 days). "Acute rejections (11 days or less) of first-sets always resulted in prolonged survival of second-sets, whereas chronic rejectors brought about accelerated destruction of second-set grafts" (Cooper, 1969c). Rejection was shown to be specific (Cooper, 1969d); histologic analysis revealed phagocytic acidophil cells at the host-graft interface, but apparently not specifically associated with blood vessels (Cooper, 1968,a). Multinucleate giant cells were found in all grafted animals and their contribution, if any, to rejection of grafts is not clear (Cooper, 1969c). No humoral factors seem to be involved. *In vitro* studies of interactions of coelomocytes of different species suggest cytotoxic effects, but the specificity of these reactions and their relation to *in vivo* phenomena remain to be demonstrated (Cooper, 1973).

It is clear that annelids can distinguish among "self," closely related "nonself" (allograft), and distantly related "nonself" (xenograft) tissues. Memory is demonstrable but somewhat capricious over extended periods. Specificity of rejection is high, but not absolute, and may vary with species. Most second-set xenografts are destroyed by accelerated reactions, but a significant number show the opposite effect – prolonged survival. The conclusion that seems most obvious is that graft rejection in annelids may well have quite a different physiological basis than in vertebrates. Any inferences about the immunological basis of this reaction should be conservatively drawn.

In other phyla with quite different morphological and physiological characteristics, even more caution may be warranted. For example, it has been reported that mollusks *(Biomphalaria glabrata)* accept allografts but reject xenografts (Tripp, 1961). Since these "grafts" were implantations of bits of foot tissue into the cephalopedal sinus, it is probably not reasonable to use the word graft in this context and a conservative interpretation of the data seems wise. It is clear that "self" can be distinguished from "nonself" but how this is done is not understood.

Recent transplantation experiments in echinoderms are particularly interesting (Hildemann and Dix, 1972). Repeat integumentary allografts on *Cucumaria tricolor* were rejected. First-set integumentary allograft rejection was still under way when a second-set response was recorded. This shows at least short-term memory among an advanced group of invertebrates. Leukocytes of several types were found invading grafts undergoing rejection. The role of these cells is unknown, but the histological similarity to vertebrate skin graft reaction is striking.

Similarly transplantation of various parts of insects from individual to another has been done in all conceivable ways (Salt, 1970). But is this "transplantation" in the conventional sense of immunologists? In many cases these "grafts" are accepted, in others, they are not. What conclusions, if any, can be drawn about the immune capabilities of insects? They too can tell "self" from "nonself" in most cases, but the cellular basis for this does not seem comparable to anything in vertebrate systems.

Thus it seems prudent to reserve judgment for now and proceed instead to obtain more detailed data on the role cells play in recognition of foreignness and whether humor factors aid them in that recognition. Also this search should range systematically over a greater diversity of invertebrate phyla so that a more general concept of immune mechanisms may emerge. In short, we do not yet know what it means if a graft is rejected or not by an invertebrate recipient.

B. What does it mean when injected immunogen is (or is not) cleared from the hemolymph of an invertebrate?

The rate of clearance of injected immunogen from the circulation has been employed as a measure of the immune status of vertebrates. It is generally assumed that the 3-phase elimination curve characteristic of removal of colloids is a result of: 1) a brief phase of rapid equilibration by diffusion of immunogen into extravascular spaces, 2) a period of metabolic decay characterized by a half-life distinctive for each substance tested, and 3) a phase of immune elimination due to presence of specific circulating antibody. The total period for primary response approximates 7 days, but on secondary response the period is only 4 days or less (Davis et al, 1968).

However, it is not always necessary for specific antibody to be present in order to have "immune clearance." In a series of three critical studies, Nelstrop, Taylor and Collard (1968 a,b,c) examined the kinetics of phage elimination and the mechanisms involved. Rabbits receiving primary injections of coliphage T1 or T2 cleared it from the blood in an initial rapid phase and a slower second phase. Secondary stimulation resulted in more rapid elimination of phage from the blood, also in a biphasic way. These clearance phenomena were specific and occurred in the absence of detectable circulating antibody (Nelstrop et al, 1968a). Subsequently these authors demonstrated that "immune" macrophages remove phage (T1) from an *in vitro* system more rapidly than "nonimmune" macrophages. This *in vitro* clearance occurs in 2 phases, similar to that occurring *in vivo* (Nelstrop et al, 1968b).

Accelerated clearance of injected bacteriophage from hemolymph has been sought by a number of investigators as an indicator of immune response of invertebrates. Variable results have been obtained. Bernheimer et al, (1952) found that *Samia cecropia* caterpillars removed 90% or more of injected T2 phage from the hemolymph between 2 and 10 hours following injection. They concluded that elements in the hemolymph were not responsible for that removal (Bernheimer et al, 1952). Secondary clearance of T1 phage in crabs *(Carcinus maenas)* was more rapid in 2 out of 3 animals tested; no neutralizing factor was found in hemolymph (Taylor et al, 1964). Nelstrop et al, (1968c) found enhanced clearance of T1 phage from the hemolymph on secondary injection (and in a single crab given a tertiary injection). Crabs given T1 phage initially and challenged with T6 phage cleared the latter at a rate similar to the primary rate; no neutralizing activity was found in the hemolymph of any crab used (Nelstrop, 1968c). Sipunculids *(D. zostericolum)* injected with T4 phage cleared at equal rates following primary or secondary injections; coelomic fluid contained no neutralizing factor (Cushing, 1967).

Studies of phage clearance in mollusks have not yielded to definitive answers. Oysters *(C. virginica)* injected with T2 phage seem to clear virus from the hemolymph more rapidly after secondary injections, but the differences were not striking and may have lacked specificity (Acton and Evans, 1968). No neutralizing substance was detected in the hemolymph and it does not seem that any one tissue accumulated phage in an active way (Acton and Evans, 1968). *Staphylococcus aureus* phage 80, when injected into *C. virginica* is eliminated from the tissues to the environment; few virions are taken up by circulating leukocytes and hemolymph from normal or immune animals has little or no effect on the virus (J. Feng, 1966). In land snails *(Achatina)* injected with T1 phage primary clearance took many weeks and secondary clearance proceeded even more slowly, no neutralizing substance was detected in the hemolymph of any snail used (Nelstrop et al, 1968c).

It seems clear that of the invertebrates tested thus far, none is capable of forming a specific phage neutralizing humoral factor. This is consistent with their apparent general inability to form antibody-like materials. But even in the absence of antibody or analagous material, the mechanism of cellular removal of virus in body fluid needs detailed study. Are virions phagocytosed as they are in vertebrates? Do they adhere nonspecifically to any other selected cells in the body? Or are there specific "activated immunocytes" that clear them from the hemolymph? How are they eliminated from the body? The answers to these questions have interest for both basic and applied biology.

V. CONCLUSIONS AND NEW DIRECTIONS

There are several interesting problems that emerge from consideration of the above evidence. Some of the most obvious include:

1. What is the role of lysozyme in normal animals? Although material with lysozyme activity is undoubtedly secreted by mollusks there is no evidence concerning its function. Does it inactivate potentially infectious microbes? What is its action spectrum? Or does it help in feeding and have no defensive function? Is it possible that there are other antimicrobial substances in the mucus of mollusks? Perhaps we have been looking in the wrong place; we should have been seeking something analogous to IgA rather than IgG.

2. Do blood lysozyme levels fluctuate in response to infection? The findings of Feng and Canzonier are clearly suggestive that different types of naturally occurring infections can affect hemolymph levels of lysozyme and other proteins. Do these reflect the disease state in a specific way or are they nonspecific reflections of pathology? What is the role of environment in varying hemolymph protein concentrations? Little attention has been paid to seasonal variation of hemolymph composition and yet this must be firmly established before we can understand other variations resulting from disease.

3. Can mollusks (or other invertebrates) mobilize specifically activated amocbocytes? The current emphasis on cellular immunity in vertebrates has led to sophisticated (but still incomplete) concepts of cell cell interaction. Similar questions need to be asked of invertebrates. Are there specific receptors on invertebrate amoebocytes? How many cell types are involved in reactions to foreign substances? Do these cells interact directly or indirectly? Are

lymphokines formed and how can they be demonstrated? In short, can we extend the emerging concepts of chemical mediation of cellular immunity to invertebrates?

4. Is it possible that invertebrates can have nonspecific humoral factors that may fulfill the role of antibody? We know that invertebrates can distinguish with considerable precision between "self" and "nonself." We know also that they do not make these distinctions via specific proteins analogous to vertebrate immunoglobulins. Is it possible that an "all purpose" protein coats all foreign substances and thus "tags" them for further attention by amoebocytes? There are tantalizing hints that such a possibility exists (Salt, 1970) and the question needs to be asked. It is interesting by itself and is part of the more general notion that parasites in normal hosts may change their surface properties by acquiring host substances or synthesizing ersatz host materials to camouflage themselves.

These questions are phrased in a general way because they are of general basic interest to biologists. Because of the very large amount of basic data available and summarized above, mollusks will make particularly valuable experimental systems for answering these questions.

REFERENCES

Anderson, R.S., N.K.B. Day and R.A. Good. 1972. Specific hemagglutinin and a modulator of complement in cockroach hemolymph. *Infection and Immunity* *5*: 55-59.

Anderson, R.S., B. Golmes and R.A. Good. 1973. *In vitro* bactericidal capacity of *Blaberus craniifer* hemocytes.

Acton, R.T. 1970. Immunobiological and immunochemical studies of the oyster *Crassostrea virginica.* MSS Information Corporation, N.Y. pp. 1-134.

Acton, R.T., J.C. Bennett, E.E. Evans and R.E. Schronhenhoher. 1969. Physical and chemical characterization of an oyster hemagglutinin. *J. Biol. Chem. 244*: 4128-4135.

Acton, R.T. and E.E. Evans. 1968. Bacteriophage clearance in the oyster *(Crassostrea virginica). J. Bact. 95*: 1260-1266.

Bang, F.B. 1966. Serological response in a marine worm, *Sipunculus nudus. J. Immunol. 96*: 960-972.

Bayne, C.J. 1973. Internal defense mechanisms of *Octopus dolfleini. Mal. Rev. 6*: 13-17.

Bayne, C.J. and J.B. Kime. 1970. *In vivo* removal of bacteria from the hemolymph of the land snail *Helix pomatia* (Pulmonata: Sylommatophora). *Malacol. Rev. 3*: 103-113.

Benex, J. and L. Lamy. 1959. Immobilisation des miracidiums de *Schistosoma mansoni* par des extraits de planoches. *Bull. Soc. path. erot. 52*: 188-193.

Bernheimer, A.W. 1952. Hemagglutinins in caterpillar bloods. *Science 115*: 150-151.

Bernheimer, A.W., E. Caspani and A. D. Kaiser. 1952. Studies on antibody formation in caterpillars. *J. Exp. Zool. 119*: 23-35.

Boyd, W.C., R. Brown and L.G. Boyd. 1966. Agglutinins for human erythrocytes in mollusks. *J. Immunol. 96*: 301-303.

Briggs, J.D. 1958. Humoral immunity in lepidopterous larvae. *J. Exp. Zool. 138*: 155-188.

Cheng, T.C. 1969. An electrophoretic analysis of hemolymph proteins of the snail *Helisoma duryi normale* experimentally challenged with bacteria. *J. Invert. Pathol. 14*: 60-81.

Cheng, T.C. and B.G. Sanders. 1962. Internal defense mechanisms in mollusks and an electrophoretic analysis of a naturally occurring serum hemagglutinin in *Vivparus malleatus* Reeve. *Proc. Penn. Acad. Sci. 36*: 72-83.

Cheng, T.C., C.N. Shuster and A.H. Anderson. 1966a. A comparative study of the susceptibility and response of eight species of marine pelecypods to the trematode *Himasthala quissentensis. Trans. Amer. Micros. Soc. 85*: 284-295.

Cheng, T.C., C.N. Shuster and A.H. Anderson. 1966b. Effects of plasma and tissue extracts of marine pelecypods on the cercariae of *Himasthala quissentensis. Exp. Parasitol. 19*. 9-14.

Chipman, D.M. and N. Sharon. 1969. Mechanism of lysozyme action. *Science 165*: 454-465.

Cohen, E., A.W. Rowe and F.C. Wissler. 1965. Heteroagglutinins of the horseshoe crab *Limulus polyphemus. Life Sci. 4*: 2009-2016.

Cooper, E.L. 1968a. Transplantation immunity in annelid worms. I. Rejection of xenografts exchanged between *Lumbricus terrestris* and *Eisenia foetida. Transplantation 6*: 322-337.

Cooper, E.L. 1968b. Multinucleate giant cells, granulomata, and "myoblastomas" in annelid worms. *J. Invert. Pathol. 11*: 123-131.

Cooper, E.L. 1969a. Chronic allograft rejection in *Lumbricus terrestris. J. Exp. Zool. 171*: 69-74.

Cooper, E.L. 1969b. Allograft rejection in *Eisenia foetida. Transplantation 8*: 220-223.

Cooper, E.L. 1969c. Neoplasia and transplantation immunity in annelids. *Nat. Cancer Inst. Monogs 31*: 655-669.

Cooper, E.L. 1969d. Specific tissue graft rejection in earthworms. *Science 166*: 1414-1415.

Cooper, E.L. 1973. Earthworm coelomocytes: role in understanding the evolution of cellular immunity. I. Formation of monolayers and cytotoxicity. Proc. 3rd Internatl. Colloq. Invert. Tissue Culture. J. Rehacek, D. Blaskovic and W.F. Hink, Eds. Publishing House of the Slovak Acad. Sci., Bratislava.

Cornick, J.W. and J.E. Stewart. 1973. Partial characterization of a natural agglutinin in the hemolymph of the lobster, *Homarus americanus. J. Invert. Pathol. 21*: 255-262.

Cushing, J.E. 1967. Invertebrates, immunology and evolution. *Fed. Proc. 26*: 1666-1670.

Cushing, J.E., N.L. Calaprice and G. Trump. 1963. Blood group reactive substances in some marine invertebrates. *Biol. Bull. 125*: 69-80.

Cushing, J.E., E.E. Evans and M.L. Evans. 1971. Induced bactericidal responses of abalones. *J. Invert. Pathol. 17*: 446-448.

Crespi, H.L., S.E. Mandeville and J.J. Katz. 1962. The action of lysozyme on several blue-green algae. *Biochem. Biophys. Res. Comm. 9*: 569-572.

Davis, B.D., R. Dulbecco, H.N. Eisen, H.S. Ginsberg and W. Barry Wood, Jr. 1967. Principles of Microbiology and Immunology. Harper & Row, New York. 853 pp.

Day, N.K.B., H. Gewurz, R. Johannsen, J. Finstad and R.A. Good. 1970. Complement and complement-like activity in lower vertebrates and invertebrates. *J. Exp. Med. 132*: 941-950.

Dusanic, D.G. and R.M. Lewert. 1963. Alterations of proteins and free amino acids of *Australorbis glabratus* hemolymph after exposure to *Schistosoma mansoni* miracidia. *J. Infect. Dis. 112*: 243-246.

Eble, A.F. and M.R. Tripp. 1968. Enzyme histochemistry of phagosomes in oyster leucocytes. *Bull. N.J. Acad. Sci. B. 13*: 93.

Evans, E.E. 1969. Induced bactericidal response in a sipunculid worm. *Nature 222*: 695.

Evans, E.E., B. Painter, M.D. Evans, P. Weinheimer and R.T. Acton. 1968. An induced bactericidin in the spiny lobster, *Panulirus argus. Proc. Soc. Exp. Biol. Med. 128*: 394-398.

Evans, E.E., P.F. Weinheimer, B. Painter, R.T. Acton and M.L. Evans. 1969. Secondary and tertiary responses of the induced bactericidin from the West Indian spiny lobster, *Panulirus argus. J. Bacteriol. 98*: 943-946.

Feir, D. and M.A. Walz. 1964. An agglutinating factor in insect hemolymph. *Ann. Entomol. Soc. Amer. 57*: 388.

Feng, J. 1966. The fate of a virus, *Staphylococcus aureus* phage 80, injected into the oyster, *Crassostrea virginica. J. Invert. Pathol. 8*: 496-504.

Feng, S.Y. 1966. Experimental bacterial infections in the oyster *Crassostrea virginica. J. Invert. Pathol. 8*: 505-511.

Feng, S.Y. 1974. Lysozyme-like activities in the hemolymph of *Crassostrea virginica. Contemporary Topics in Immunobiol. 4*: 225-231.

Feng, S.Y. and W.J. Canzonier. 1970. Humoral responses in the American oyster *(Crassostrea virginica)* infected with *Bucephalus* sp. and *Minchinia nelsoni. In*: A Symposium on Diseases of Fishes and Shellfishes, S.F. Snieszko, ed. Amer. Fis. Soc. Spec. Publ. no. 5. Washington, D.C. 526 pp.

Feng, S.Y. and L.A. Stauber. 1968. Experimental hexamitiosis in the oyster *Crassostrea virginica. J. Invert. Pathol. 10*: 94-110.

Fernandez-Moran, H., J.J. Machalonis and G.M. Edelman. 1968. Electron microscopy of a hemagglutinin from *Limulus polyphemus. J. Mol. Biol. 32*: 467-469.

Finstad, C.L., G.W. Litman, J. Finstad and R.A. Good. 1972. The evolution of the immune response. XIII. The characterization of purified erythrocyte agglutinins from two invertebrate species. *J. Immunol. 108*: 1704-1711.

Fletcher, T.C. and A. White. 1973. Lysozyme activity in the plaice *(Pleuronectes platessa* L.). *Experientia 29*: 1283-1285.

Gingrich, R.E. 1964. Acquired humoral immune response of the large milkweed bug, *Oncopetus fasciatus* (Dallas), to injected materials. *J. Insect Physiol. 10*: 179-194.

Gotz, P. 1969. Die Einkapselung von Parasiten in der Hamolymphe von Chironomus-Larven (Diptera) *Zool. Anz., Suppl. -BD33*: 610-617.

Gress, F.M. and T.C. Cheng. 1973. Alterations in total serum proteins and protein fractions in *Biomphalaria glabrata* parasitized by *Schistosoma mansoni. J. Invert. Pathol. 22*: 382-390.

Hammarstrom, S. and E.A. Kabat. 1969. Purification and characterization of a blood-group A reactive hemagglutinin from the snail *Helix pomatia* and a study of its combining site. *Biochemistry 8*: 2696-2705.

Heise, E.R. and W.N. Myrvik. 1967. Secretion of lysozyme by rabbit alveolar macrophages *in vitro. J. Reticuloendothelial Soc. 4*: 510-523.

Hildemann, W.H. and T.G. Dix. 1972. Transplantation reactions of tropical Australian echinoderms. *Transplantation 15*: 624-633.

Hildemann, W.H. and G.H. Theones. 1969. Immunological responses of Pacific hagfish. I. Skin transplantation immunity. *Transplantation 7*: 506-521.

Hirsch, J.G. 1959. Antimicrobial factors in tissues and phagocytic cells. *Bact. Rev. 24*: 133-140.

Inouye, M. and A. Tsugita. 1966. The amino acid sequency of T_4 bacteriophage lysozyme. *J. Mol. Biol. 22*: 193-196.

Janoff, A. and E. Hawrylko. 1964. Lysosomal enzymes in invertebrate leucocytes. *J. Cell. Comp. Physiol. 63*: 267-271.

Jenkin, C.R. and D. Rowley. 1970. Immunity in invertebrates. The purification of a haemagglutinin to rat and rabbit erythrocytes from the haemolymph of the murray mussel *(Velesunio ambiguus). Aust. J. Exp. Biol. Med. Sci. 48*: 129-137.

Johannsen, R., R.S. Anderson, R.A. Good and N.K. Day. 1973. A comparative study of the bactericidal activity of horseshoe crab *(Limulus polyphemus)* hemolymph and vertebrate serum. *J. Invert. Pathol. 22*: 372-376.

Johnson, H. 1964. Human blood group A_1 specific agglutinin of the butter clam *Saxidomus giganteus. Science 146*: 548-549.

Johnson, P.T. and F.A. Chapman. 1970. Comparative studies on the *in vitro* response of bacteria to invertebrate body fluids. II. *Aplysia californica* (sea hare) and *Ciona intestinalis* (tunicate). *J. Invest. Pathol. 16*: 259-267.

Jolles, P. and S. Luili. 1960. Purification et etude comparie de nouveaux lysozymes; extraits du poumon de poule et de *Nephthys bombergi. Biochim. Biophys. Acta 39*: 212-217.

Judge, J.R. 1966. Inhibition of effects of leukemongenic viruses in mice by extracts of *Mercenaria mercenaria. Proc. Soc. Exp. Biol. Med. 123*: 299-302.

Krassner, S.M. and B. Flory. 1970. Antibacterial factors in the sipunculid worms *Golfingia gouldii* and *Dendrostomum pyroides. J. Invert. Pathol. 16*: 331-336.

Landsteiner, K. 1945. The specificity of serological reactions. Harvard Univ. Press, Cambridge, Mass.

Lee, F.O. and T.C. Cheng. 1972. *Schistosoma mansoni:* alterations in total protein and hemoglobin in the hemolymph of infected *Biomphalara glabrata. Exp. Parasitol 31*: 203-216.

Li, C.P. 1960a. Antimicrobial effect of abalone juice. *Proc. Soc. Exp. Biol. Med. 103*: 522-524.

Li, C.P. 1960b. Antimicrobial activity of certain marine fauna. *Proc. Soc. Exp. Biol. Med. 104*: 366-368.

Li, C.P., B. Prescott, B. Eddy, G. Caldes, W.R. Green, E.C. Martino and A.M. Young. 1965. Antiviral activity of paolins from clams. *Ann. N.Y. Acad. Sci. 130*: 374-382.

Li, C.P., B. Prescott and W.G. Johnes. 1962a. Antiviral activity of a fraction of abalone juice. *Proc. Soc. Exp. Biol. Med. 109*: 534-538.

Li, C.P., B. Prescott, W. G. Johnes and E.C. Martino. 1962b. Antimicrobial agents from mollusks. *Trans. N.Y. Acad. Sci. (Sec. II), 24*: 504-509.

Li, M.F. and C. Flemming. 1967. Hemagglutinins from oyster hemolymph. *Can. J. Zool. 45*: 1225-1234.

Lillie, F.R. 1913. Studies of fertilization. V. The behavior of the spermatozoa of *Nereis* and *Arbacia* with special reference to egg-extractives. *J. Exp. Zool. 14*: 515 571.

Malke, H. 1965. Uher das Vorkommen von Lysozym in Insekten. *Zeit. Allg.Mikrobiol. 5*: 42-47.

Marchalonis, J.J. and G.M. Edelman. 1968. Isolation and characterization of a hemagglutinin from *Limulus polyphemus. J. Mol. Biol. 32*: 453-465.

Michelson, E.M. 1963. Development and specificity of miracidial immobilizing substances in extracts of the snail *Australorbis glabratus* exposed to various agents. *Ann. N.Y. Acad. Sci. 113*: 486-491.

Michelson, E.M. 1964. Miracidia-immobilizing substances in extracts prepared from snails infected with *Schistosoma mansoni. Amer. J. Trop. Med. Hyg. 13*: 36-42.

McDade, J.E. and M.R. Tripp. 1967a. Mechanism of agglutination of red blood cells by oyster hemolymph. *J. Invert. Pathol. 9*: 523-530.

McDade, J.E. and M.R. Tripp. 1967b. Lysozyme in the hemolymph of the oyster, *Crassostrea virginica. J. Invertebrate Pathol. 9*: 531-535.

McDade, J.E. and M.R. Tripp. 1967c. Lysozyme in oyster mantle mucus. *J. Invert. Pathl. 9*: 581-582.

McKay, D. and C.R. Jenkin. 1969. Immunity in the invertebrates. II. Adaptive immunity in the crayfish *(Parachaeraps bicarinatus). Immunology 17*: 127-137.

McKay, D. and C.R. Jenkin. 1970a. Immunity in the invertebrates. The fate and distribution of bacteria in normal and immunised crayfish *(Parachaeraps bicarinatus). Aust. J. Exp. Biol. Med. Sci. 48*: 599-607.

McKay, D. and C.R. Jenkin. 1970b. Immunity in the invertebrates. Correlation of the phagocytic activity of haemocytes with resistance to infection in the crayfish *(Parachaecaps bicarinatus)*. Aust. J. Exp. Biol. Med. Sci. 48*: 609-617.*

McKay, D. and C.R. Jenkin and D. Rowley. 1969. Immunity in the invertebrates. I. Studies on the naturally occurring haemagglutinins in the fluid from invertebrates. *Aust. J. Exp. Biol. Med. Sci. 47*: 125-134.

Messner, V.B. and W. Mohrig. 1969. Zum lysozym - vorkommen bei muscheln *(Anadonta anatina* (L.)). *Zool. Jb. Physiol. 74*: 427-435.

Miller, V.H., R.S. Ballback, G.B. Pauley and S.M. Krassner. 1972. A preliminary physiochemical characterization of an agglutinin found in the hemolymph of the crayfish *Procambarus clarkii. J. Invert. Pathol 19*: 83-93.

Mohrig, W. and B. Messner. 1968. Immun-reaktionen bei Insekten. I. Lysozym als grundlegendes antibakteriellie Faktor in humoralen Abwehrhechanismus der Insekten. *Biol. Zbl. 87:* 439-470.

Nelstrop. A.E., G. Taylor and P. Collard. 1968a. Studies on phagocytosis. I. Antigen clearance studies in rabbits. *Immunology 14*: 325-337.

Nelstrop, A.E., G. Taylor and P. Collard. 1968b. II. *In vitro* phagocytosis by macrophages. *Immunology 14*: 339-346.

Nelstrop, A.E., G. Taylor and P. Collard. 1968c. Studies on phagocytosis. III. Antigen clearance studies in invertebrates and poikilothermic vertebrates. *Immunology 14*: 347-356.

Pauley, G.B., G.A. Granger and S.M. Krassner. 1971. Characterization of a natural agglutinin present in the hemolymph of the California sea hare, *Aplysia Californica. J. Invert. Pathol. 18*: 207-218.

Phillips, J.H. 1960. Antibody-like materials of marine invertebrates. *Ann. N.Y. Acad. Sci. 90*: 760-769.

Pan, C-T. 1965. Studies on the host-parasite relationship between *Schistosoma mansoni* and the snail *Australorbis glabratus. Amer. J. Trop. Med. Hyg. 14*: 931-976.

Phillips, J.H. and B.J. Yardley. 1960. Detection in invertebrates of inducible, reactive materials resembling antibody. *Nature 188*: 728-730.

Prescott, B. and C.P. Li. 1960. Abalone juice. Fractionation and antibacterial spectrum. *Proc. Soc. Exp. Biol. Med. 105*: 498-500.

Prescott, B., C.P. Li, G. Caldes and E.C. Martino. 1966. Chemical studies of paolin II, an antiviral substance from oysters. *Proc. Soc. Exp. Biol. Med. 123*: 460-464.

Prowse, R.H. and N.N. Tait. 1969. *In vitro* phagocytosis by amoebocytes from the haemolymph of *Helix aspersa* (Muller). I. Evidence for opsonic factor(s) in serum. *Immunology 17*: 437-441.

Rabin, H. and F.B. Bang. 1964. *In vitro* studies of the antibacterial activity of *Golfingia gouldii* (Pourtalis) coelomic fluid. *J. Insect Pathol. 6*: 457-465.

Reade, P.C. 1968. Phagocytosis in invertebrates. *Aust. J. Exp. Biol. Med. Sci. 46*: 219-229.

Salt, G. 1970. The Cellular Defense Reactions of Insects. University Printing House, Cambridge. 118 pp.

Salton, M.R.J. 1957. The properties of lysozyme and its action on microorganisms. *Bact. Rev. 21*: 82-99.

Schmeer, M.R. and G. Beery. 1965a. Mercenene: a preliminary investigation of the cytological effects of this antitumor agent extracted from *Mercenaria mercenaria* on the Krebs-2 carcinoma. *Biol. Bull. 129*: 420.

Schmeer, M.R. and G. Berry. 1965b. Mercenene: Growth-inhibitor extracted from *Mercenaria compechiensis. In vivo* and *in vitro* activity. *Biol. Bull. 129*: 420-421.

Schmeer, M.R., D. Horton and A. Tanimura. 1966. Mercenene, a tumor inhibitor from *Mercenaria:* purification and characterization studies. *Life Sciences 5*: 1169-1178.

Schmeer, M.R. and C.V. Huala. 1965. Mercenene: *in vivo* effects of mollusk extracts on the sarcoma 180. *Ann. N.Y. Acad. Sci. 118*: 603-610.

Schubert, V.I. and B. Messner. 1971. Unterschrengen uber das Vorkommen von Lysozym bei Anneliden. *Zool. Johrb. Physiol. 76*: 36-50.

Seaman, G.R. and N.L. Robert. 1968. Immunological response of male cockroaches to injection of *Tetrahymena pyriformis. Science 161*: 1359-1361.

Sindermann, C.J. and D. F. Mairs. 1959. A major blood group system in Atlantic sea herring. *Copeia 3*: 228-232.

Skarnes, R.C. and D.W. Watson. 1957. Antimicrobial factors of normal tissues and fluids. *Bact. Rev. 21*: 273-294.

Stauber, L.A. 1950. The fate of india ink injected intracardially into the oyster, *Ostrea virginica* Gmelin. *Biol. Bull. 98*: 227-241.

Stephens, J.M. 1959. Immune responses of some insects to some bacterial antigens. *Can. J. Microbiol. 5*: 203-228.

Stephens, J.M. 1962. Influence of active immunization on melanization of the blood of wax moth larvae. *Can. J. Microbiol. 8*: 597-602.

Stephens, J.M. 1963a. Bactericidal activity of hemolymph of some normal insects. *J. Insect. Pathol. 5*: 61-65.

Stephens, J.M. 1963b. Protective effects of several immunizing preparations that produce active immunity in *Galleria mellonella* (Linnaeus). *J. Insect. Pathol. 5*: 129-130.

Stephens, J.M. and J.H. Marshall. 1962. Some properties of an immune factor isolated from the blood of actively immunized wax moth larvae. *Can. J. Microbiol. 8*: 719-725.

Stewart, J.E. and B.M. Zwickar. 1972. Natural and induced bactericidal activities in the hemolymph of the lobster, *Homarus americanus*: products of hemocyte-plasma interaction. *Can. J. Microbiol. 18*: 1499-1509.

Taylor, A.E., G. Taylor and P. Collard. 1964. Secondary immune response to bacteriophage T1 in the shore crab, *Carcinus maenas. Nature 263*: 775.

Triplett, E.L., J.E. Cushing and G.L. Durall. 1958. Observations on some immune reactions of the sipunculid worm *Dendeostomum zostericolum. Amer. Natural. 92*: 287-293.

Tripp, M.R. 1960. Mechanisms of removal of injected microorganisms from the American oyster, *Crassostrea virginica. Biol. Bull 119*: 273-282.

Tripp, M.R. 1961. The fate of foreign materials experimentally introduced into the snail *Australorbis glabratus. J. Parasitol. 47*: 745-751.

Tripp, M.R. 1966. Hemagglutinin in the blood of the oyster *Crassostrea virginica. J. Invert. Pathol. 8*: 478-484.

Tripp, M.R. 1970. Defense mechanisms of mollusks. *J. Reticuloendothelial Soc. 7*: 173-182.

Tyler, A. 1046. Natural heteroagglutinins in the body-fluids and seminal fluids of various invertebrates. *Biol. Bull. 90*: 213-219.

Tyler, A. And C.B. Metz. 1945. Natural heteroagglutinins in the serum of the spiny lobster, *Panulirus interruptus.* I. Taxonomic range of activity, electrophoretic and immunizing properties. *J. Exp. Zool. 100*: 387-406.

Tyler, A. and B.T. Scheer. 1945. Natural heteroagglutinins in the serum of the spiny lobster, *Panulirus interruptus.* II. Chemical and antigenic relations to blood proteins. *Biol. Bull. 89*: 193-200.

Ublenbruck, G., G.I. Pardoe, O. Prokop and I. Ishiyama. 1972. The serological specificity of snail agglutinins (protectins). *Anim. Blood Grps. Biochem. Genet. 3*: 125-139.

Weinheimer, P.F., R.T. Acton, S. Sawyer and E.E. Evans. 1969. Specificity of the induced bactericidin of the West Indian spiny lobster, *Panulirus argus. J. Bacteriol. 98*: 947-948.

HEMOLYMPH IN INSECTS AND ARACHNIDS - SOME BIOCHEMICAL FEATURES

G. R. Wyatt

Department of Biology
Queen's University
Kingston, Ontario, Canada

The hemolymph of insects is a rich and varied fluid, the study of which has provided many surprises. In quantity, it may represent 20-40% of an insect's weight. Pumped by a simple tubular heart, it bathes all of the organs. Although the cellular tissues are covered by a thin secreted layer, the basement membrane, which consists of mucopolysaccharide and sometimes fibres of collagen, there is little evidence that this provides any permeability barrier (except for macromolecules) between the hemolymph and the cells (Ashhurst, 1968). Thus, the hemolymph supplies the tissues with nutrients, removes their waste products and generally provides their buffer from the outside world. Changes in the composition of hemolymph during ontogeny of a species can provide clues to significant processes in developmental physiology. Similarly, changes in the hemolymph after wounding, or during infection or parasitism, may reflect the nature of the pathological processes and defensive reactions of the host.

Knowledge of the chemical composition of body fluids is also important for the development of Ringer solutions (Lockwood, 1961) and culture media congenial to the cells of different organisms. An early successful insect cell culture medium was based upon analyses of silkworm hemolymph (S. S. Wyatt, 1956), and more recent refinement and extension of this principle has yielded media suitable for culturing tissue from many insect species (Aizawa, 1971; Brooks and Kurtti, 1971).

Unfortunately, most of the systematic analysis of hemolymph was done many years ago by techniques much less sensitive and reliable than those now available. Many estimates of "sugar" in the hemolymph were reported on the basis of measurements of reducing power, before it was discovered that the main sugar in most insect species is the non reducing disaccharide trehalose (Wyatt and Kalf, 1957). The most extensive quantitative data on amino acids in insect hemolymph to this day are those obtained by Duchateau and Florkin (1958) by means of microbiological assay; very few analyses have been reported on this material with the chromatographic amino acid analyzer. Similarly, Duchateau et al. (1953) published an extensive tabulation of the inorganic cation composition of the hemolymph of 63 species of insect, based on traditional methods of chemical analysis, but there has been as yet little application of the far simpler and more sensitive newer techniques, such as atomic absorption spectroscopy (cf. Jungreis et al., 1973). These older analyses have been immensely valuable, but they now need supplementing by systematic study of hemolymph with modern techniques.

In this paper, I shall draw attention to some of the more significant biochemical features of insect and arachnid hemolymph, as well as some areas of persisting ignorance. More details are available in reviews (Buck, 1953; Wyatt, 1961; Florkin and Jeuniaux, 1964; Jeuniaux, 1971; Jeuniaux and Florkin, 1974).

INORGANIC COMPONENTS

Whereas the predominant inorganic ions in the blood plasma of most animals are sodium and chloride, supposedly in reflection of the composition of the oceans in which they evolved, this is by no means always true with the insects. The inorganic composition of insect hemolymph shows wide variations, as illustrated by the selected data in Table I. Several generalizations may be made (Sutcliffe, 1963; Florkin and Jeuniaux, 1964; Jeuniaux, 1971). For the most primitive insects (Apterygota) the available incomplete data indicate hemolymph like the blood of other animals, with high sodium and low potassium. The Exopterygota (insects having incomplete metamorphosis) retain high sodium and chloride, but magnesium is higher than in the plasma of vertebrates. The more advanced insects having complete metamorphosis (Endopterygota) exhibit wide variations in the proportions of sodium and potassium, and generally have elevated magnesium and reduced chloride. High potassium and magnesium, along with low sodium, may be a pattern that reflects the evolution of insects in association with flowering plants, which contain much potassium and magnesium and little sodium. The extreme of this pattern of composition is exhibited by leaf-eating forms, such as *Bombyx* and *Hyalophora* larvae and the larvae and adults of the chrysomelid beetle, *Timarcha*.

The hemolymph composition of phytophagous insects, however, does not represent mere submission to the impact of their dietary, for the characteristic proportions of ions are maintained, within rather broad limits, by regulatory processes. Thus, cecropia silkmoth larvae reared on an artificial diet containing magnesium at only 7.5 mmol/kg, instead of cherry foliage with 61 mmol/kg, still maintained their hemolymph magnesium above 30 mM (Jungreis et al., 1973). Similarly, this diet contained potassium at only one quarter of the level found in foliage, but the potassium content of hemolymph was virtually the same on the two regimens.

Developmental changes are illustrated by the data on the cecropia silkmoth, the silkworm (*Bombyx*) and the honeybee (*Apis*). During metamorphosis of each species magnesium declines, and in the bee sodium also rises. In cecropia, during the transition from larva to pupa, magnesium declines also within the fat body cells, a change which, it is suggested, may play a role in modulating the synthesis of trehalose during development (Jungreis and Wyatt, 1974).

The Coleoptera show wide variation in hemolymph cation composition, as illustrated by two arbitrarily selected species in Table I; these differences have been interpreted as related to evolutionary progress within this vast Order (Jeuniaux, 1971). Larvae of several Trichoptera (caddis flies) are all high in sodium and low in potasssium, while magnesium varies extraordinarily, ranging from 10 to 200 meq/l (8 to 65% of the total cations) (Beaujot et al., 1970). In larvae of the dipteran *Drosophila melanogaster*, no one inorganic ion predominates (Table I).

TABLE I

INORGANIC COMPONENTS OF THE HEMOLYMPH OF SOME REPRESENTATIVE INSECTS

APTERYGOTA	Na	K	Ca meq/1	Mg	Cl	P_i
Petrobius maritimus[a]	208	5.8				
EXOPTERYGOTA						
Odonata						
Aeschna grandis larva[b,d]	145	9	15	15	110	6
Dictyoptera						
Periplaneta americana adult[c]	157	7.7	7.9	10	144	
Orthoptera						
Locusta migratoria adult[d]	67	9.0	15	27	98	
Dermaptera						
Anisolabis littorea adult[e]	193	5	15	11	117	5
ENDOPTERYGOTA						
Lepidoptera						
Galleria mellonella larva[d]	26	36	24	33		
Bombyx mori larva (stage V day 0)[f,g]	3.4	46	21	87	20	3
B. mori pupa (pharate adult)[f,g]	3.6	54	16	57	24	6
B. mori adult[f]	7.7	33	12	40		
Hyalophora cecropia larva[h]	2	39	12	100	6	4
H. cecropia pupa[h]	<2	38	10	38	18	12
Hymenoptera						
Apis mellifica larva[d,i]	11	31	18	21	33	10
A. mellifica adults[j]	47	27	18	1		
Diptera						
Drosophila melanogaster larva[k]	56	40	16	42	42	4
Coleoptera						
Oryctes boas larva[l]	15	10	21	160		
O. boas adult[l]	69	14	14	58		
Timarcha tenebriocasa larva[l]	1.6	47	72	158		
T. tenebriocasa adult[l]	1.4	55	27	234		

[a]Lockwood and Croghan (1959)

TABLE I
(Cont'd)

[b]Sutcliffe (1962)
[c]Van Asperen and Van Esch (1956)
[d]Duchateau et al. (1953)
[e]Leader and Bedford (1972a)
[f]Naoumoff and Jeuniaux (1970)
[g]Bialaszewicz and Landau (1938)
[h]Jungreis et al. (1973)
[i]Bishop et al. (1925)
[j]Florkin and Jeuniaux (1963)
[k]Begg and Cruickshank (1962)
[l]Jeuniaux (1971)

Although advanced insects frequently have relatively high potassium in their hemolymph, the levels are still far lower than those within the cells, permitting maintenance of a membrane potential. It has recently been observed that potassium levels about 70 mM inhibited the growth of lepidopteran cells in culture (T. Kurtti and M. Brooks, personal communication). An important mechanism in regulating hemolymph potassium levels is active transport through the mid-gut epithelium toward the lumen (Harvey and Nedergaard, 1964).

The low chloride level in the hemolymph of higher insects requires that substantial amounts of other anions contribute to the maintenance of electroneutrality. Some inorganic phosphate is always present, but still more abundant may be an organic phosphate fraction comprising several components (Wyatt, et al., 1963). In silkmoths, the total phosphates may provide 30% or more of the necessary anion titre. Other contributions are bicarbonate (on which few data are available) and certain organic acids (see below). At the pH of hemolymph, (usually 6.0-7.0) the abundant free amino acids generally contribute as many cations as anions.

ORGANIC COMPONENTS OF LOW MOLECULAR WEIGHT

Insect hemolymph is particularly rich in a variety of organic solutes, which distinguish it from the plasma of vertebrates. This is illustrated by some data on the cecropia silkmoth in Table II.

A remarkably high level of free amino acids, first noted 65 years ago in silkworm hemolymph, has been observed in many insects and even proposed as a biochemical character of the class Insecta (Duchateau and Florkin, 1958). In primitive species this characteristic is not highly developed, but among the higher insects the total free amino acids frequently approximate 100 mM and may range up to 200 mM, accounting for 20-50% of the total osmotic pressure (Sutcliffe, 1963). Changes in free amino acid levels during starvation and water deprivation indicate that they may be important in regulating hemolymph osmolarity. The amino acids also serve their appropriate biochemical functions. Thus they are precursors for synthesis of proteins, including the hemolymph proteins and special products such as silk. Tryosine and certain derivatives such as tyrosine phosphate (Mitchell and Luna, 1964) and β-alanyl-L-tyrosine (Bodnaryk and Levenbook, 1969) serve as precursors for quinones required for tanning the cuticle after molting. Proline can be a significant metabolic source for flight energy in certain adult insects, notably the tsetse fly, which contains very little available carbohydrate (Bursell, 1963).

Along with the 20 amino acids of proteins, hemolymph often contains others, such as ornithine, α- and γ-amino-butyric acids, and β-alanine. Particularly puzzling is the presence of D-alanine and D-serine in milkweed bugs and silkworms, respectively (Corrigan, 1969).

In addition to free amino acids, peptides are often found in insect hemolymph, sometimes in remarkable amount and diversity (Mitchell and Simmons, 1962). The hemolymph of blowfly larvae (*Phormia regina*) contains at least 19 small peptides, from which 9 di- and tri-peptides have been identified, all with C-terminal lysine or histidine (Bodnaryk and Levenbook, 1968). Their biological function is quite unknown.

TABLE II

INSECT HEMOLYMPH:
SOME ORGANIC COMPONENTS IN THE CECROPIA SILKMOTH

	Mature larva		Diapause pupa
		mmol/l	
Free amino acids[a,b]	70		100
Citrate[c]	8		12
Organic phosphates[d]	26		31
Trehalose[e]	45		8-16
Glucose[f]	<0.5		<0.5
Glycerol[f]	0		400
		%	
Protein[a,b]	9		7
Peptides[b]	2		

[a]Wyatt (1961)
[b]Jungreis (1974)
[c]Hollis and Levenbook (1961)
[d]Wyatt et al. (1963)
[e]Jungreis and Wyatt (1972)
[f]Wyatt (1967)

There are several reports of substantial levels of organic acids other than amino acids in insect hemolymph, but there has been little systematic investigation of these (review: Wyatt, 1961). Most interesting is the demonstration of citrate in a number of species (Hollis and Levenbook, 1961), which is reported in the *Bombyx* silkworm larvae at the remarkable level of 32 mM (96 meq/l if fully ionized). Not only does this make up a major proportion of the anion titre, but since citrate strongly chelates magnesium and calcium ions, it would greatly reduce the activities of the latter in the hemolymph.

The predominant sugar in insect hemolymph is generally trehalose, which has been reported at levels up to about 60 mg/ml (180 mM) (review: Wyatt, 1967). In some insects, trehalose is found only at low levels (the tsetse fly; Geigy et al., 1959) or is apparently totally lacking (the New Zealand earwig; *Anisolabis littorea*; Leader and Bedford, 1972b). Trehalose levels in hemolymph are subject to homeostatic regulation and may change characteristically between developmental stages of a species; the mechanisms involved in this regulation, which differ greatly from those used for blood sugar regulation by vertebrates, have recently received some attention (Murphy and Wyatt, 1965; Friedman, 1967; Jungreis and Wyatt, 1972, 1974).. Glucose levels in hemolymph are generally low as a consequence of continual conversion to trehalose by the fat body. This facilitates absorption of glucose through the intestinal wall by passive diffusion, which is important because the insect gut lacks the capacity for active absorption of sugars. A few species which have relatively high hemolymph glucose (aphids, honey bees) feed upon particularly sugar-rich fluids.

An impressive biochemical adaptation of some insects that overwinter in cold climates is the conversion of most of their carbohydrate reserves into glycerol during the winter months. In pupae of the cecropia silkmoth, glycerol may attain a level of 5% as a consequence of a metabolic pathway that is initiated with the onset of diapause and further enhanced by exposure to low temperature (Wyatt, 1967). Still higher levels of glycerol (up to 25% of body weight) are produced in some other species. Glycerol clearly has a role in protecting insects against injury by freezing, and yet is not the sole factor in this, for some species show remarkable cold-hardiness without production of glycerol (Salt, 1961).

It is certain that many unexpected compounds remain to be discovered in insect hemolymph, and some of these will undoubtedly occur at high levels and have important biological roles. The free amino acids, despite their abundance, only account for about one half of the non-protein nitrogen of hemolymph (Buck, 1953), and a tally of the remainder that approaches completeness has never been achieved for any species.

PROTEINS

The level of protein in insect hemolymph ranges from about 1 to 10%. The protein level is commonly low during most of larval development, but rises steeply during the last instar just before metamorphosis. In the *Bombyx* silkworm, for example, the hemolymph contains about 1.5% protein at the beginning of the fifth stage and 5%

at its conclusion (Wyatt et al., 1956). Since the advent of polyacrylamide gel electrophoresis ten years ago, analysis of hemolymph proteins by this technique has been a favourite occupation of insect physiologists. Elegant patterns are obtained, which bear no homology to the plasma proteins of vertebrates and differ characteristically with species and stage of the donor insect (*Bombyx*: Nakasone and Kobayashi, 1965; *Ostrinia*: Chippendale and Beck, 1966; *Phormia*: Chen and Levenbook, 1966; *Calliphora*: Kinnear et al., 1970; *Tenebrio* and *Galleria*: Lavendure, 1970; *Dysdercus*: Hollweg, 1970; *Lepidoptera*: Whitmore and Gilbert, 1974; and many other papers). Frequently, several of the major components are of high molecular weight, so that gels of relatively low concentration (4 to 5%) are necessary for optimal resolution.

Only quite recently have individual hemolymph proteins begun to be identified and characterized chemically and functionally. A few of these are listed in Table III. The name calliphorin was given to a protein which can make up as much as 60% of the total soluble protein of mature larvae of the blowfly *Calliphora erythrocephala* (Munn and Greville, 1969). The purified protein ($S_{20,w}$19.4S, molecular weight 530,000) consists of probably six subunits, and has the highest tyrosine content of any protein whose composition is known, along with an almost equally high content of phenylalanine (Munn et al., 1971). Related Diptera contain closely related proteins, and other insects examined by Munn and Greville (*Tenebrio, Pieris*) also yielded macromolecular components with similar appearance in the electron microscope and slightly lower sedimentation constants (16-17S). These are believed to represent a means of storage of amino acids to be recycled in metamorphosis. Specific storage proteins are apparently of general occurrence in insect metamorphosis, and are laid down in the fat body cells in the form of the protein granules ("albuminoid spheres") that have been known to histologists since the nineteenth century. Relatively recently, in Lepidoptera (Locke and Collins, 1968) and in Diptera (Martin et al., 1970), evidence has been obtained for output of protein from the fat body into the hemolymph during the last larval instar, followed by its resorption into the fat body at the onset of metamorphosis. A specific hemolymph protein is deposited in partially crystalline protein granules that appear abundantly in the fat body cells during the larval-pupal transformation (Bhakthan and Gilbert, 1972; Tojo and Wyatt, unpublished data).

A second class of hemolymph protein having the role of a nutrient reserve is the vitellogenins, or egg yolk proteins (Pan et al., 1969). These are generally found in female insects approaching reproductive maturity, in which they are produced in the fat body, transported in the hemolymph and absorbed by the oocytes. The vitellogenin of the cockroach *Leucophaea maderae* has been partially purified and found to have a molecular weight of 560,000 and to contain some carbohydrate and lipid (Dejmal and Brookes, 1972). Hemolymph contains several lipoproteins which can be separated in density gradients (Thomas and Gilbert, 1968, 1969) From cynthia silkmoth hemolymph, two lipoproteins capable of binding and transporting diglycerides have been purified and characterized (Chino et al., 1969). Another class of proteins very recently identified are those that bind and presumably transport the lipophilic juvenile hormone (Emmerich and Hartmann, 1973; Kramer et al., 1974). Hemolymph often contains chromoproteins, and one of these, a blue bile pigment protein from larval tobacco hornworm hemolymph, has recently been highly purified, characterized and named insectocyanin (P. Cherbas and C. M. Williams, personal communication).

TABLE III

SOME INSECT HEMOLYMPH PROTEINS

Protein	Insect	Mol. wt. x 10^{-3}	Author
Calliphorin	*Calliphora erythrocephala*	530	Munn et al. (1971)
Storage protein	*Hyalophora cecropia*	250	Tojo & Wyatt (1974)
Vitellogenin	*Leucophaea maderae*	560	Dejmal & Brookes (1972) Engelmann & Friedel (1974)
Liproteins	*Hyalophora cecropia*	100, 200	Thomas & Gilbert (1968)
Diglyceride-carrying lipoproteins	*Samia cynthia*	500, 700	Chino et al. (1969)
JH-binding protein	*Locusta migratoria*	220	Emmerich & Hartmann (1973)
JH-binding protein	*Manduca sexta*	34	Kramer et al. (1974)

It is notable that most of these proteins have high molecular weights, which would minimize their osmotic contribution. Several of the identified proteins have been shown to be produced in the fat body, and this tissue is probably the general site of synthesis of the major proteins of hemolymph, as originally suggested by Shigematsu (1958).

Many enzymes have been detected in insect hemolymph. Frequently, activities are relatively low and it is not clear that the enzymes have any particular role in the plasma; they may merely represent some leakage from the cellular tissues. Others, however, clearly have function in the hemolymph. Tyrosinase, or phenol oxidase, is regularly present, and is responsible for the blackening of hemolymph that bleeds from a wound. This enzyme has been extensively studied, including its mode of activation from an inactive proenzyme (Ohnishi et al., 1970; Ashida, 1971), and yet its biological significance is still not properly understood. In some insects, a trehalase has been demonstrated in hemolymph, which is normally maintained in a state of inhibition but which may be responsible for the cleavage of a certain amount of trehalose to glucose for use by tissues which lack trehalase themselves (Friedman, 1961; see also Wyatt, 1967). A number of esterases have also been observed (Laufer, 1960; Whitmore and Gilbert, 1974). Distinct from the general esterases is an enzyme specific for hydrolytic inactivation of juvenile hormone (Weirich et al., 1973). The synthesis of this enzyme in the fat body appears to be induced by juvenile hormone itself, which thus sows the seeds of its own destruction (Whitmore et al., 1974).

ROLES OF HEMOLYMPH IN DEFENSE

Upon exposure at a wound, hemolymph generally forms a clot containing both cells and coagulated protein. The cellular reactions are discussed elsewhere in this volume (Jones, 1974). The form and extent of clotting by the plasma proteins varies greatly in different species and developmental stages, and the mechanisms are not well understood (Gregoire, 1964). There is little evidence, however, for a calcium-dependent cascade of reactions like that known in vertebrate plasma. While clotting and darkening are not necessarily correlated, it is probable that the phenol oxidase reaction mentioned above may contribute to protection at a wound site. The products of the reaction include quinones, which may be toxic to invading microorganisms. Melanin formation as part of the process of encapsulation of parasites within the hemocoel is noted elsewhere (Nappi, 1974). In addition, inoculation of insects with bacteria quickly evokes the appearance of antibacterial substances in the hemolymph, which include lysozyme (Powning and Davidson, 1968) and other factors. This important, intriguing and inadequately understood phenomenon is discussed by Chadwick (1974).

Accidental or experimental wounding of the integument evokes a syndrome of physiological and biochemical responses which are readily observed in diapausing silkmoth pupae. In these pupae, integumentary injury is followed by an immediate approximate doubling of respiration, followed by further several fold elevation during several days, dependent on the size of the wound (Schneiderman and Williams, 1953; Harvey and Williams, 1961). The rate of protein synthesis, measured by incorporation of injected amino acids, rises steeply both in the epidermis and in other tissues. Levels of respiratory enzymes increase in the epidermis (Shappirio and Harvey, 1965), and the fat body releases new protein into the hemolymph (Stevenson and Wyatt, 1962, Skinner, 1963). The fat body also shows sharply elevated synthesis of RNA, which appears to precede the rise in protein synthesis (Wyatt, 1963). Another manifestation of the "injury syndrome" is elevation of blood trehalose (Wyatt, 1963). This is produced at the expense of glycogen or, in the diapausing cecropia pupa at the expense of glycerol, and increased activities of glycogen phosphorylase (Stevenson and Wyatt, 1964) and glycerol kinase (Yamashita and Wyatt, unpublished data) have been found in the fat body after injury. These various responses must depend upon a circulating "wound hormone", and presumably contribute to providing energy and materials for the repair of a wound, although it is not obvious why so extensive a physiological and biochemical arousal should be required.

Making a more obvious contribution to wound repair is the mobilization of hemocytes after injury to diapause pupae, first observed by Harvey and Williams (1961). Whereas hemolymph drawn directly from cecropia pupae in diapause contains on the average only about 10 cells per μ1, the cell count rose progressively in a series of samples taken during 3 days to more than 500 cells per μ1, due to the injury of collecting the samples. There was no evidence of mitosis (and wounding has little effect on DNA synthesis), so that the cells must have been mobilized from sites where they were sequestered. In an important recent contribution L. Cherbas (1973) observed that the cells in hemolymph collected through a slit in the cuticle of diapausing silkmoth pupae became amoeboid and adhesive when placed in culture, whereas cells in hemolymph collected by a technique which avoided contact with the integument retained the inactivity and round or spindle-shaped form that they had *in vivo*. A factor capable of activating hemocytes *in vitro* was demonstrated in hemolymph collected from a wound, and was named hemokinin. Studies on the nature and role of this factor will be of great interest, in particular its possible relationship to the biochemical consequences of wounding mentioned above.

HEMOLYMPH OF ARACHNIDS

Despite the importance of ticks as vectors of human disease, their physiology has been little studied and we have few data on the composition of tick hemolymph. The most complete available analysis, on the hemolymph of the cattle tick, *Boophilus microplus*, was carried out in relation to a study of the mechanisms of excretion of the excess water and electrolytes taken in with the blood meal (Tatchell, 1969) and is summarized in Table IV. Sodium and chloride are high and potassium and magnesium quite low. Free amino acids are very low compared with insect hemolymph. No trehalose could be detected, but glucose was quite high, and, surprisingly, a significant level of free inositol was found. Similar data obtained independently on the same species and on *Argas lagenoplastis* were used for the formulation of a culture medium for tick cells which gave better growth and survival than media previously tested (Rehacek and Brzostowski, 1969). The medium was adjusted to pH 6.8, although the pH of tick hemolymph was not recorded, and may possibly be higher than this in view of the value observed for a spider (Table IV).

TABLE IV

THE COMPOSITION OF HEMOLYMPH OF TWO ARACHNIDS

	*Boophilus microplus**	*Cupiennius salei*‡
pH		8.1-8.3
Osmotic pressure (mOsmolar)	376	
Non-aqueous matter, %		7.8
Protein (N x 6.25), %		4.7
Na, meq/1	136	223
K, meq/1	15	6.8
Ca, meq/1	8.3	8.0
Mg, meq/1	4.2	
Cl, meq/1	118	258

TABLE IV
(Cont'd)

	*Boophilus microplus**	*Cupiennius salei≠*
Cu, mg/1		58
P, acid-soluble, mM		1.1
P, inorganic, mM		1.1
N, acid-soluble, mM		60
Amino acids, mM	6.3	
Urea, mM	17	
Glucose, mM	28	70
Inositol, mM	2.2	

*Engorged female ticks (Tatchell, 1969).
≠Starved adult female spiders (Loewe et al., 1970).

Recently, systematic studies on tick physiology in progress at the U.S. Naval Medical Research Unit in Cairo have begun to yield results. Species of *Argas* and *Hyalomma* have been studied, with particular attention to changes in the composition of body fluids during the nutritive and reproductive cycles (Aboul-Nasr and Bassal, 1971; Boctor and Araman, 1971; Dolp and Hamdy, 1971; Araman, 1972 ; Araman and Said, 1972; Boctor, 1972; Hefnawy, 1972a, 1972b). Although the range of chemical components measured is as yet incomplete, the results are fully consistent with the pattern observed for *Boophilus*.

A rather extensive analysis of the hemolymph of a spider has recently been reported (Table IV). The pattern of composition is very similar to that found in ticks. In addition, copper was reported (as a component of hemocyanin) and acid-soluble phosphate at rather low levels, almost wholly inorganic. It is notable that the pH (8.1-8.3) is significantly higher than is found either in vertebrates or in insects. The hemolymph of several species of scorpion is also similar in the proprotions of its inorganic components (see Jeuniaux, 1971).

From these scanty data, one can infer that the hemolymph of certain Arachnids somewhat resembles that of primitive insects, and has not evolved the features characteristic of advanced insects, such as high potassium and magnesium, free amino acids, organic phosphates and trehalose.

Qualitative and quantitative analysis has already contributed substantially to the progress of arthropod physiology, but there are still remarkable gaps in our knowledge. Further systematic comparative studies, using modern techniques, on hemolymph from this vast and diverse group of animals will continue to be rewarding.

REFERENCES

Aboul-Nasr, A. E. and Bassal, T. T. M. (1971). *J. Med. Entomol., 8*:521-524.

Aizawa, K. (1971). In "Invertebrate Tissue Culture" (C. Vago, ed.). pp. 211-246. Academic Press, New York.

Araman, S. R. (1972). *J. Parasitol., 58*:354-357.

Araman, S. F. and Said, A. (1972). *J. Parasitol., 58*:348-353.

Ashhurst, D. E. (1968). *Ann. Rev. Ent., 13*:45-74.

Ashida, M. (1971). *Arch. Biochem. Biophys., 144*:749.

Beaujot, J., Naoumoff, M. and Jeuniaux, C. (1970). *Arch. Intern. Physiol. Biochim., 78*:111-118.

Begg, M. and Cruickshank, W. J. (1962). *Proc. Roy. Soc. Edinburgh, 68*:215-237.

Bhakthan, N. M. G. and Gilbert, L. I. (1971). *Z. Zellforsch., 124*:433-444.

Bialaszewicz, K. and Landau, C. (1938). *Acta Biol. Exptl.* (Warsaw) *12*:307-320.

Bishop, G. H., Briggs, A. P. and Ronzoni, E. (1925). *J. Biol. Chem., 66*:77-88.

Boctor, F. N. (1972). *J. Med. Entomol., 9*:201-204.

Boctor, F. N. and Araman, S. F. (1971). *J. Med. Entomol., 8*:525-528.

Bodnaryk, R. P. and Levenbook, L. (1968). *Biochem. J., 110*:771-773.

Bodnaryk, R. P. and Levenbook, L. (1969). *Comp. Biochem. Physiol., 30*:909-921.

Brooks, M. A. and Kurtti, T. J. (1971). *Ann. Rev. Ent., 16*:27-52.

Buck, J. B. (1953). In "Insect Physiology" (K. Roeder, et.) Chap. 6. J. Wiley and Sons, New York.

Bursell, E. (1963). *J. Insect Physiol., 9*:439-452.

Chadwick, J. M. (1974). In. "Mechanisms of Invertebrate Vector-Parasite Relations" (this volume).

Chen, P. S. and Levenbook, L. (1966). *J. Insect Physiol., 12*:1595-1609.

Cherbas, L. (1973). *J. Insect Physiol.*, *19*:2011-2024.

Chino, H., Murakami, S. and Harashima, K. (1969). *Biochim. Biophys. Acta*, *176*:1-26.

Chippendale, G. M. and Beck, S. D. (1966). *J. Insect Physiol.*, *12*:1629-1638.

Corrigan, J. J. (1969). *Science*, *164*:142-149.

Dolp, R. M. and Hamdy, B. H. (1971). *J. Med. Entomol.*, *8*:636-642.

Duchateau, F. and Florkin, M. (1958). *Arch. Internat. Physiol. Biochim.*, *66*:573-591.

Duchateau, G., Florkin, M. and Leclercq, J. (1953). *Arch. Internat. Physiol.*, *61*:518-549.

Emmerich, H. and Hartmann, R. (1973). *J. Insect Physiol.*, *19*:1663-1675.

Engelmann, F. and Friedel, T. (1974). *Life Sciences*, *14*:587-594.

Florkin, M. and Jeuniaux, C. (1963). *Life Sciences*, *12*:982-985.

Florkin, M. and Jeuniaux, C. (1964). In "The Physiology of Insecta" (M. Rockstein, ed.) Vol. 3, pp. 109-152. Academic Press, New York.

Friedman, S. (1961). *Arch. Biochem. Biophys.*, *93*:550-554.

Friedman, S. (1967). *J. Insect Physiol.*, *13*:397-405.

Geigy, R., Huber, M., Weinman, D. and Wyatt, G. R. (1959). *Acta Tropica* (Basel) *16*:255-262

Gregoire, C. (1964). In "The Physiology of Insecta" (M. Rockstein, ed.) Vol. 3, pp. 153-188. Academic Press, New York.

Harvey, W. R. and Nedergaard, S. (1964). *Proc. Nat. Acad. Sci.*, *51*:757-765.

Harvey, W. R. and Williams, C. M. (1961). *J. Insect Physiol.*, *7*:81-99.

Hefnawy, T. (1972a). *J. Parasitol.*, *58*:358-364.

Hefnawy, T. (1972b). *J. Parasitol.*, *58*:1197-1200.

Hollweg, G. (1970). *Naturwissenschaften*, *59*:140.

Jeuniaux, C. (1971). In "Chemical Zoology" (M. Florkin and B. Scheer, eds.). Vol. 6, pp. 63-118.

Jeuniaux, C. and Florkin, M. (1971). In "The Physiology of Insecta" (M. Rockstein, ed.). Vol. 5, Second Edition, Academic Press, New York. (in press).

Jones, J. C. (1974). In "Mechanisms of Invertebrate Vector-Parasite Relations". (this volume)

Jungreis, A. M. (1974). *J. Comp. Physiol., 88*:113-127.

Jungreis, A. M. and Wyatt, G. R. (1972). *Biol. Bull., 143*:367-391.

Jungreis, A. M. and Wyatt, G. R. (1974). *J. Exp. Zool., 187*:41-45.

Jungreis, A. M., Jatlow, P., and Wyatt, G. R. (1973). *J. Insect Physiol., 19*:225-233.

Kinnear, J. F., Martin, M. D. and Thomson, J. A. (1970). *Aust. J. Biol. Sci., 24*:275-289.

Kramer, K. J., Sanburg, L. L., Dezdy, F. J. and Low, J. H. (1974). *Proc. Nat. Acad. Sci., 71*:493-497.

Laufer, H. (1960). *Ann. N. Y. Acad. Sci., 89*:490-515.

Laverdure, A. M. (1970). *Ann. D'Endocrinologie*, Paris *31*:504-515.

Leader, J. P. and Bedford, J. J. (1972a). *J. Insect Physiol., 18*:2229-2235.

Leader, J. P. and Bedford, J. J. (1972b). *Comp. Biochem. Physiol. B., 43*:233-235.

Levenbook, L. and Hollis, W. W. (1961). *J. Insect Physiol., 6*:52-61.

Locke, M. and Collins, J. V. (1968). *J. Cell Biol., 36*:453-483.

Lockwood, A. P. M. (1961). *Comp Biochem. Physiol., 2*:241-289.

Lockwood, A. P. M. and Croghan, P. C. (1959). *Nature, 184*:370-371.

Loewe, R., Linzen, B. and von Stackelberg, W. (1970). *Z. vergl. Physiol., 66*:27-34.

Martin, M. D., Kinnear, J. F. and Thomson, J. A. (1970). *Aust. J. Biol. Sci., 24*:291-299.

Mitchell, H. K. and Lunan, K. D. (1964). *Arch. Biochem. Biophys., 106*:219-222.

Mitchell, H. K. and Simmons, J. R. (1962). In "Amino Acid Pools" (J. T. Holden, ed.). pp. 136-146. Elsevier, Amsterdam.

Munn, E. A. and Greville, G. D. (1969). *J. Insect Physiol., 15*:1935-1950.

Munn, E. A., Feinstein, A. and Greville, G. D. (1971). *Biochem. J.*, *124*:367-374.

Murphy, T. A. and Wyatt, G. R. (1965). *J. Biol. Chem.*, *240*:1500-1508.

Nakasone, S., and Kobayashi, M. (1965). *J. Sericul. Sci. Jap.*, *34*:257-262.

Naoumoff, E. C. and Jeuniaux, M. (1970). *Arch. Intern. Physiol. Biochim.*, *78*:357-365.

Nappi, A. J. (1974). In "Mechanisms of Invertebrate Vector-Parasite Relations". (this volume)

Ohnishi, E., Dohke, K. and Ashida, M. (1970). *Arch. Biochem. Biophys.*, *139*:143-148.

Pan, M. L., Bell, W. J. and Telfer, W. H. (1969). *Science*, *165*:393-394.

Powning, R. F. and Davidson, W. J. (1973). *Comp. Biochem. Physiol.*, *45*:669-681.

Rehacek, J. and Brzostowski, H. W. (1969). *J. Insect Physiol.*, *15*:1431-1436.

Salt, R. W. (1961). *Ann. Rev. Entomol.*, *6*:55-74.

Schneiderman, H. A. and Williams, C. M. (1953). *Biol. Bull.*, *105*:320-334.

Shappirio, D. G. and Harvey, W. R. (1965). *J. Insect Physiol.*, *11*:305-327.

Shigematsu, H. (1958). *Nature*, *182*:880-882.

Skinner, D. M. (1963). *Biol. Bull.*, *125*:165-176.

Stevenson, E. and Wyatt, G. R. (1962). *Arch. Biochem. Biophys.*, *99*:65-71.

Sutcliffe, D. W. (1962). *J. Exp. Biol.*, *39*:325-343.

Sutcliffe, D. W. (1963). *Comp. Biochem. Physiol.*, *9*:121-135.

Tatchell, R. J. (1969). *J. Insect Physiol.*, *15*:1421-1430.

Thomas, K. K. and Gilbert, L. I. (1968). *Arch. Biochem. Biophys.*, *127*:512-521.

Thomas, K. K. and Gilbert, L. I. (1969). *Physiol. Chem. Physics*, *1*:293-311.

Van Asperen, K. and Van Esch, I. (1956). *Arch. Neerl. Zool.*, *11*:342-360.

Weirich, G., Wren, J. and Siddall, J. B. (1973). *Insect Biochem.*, *3*:397-407.

Whitmore, E. and Gilbert, L. I. (1974). *Comp. Biochem. Physiol., 47B*:63-78.

Whitmore, D., Gilbert, L. I. and Ittycheriah, P. I. (1974). *Mol. Cell. Endocrinol., 1*:37-54.

Wyatt, G. R. (1961). *Ann. Rev. Entomol., 6*:75-102.

Wyatt, G. R. (1963). In "Insect Physiology" (V. J. Brookes, ed.). 23rd Biology Symposium, pp. 23-41. Oregon State University, Corvallis, Oregon.

Wyatt, G. R. (1967). *Adv. Insect Physiol., 4*:281-354.

Wyatt, G. R. and Kalf, G. F. (1957). *J. Gen. Physiol., 40*:833-847.

Wyatt, G. R., Loughheed, T. C. and Wyatt, S. S. (1956). *J. gen. Physiol., 39*:853-868.

Wyatt, G. R., Kropf, R. B. and Carey, F. G. (1963). *J. Insect Physiol., 9*:137-152.

Wyatt, S. S. (1956). *J. Gen. Physiol., 39*:841-852.

HEMOLYMPH CHANGES WITH INFECTION OR INDUCED IMMUNITY IN INSECTS AND TICKS

June Stephens Chadwick

Department of Microbiology and Immunology
Queen's University
Kingston, Ontario, Canada

The varying relationships between parasites and hosts have long been recognized, whether the host be an insect, a tick, or some higher form of life. The mere fact of parasitism in itself implies a mechanism of offense on the part of the parasite and, very likely, defensive effort to ward off the parasite, on the part of the host. The host-parasite interaction may be manifested in various organs and tissues, including hemolymph. When the parasite is an infectious agent the same criteria undoubtedly apply, except that the concept of possible resistance or immunity to the infectious agent must also be taken into account. If the parasite is an infectious agent for an insect host, very little immune response can be expected. That is, most insects and ticks do not, in due course, generally recover from an infectious agent as does man from many of the infections caused by microorganisms. When immunity against a pathogen or potential pathogen has been achieved in an insect it has most often been acquired artificially than naturally.

It is not uncommon to use the blood of vertebrates as an index of health or disease. Similarly, it is not inconceivable that the hemolymph of insects or ticks may reveal useful information concerning the physiological condition of that arthropod during infection and ultimately show changes that reveal a state of immunity.

If we are to consider the changes that infection may bring about in the hemolymph of arthropods such as insects and ticks, it is probably best to qualify what is meant by infection. To quote Steinhaus (1963) "The infectious diseases of insects are usually thought of in terms of the type of etiological agent involved." Thus as in mammals infection in insects and ticks may be brought about by viruses, rickettsiae, bacteria, fungi, protozoa, etc. When such agents have been studied with respect to the accompanying changes that infection or induced immunity has wrought in the host hemolymph, reports in the literature on viruses and bacteria tend to exceed those on all other infectious agents.

Information dealing with infection in ticks is sparse but there are many excellent treatises on infection in insects (Burges and Hussey, 1971; Steinhaus, 1963; Steinhaus, 1969), and reviews on the state of insect immunity (Briggs, 1964; Chadwick, 1967; Heimpel and Harshbarger, 1965). Many of the articles reported in the above-mentioned reviews tend to deal with the over-all pathology, for example the killing dose, time of death, gross appearance, etc. rather than actually dealing with changes that occur in the hemolymph. It is perhaps within the last decade that more investigators have become concerned with the mechanisms of vector-parasite relationships, or more specifically with the changes in hemolymph that accompany infection. One of the most comprehensive reviews that deals with defense mechanisms against microorganisms and parasitoids is the recent work of Whitcomb et al., (1974).

The reason for the less specific approach and lack of information on hemolymph changes may well be that it is only within the last decade that much information has become available as to hemolymph composition and function in various insect species, and to a lesser degree in ticks. Improved and more sophisticated techniques for the separation of components of mammalian sera probably led to increased studies of hemolymph proteins and the eventual comparison of those in normal and infected insects or ticks.

CHANGES IN INSECT HEMOLYMPH FOLLOWING INFECTION

Infections Due to Bacteria. The association of bacteria with insects has been appreciated since Pasteur's time and many species of pathogenic bacteria have been described. Lysenko (1964) stated that bacteria and other pathogens should be classified according to their relationship to the host. Fundamental knowledge of the relationship according to Lysenko is sorely needed so future work will be less empirical. It is true that insofar as bacteria are concerned host-parasite relationships are still poorly understood.

Bucher (1960) proposed four groupings to include all pathogenic bacterial species; (1) obligate pathogens; (2) crystalliferous spore formers; (3) facultative pathogens and (4) potential pathogens. *B. popilliae* and *B. lentimorbus* fall into group 1 and *B. thuringiensis* into group 2, and here is where the attention has been focused. It is unfortunate that these examples (or their variants) are practically the sole member of their group. A number of the potential pathogens such as *Pseudomonas* or *Proteus* species frequently serve as laboratory models for studies on infection or acquired immunity. It is probably still safe to state that even so, our knowledge about the host-parasite relationship between these potential pathogens and their various hosts remains extremely limited. Effects on a specific tissue such as hemolymph are indeed not well-defined. One of the most comprehensive studies on the pathogenesis of potential pathogens is the series of papers by Lysenko and associates (Lysenko, 1963 a,b,c,d; Lysenko and Kǔcera, 1968) dealing with the pathogenicity of *Pseudomonas aeruginosa*.

While phagocytosis of foreign particles is dealt with elsewhere, it is probably fair to include here one of the most obvious and most easily recognizable of hemolymph changes, the occurrence of bacteria in the hemolymph. This undoubtedly is a change from the normal. It is probable that bacteremia or septicemia occur as one phase of almost any bacterial infection. The same may not be true where a toxin is the most important virulence factor of the pathogen. There are indeed a number of reports of bacterial diseases in insects where detectable numbers of microorganisms in the hemolymph is a salient sign of infection.

Another change which occurs is a difference in the total hemocyte count. Again this may occur independently of phagocytosis so will be reported in a few instances here. Some authors have actually attempted to classify the types of blood cells in which significant changes in numbers have occurred.

Milky Disease Organisms. The comprehensive work of Dutky (1940) showing that milky disease is an effective means of control of the Japanese beetle has most certainly led to some interesting investigations on that insect-pathogen relationship. Many of these investigations understandably have been directed at methods of improving the control from an economic viewpoint rather than being in-depth basic investigations on the killing mechanism. It is well established that during the 7-21 day developmental period of milky disease in the Japanese beetle that about 5 x 10^{10} spores accumulate per ml of hemolymph (Dulmage and Rhodes, 1971, p. 528). Further to this some precise and extremely useful determinations have been made of differences between the hemolymph of normal and diseased larvae.

Schwartz and Townshend (1968) studied the effects of milky disease on hemolymph coagulation and on the numbers of hemocytes in the Japanese beetle. They reported no difference in coagulation time between normal and disease larvae. In the same investigation they reported a 1-4 fold increase in total hemocyte numbers in diseased larvae after 98-day degrees. The average counts were not significantly different in healthy and diseased larvae over the total 161-day degree test period. This apparently indicates that the total counts were significantly different at specific times even though the averages did not differ over a long time period. These authors speculate that the occasional increases in numbers may be caused by a response needed to offset the drain of hemolymph nutrients or by an attempt to absorb the toxins coincident with the large number of vegetative forms in the hemolymph. Presumably the return to lower numbers could reflect inability on the part of the host insect to overcome these detrimental effects. It is well known that there is little chance of the Japanese beetle overcoming a septicemic condition with *B. popilliae*.

Insofar as the author is aware the literature is devoid of references on attempts to induce immunity against either *B. popilliae* or *B. lentimorbus*. It would be interesting to know whether induced immunity would totally counteract these shifts in hemocyte numbers.

Weiner et al., (1966) made a study of the oxygen concentration in larvae of the Japanese beetle infected with *B. popilliae*. They found that the concentration of oxygen was significantly lower early in the infection. At that time apparently only vegetative cells of *B. popilliae* were present in the hemolymph. If both vegetative cells and spores were present the oxygen level approached that of a control group. These authors also investigated the hemolymph pH and found that it was not significantly different with infection. It was in the order of 7.0 ± 0.1. They concluded that pH was not a factor in the variation of concentration of dissolved oxygen in larvae. This study and the study of Schwartz and Townshend indicate that a variety of physical changes can occur in hemolymph as a result of an acute infection such as milky disease. No author clearly indicates the condition of the hemolymph just prior to the death of the larvae.

Bennett and Shotwell (1970) reported that *P. japonica* affected by milky disease showed changes in the content of the major protein of the larval hemolymph. The composition of hemolymph in both normal and infected larvae was first investigated by Bennett et al. in 1968. It was then noted that one protein disappeared with disease and that the major protein also decreased in concentration. The protein present in greatest concentration exhibited the greatest electrophoretic mobility and had an iso-electric point of 5.85. At that time it was known that the protein had both lipid and carbohydrate moieties. It is now known that this major protein is a lipoglycoprotein. Bennett and Shotwell (1970) in addition to noting depletion of this protein in diseased larvae also characterized the lipids and amino acids of the protein. They found that the relative concentration of palmitic acid was greatly decreased in the diseased insects though the amino acid composition of the lipoglycoprotein was similar in both groups of insects. Studies such as this might well throw light on mechanisms of pathogenicity and host resistance. Unfortunately such detailed studies are rare.

Bacillus thuringiensis. One of the best known and probably the most ubiquitously studied of all insect pathogens is *Bacillus thuringiensis* often referred to as a crystalliferous bacillus. There have been many excellent reviews on the developmental studies on this organism which led to its use as a biological control agent for insect pests. Perhaps more is known of its mechanism of pathogenicity than of any other invertebrate bacterial pathogen. A detailed discussion of the organism and its toxicity is given in the review by Heimpel (1967), and a discussion of its use as a microbial insecticide by Angus (1968). Because it has been well established that varieties of this organism produce at least four toxins and because it is largely virulent due to the toxins, it may be somewhat out of context to discuss the hemolymph changes that it induces under the heading of changes due to infection. However, it would be unfortunate and remiss to exclude this organism not only because of its importance, but also because of some precise and esoteric work illustrating that it is possible with this concentrated type of investigation to elucidate a basic mechanism of virulence.

There are six types of crystalliferous bacteria varying in pathogenicity for different insects. According to Tanada (1971), pathogenicity is associated with the quantity and quality of the toxins produced by a given strain and with certain characteristics of the host insect.

Insects susceptible to *B. thuringiensis* belong to the Lepidoptera and have a gut pH in the average of pH 9 - 10.5. But not all Lepidopterous insects whose gut contents have a pH in the above range are susceptible to this organism. The toxic activity associated with the parasporal body has various effects on tissues of susceptible species such as *Bombyx mori* (Heimpel and Angus, 1959). One such effect that occurs in susceptible insects with an alkaline gut pH begins with altered permeability of the mid gut epithelium followed by penetration of the hemocoel by the highly alkaline gut juices leading to increased pH of the hemolymph (Fast and Angus, 1965). These authors found there was a slower rise in hemolymph glucose in infected larvae compared to the control, apparently due to a deficient active transport system. As a result of the change in the hemolymph pH, first paralysis and later death of the insect ensues.

There are numerous manifestations of toxic effects of this organism which have been elucidated but it is only within the province of this discussion to mention those directly related to changes in character or composition of hemolymph. It is probably relevant to an understanding of the mechanism to consider that there are certain susceptible species with high gut pH which do not react as described. One such example is *Malacosoma* spp. in which there is no sudden alteration in the hemolymph even though the insect eventually dies with generalized paralysis.

In a preliminary investigation Faust (1968) studied the *in vitro* chemical reaction of the δ-endotoxin of *B. thuringiensis* var. dendrolimus with other proteins and demonstrated that it had a strong affinity with proteins whose iso-electric points are in the alkaline range of pH. He speculated at this time that the mechanism by which the toxin caused the breakdown of cellular function and killed the host was probably related to this interaction at alkaline pH. Further to this, Faust and Dougherty (1969) studied the effects of the δ-endotoxin of the same strain on the hemolymph of *B. mori*. While they did not discount the fact that the first tissue affected in the silkworm was in the mid gut region, they did show that interaction exists between the hemolymph and toxin. The nature of interaction was not known. Whether this is an example of hemolymph activity not directly related to mode of action of the toxic portion of the organism remains to be seen. This finding if confirmed should forestall a premature conclusion that one observed hemolymph change or complexing of material, for example, is enough to explain the mechanism of action of a given infectious agent.

There are no reports of efforts to induce immunity against *B. thuringiensis* and perhaps no apparent practical reason to do so. However, resistance likely will occur. In fact, Harvey and Howell (1965) reported an instance of an insect, the housefly, becoming resistant to *B. thuringiensis* after culture for 50 generations on media treated with the bacillus. The action here of course is due to a thermostable toxin. It is not impossible that resistance to the toxic crystal may later be demonstrable. It may seem a bit incongruous, but not impossible that information on induced resistance to this organism may well lead indirectly to an understanding of the mechanism of pathogenicity in areas as yet unexplained. On the whole it is apparent that the mechanism of action of *B. thuringiensis*, is probably understood much better than that of many parasites, but much still remains to be investigated.

Non-crystalliferous Bacteria. Bacillus cereus sometimes regarded as the parent species of *B. thuringiensis* but differing in its lack of formation of a protein parasporal inclusion body, is considered generally to be a ubiquitous, saprophytic organism. On the contrary it can be associated with disease in invertebrates or higher animals. The insects which may be affected by this organism have been listed by Heimpel and Angus (1963). Its host range like that of *B. thuringiensis* is dictated by the gut pH of the insect species. Virulence is probably due to toxic phospholipase C which is active only at alkaline pH (Heimpel, 1955). It is interesting that in contrast to insects affected by *B. thuringiensis*, those infected by *B. cereus* show little pH change in their hemolymph (Heimpel, 1955).

Further investigations on the pathogenicity of *B. cereus* for insects were made by Lysenko (1972). He actually found changes in the hemolymph proteins of *G. mellonella* after the injection of larvae with known lethal doses of phospholipase C. There was a decrease in two proteins in the diseased hemolymph. The hemolymph investigated actually came from dead larvae since the toxin killed the insects within one hour after injection. Even so, presumably the hemolymph proteins were representative of changes induced by the phospholipase C.

Pseudomonas aeruginosa. The precise mechanism of pathogenicity of *Pseudomonas aeruginosa* for insects in general and particularly for many susceptible species of Lepidoptera is still not totally elucidated. Neither is the actual mechanism of pathogenicity of this organism for mammals totally understood (Liu et al., 1961; Meinke and Beck, 1970). Though there have been several investigations on its pathogenic mechanism in insects (Lysenko, 1963 a,b,c,d, 1964; Kučera and Lysenko, 1968; Lysenko and Kučera, 1968), such investigations have been fewer in number than the great many carried out on induced immunity to potential pathogens, largely employing *P. aeruginosa*. Beyond a thorough knowledge of the high virulence of the organism (Chadwick, 1971) and the fact that protection against it can be achieved with prior injection of vaccine or endotoxin (Chadwick, 1968), a great deal is still lacking in understanding the mechanism of virulence and of immunity. It is understandable therefore that the hemolymph changes that are a consequence of *Pseudomonas* infection remain largely unknown.

The most concentrated efforts to determine the mechanism of pathogenicity have been those of Lysenko and his co-workers. Lysenko (1963a, b) first described a toxic substance from *P. aeruginosa* which had proper biochemical characteristics to suggest it was a toxin. The toxic substances were believed to be proteinases. Lysenko and Kučera (1968) studied their toxicity and investigated the changes they produced in the hemolymph of *Galleria mellonella* larvae under *in vitro* conditions. The toxins had specific affinity for particular protein fractions. Over a 16-hour time interval it was assumed that 2 major protein fractions could be totally digested *in vitro* by 80 LD_{50} of the most potent toxins per milligram of hemolymph. It is very difficult to relate this to the *in vivo* situation occurring in larvae affected with *P. aeruginosa*.

Lysenko and Kučera (1968) attempted to understand the mechanism of toxicity by studying *in vitro* changes in the hemolymph when it was treated with various toxins. The toxic enzyme attacked only 2 of 18 proteins in *Galleria* hemolymph. The above finding was substantiated by the fact that Lysenko (1966) had reported similar *in vivo* results. This undoubtedly lends more weight to the fact that protein degradation is a major function of the toxin. Even more conclusive would be parallel *in vivo* and *in vitro* studies of the toxic effects on hemolymph. Burges and Hussey (1971) suggest two possible causes of death in wax moth larvae: (1) the insect is depleted of necessary proteins by virtue of the destructive effects of the toxin or, (2) the breakdown products liberated from the proteins may be lethally toxic to the insect. The final mode of action however is still unknown.

Insofar as other *in vivo* hemolymph changes are concerned, Madziara-Borusiewicz and Lysenko (1971) reported that the toxins of *P. aeruginosa* had an effect on the hemocytes. It is well known that the hemolymph of *G. mellonella* larvae rapidly melanizes but apparently the effect of the *Pseudomonas* toxin was to increase the rate of melanization. Yet another reported effect was the destruction of some hemocytes as a result of degenerate nuclei.

Lysenko and co-workers have probably investigated in greatest depth the mechanism of pathogenicity of insect pathogens. There are however unanswered questions about this organism. It would be interesting to know if the same effects and hemolymph changes occur in other susceptible insect hosts.

Infections Due to Viruses. Insect viruses particularly the nuclear polyhedrosis (NPV) viruses have received a considerable amount of attention and there is reasonably prolific literature on the mechanisms of their pathogenicity for insects. Some of the reports actually deal with hemolymph changes brought about by these agents. Because of the depth of information which has evolved on the NPV it has become possible to use some of them to control pests. The nature of the nuclear-polyhedrosis viruses has been reviewed by Bergold (1963), the type of infections they cause discussed by Aizawa (1963) and their use as microbial control agents by Stairs (1971). Useful information also exists on the cytoplasmic polyhedrosis viruses and to a lesser extent reports are now filtering into the literature on the granulosis viruses. The reader is referred to the previous descriptive articles and also to Smith (1963) and Huger (1963) for accounts of the latter two types of insect viruses mentioned.

It would be reasonable to suppose that viral infections would simulate those due to bacteria in causing physico-chemical and biochemical changes in the hemolymph of infected insects or ticks. Over the last decade reports of biochemical changes predominate. It is generally assumed that insect viruses must stimulate the production of nucleic acid and protein in infected cells as to animal viruses. Bergold (1963) stated that for insects there were very few references on the alterations in the amount of nucleic acid or protein during infection. References to such changes have been made since Bergold's assessment of the situation. The remainder of this section will largely deal with some of these reports.

Nucleopolyhedrosis Viruses (NPV). It is generally accepted that nucleopolyhedrosis viruses multiply in hemocytes but there have actually been few references to cell numbers during infection. Morris (1962) found that 8-9 days after inoculation with NPV that there were fewer blood cells in *Lambdina fiscellaria*, the western oak looper. Smith and Xeros (1954) had reported an increase in hemocytes in larvae of *Tipula paludosa* infected with NPV. More recently, to investigate whether or not cell numbers increased or decreased during infection by NPV, Shapiro (1967) carried out a carefully controlled experiment using *Galleria mellonella* larvae. The total hemocyte numbers in the wax moth larvae were determined during the NPV infection and in both untreated and starved controls. In uninfected controls the total hemocyte counts increased until just prior to pupation. In the starved larvae the total hemocyte count decreased steadily over the experimental period of 14 days and the differences were significant after day one. In larvae inoculated with NPV, the counts did not change significantly until the tenth day. After this, counts in infected larvae decrease significantly. It is interesting that during the first five days of infection the total hemocyte count in infected larvae was actually greater than in larvae suffering from starvation though the count in untreated controls was greater than in infected larvae. It is apparent that infection with NPV produces a pattern in hemocyte numbers that is even different from that produced by starvation but that a decrease in numbers will occur before death of wax moth larvae. Unfortunately there are few reports of this type for other insect species infected with NPV.

In another study of NPV infection in wax moth larvae (Smirnoff 1971) compared the concentrations of potassium and sulfur in healthy and infected larval tissue. His investigation was made because quantities of elements such as potassium and sulfur are normally stable and variation could be expected only if the physiological condition of the host altered. It is unfortunate the concentrations in hemolymph were not investigated. Smirnoff found that the amount of potassium decreased by one-half in infected larvae though it normally increases constantly during development of healthy larvae. Sulfur concentrations were found to increase through the various instars in healthy larvae but the accumulations were greater in infected larvae. The author contended that quantitative analyses such as this might be useful to determine both the presence and evolution of viral infection. There are few investigations of this type reported for NPV infections and almost none in hemolymph. It may well be that determinations of other elements in conjunction with the two reported here could be even more useful if extended to hemolymph. Wyatt (1974) has discussed the variation in sodium:potassium:ratios as well as comparing the concentrations of calcium and magnesium in vertebrate and invertebrate sera. Perhaps investigations on possible changes in such levels due to viral infections would also aid in tracing the course of the disease.

Most of the biochemical investigations on changes associated with NPV in insects have been directed at amino acid or protein levels and occasionally to lipid content. An early investigation of this type was made by Martigoni and Milstead (1964) when they studied the changes in both total solids and total protein in hemolymph of the infected variegated cutworm, *Peridroma saucia*. The senior author (Martigoni, 1964) had stated elsewhere that the information on protein metabolism in diseased insects was sorely lacking. Their study was aimed at earlier stages of the NPV infection at a point where protein values are not spuriously high due to tremendous release of viral inclusion bodies into the blood just prior to death of diseased insects. Martigoni and Milstead found an actual decrease in total solids and total proteins in the blood of diseased sixth instar larvae. They concluded that hypoproteinemia is a part of the NPV syndrome and suggested this could be true in other Lepidoptera. It was also suggested that hypoproteinemia would be greatest whenever there were marked cytopathological changes in the fat body. The methods employed here included refractometry, specific gravity determinations and protein analysis and a survey of this type made on a variety of insects showing abnormalities in the fat body could be quite informative.

A similar type of investigation was made by van der Geest and Craig (1967). They also found a decrease in total solid content in diseased hemolymph of the variegated cutworm and attributed it partly to a decrease in hemolymph protein concentration and in free amino acids though there must have been other contributing factors. They suggested that decreased food intake during disease as well as synthesis of virus might have caused a depletion of reserve carbohydrates. Experiments to confirm this would be most interesting. These authors also indicated that lysine, valine, leucine, isoleucine, phenylalanine, threonine, serine, proline and tyrosine decreased in infected larvae. Because these amino acids are important components of inclusion body proteins, it was suggested that amino acids for viral protein might be partially derived from the hemolymph. In a later investigation by van der Geest and Wassink (1969) it was shown by gel electrophoresis that a decrease of almost all proteins occurred in the hemolymph of the cabbage armyworm infected with NPV. The authors pointed out the difficulty in comparison of healthy and diseased protein patterns due to large variations in normal individuals.

Shapiro and Ignoffo (1970) using refractometry, the Lowry et al. (1951) method of protein determination and ultraviolet absorptiometry analysed protein and free amino acid changes in the hemolymph of *Heliothis zea* larvae during nucleopolyhedrosis. In 24 hours after infection it was possible to detect both hyperproteinemia and hyperamino acidemia in larve of the bollworm. However, after 96 hours the hemolymph of the infected larvae showed a decrease in amino acids as well as decrease in proteins.

Young and Scott (1970) investigated hemolymph proteins in the cabbage looper, *Trichoplusia ni* during an NPV infection. The proteins were analysed by immunoelectrophoresis and again it was found that there was a decrease in the protein concentration of the hemolymph. The decrease was due to the loss of two normal hemolymph protein components as well as to a reduced concentration of the other major components of the hemolymph. It was interesting that these authors also reported that the hemolymph from infected cabbage looper larvae contained some proteins not present in normal hemolymph. Previous authors had not reported this probably because the techniques used were not suitable to demonstrate this type of change. Young and Scott found that these new proteins cross-reacted with antisera from both diseased and healthy larvae. They suggested that on this basis the new proteins could not be serologically related to those in the virus. It would however seem fallacious to state that they were new proteins if they reacted with antisera from healthy larvae. The authors counteract this argument by suggesting that they are probably present in normal insects in too minute quantities to be detectable by the technique used. The significance of these new (or at least increased) proteins in the process of the NPV infection is unknown.

Young and Lovell (1971) carried out further investigation to characterize the hemolymph proteins of infected fifth instar larvae of the cabbage looper using acrylamide gel electrophoresis. Again it was found that throughout the disease, protein concentration of hemolymph was lower than in healthy larvae. The sharpest drop in concentration was on the fourth day after infection while at a comparable time protein concentration was increasing in healthy larvae. By the use of seven per cent acrylamide gels it was shown that several major proteins of high molecular weight actually decreased though the change in pattern was not very apparent in the first 2 days of the infection. Seventeen protein bands were detected in both types of hemolymph. The proteins were mainly glycoprotein and lipoprotein. Two bands of alkaline phosphatase were present in normal hemolymph. No changes were apparent in enzymes such as esterases, malic dehydrogenase or aminopeptidase. As a result of this investigation it would appear that the decrease in protein concentration can be expected to be more pronounced as the disease progresses. The authors believe that after the fat body is infected with NPV, normal protein synthesis and release may be markedly reduced, or perhaps even eliminated.

Watanabe (1967) compared the hemolymph proteins of normal and NPV-infected larvae of the fall webworm, *Hyphantria cunea*. Separation was made by agarose-gel electrophoresis. The hemolymph of normal larvae showed three major fractions migrating toward the anode and in the early stages of disease the pattern was the same. In later stages of the disease there were less of the medially migrating fraction and the fraction with lower mobility, but no apparent change in the faster migrating fraction. Again we have an instance of heavily reduced protein concentration as the disease progressed. Once again the protein seen in diseased hemolymph is believed to be a reflection of lesions in the fat body.

Another electrophoretic study was that of Tanada and Watanabe (1971) in which disc electrophoretic patterns of hemolymph proteins of NPV-infected larvae of the armyworm *Pseudaletia unipuncta* were compared to controls and those of larvae infected with granulosis virus. Late fourth instar larvae were used and at least 13 bands were found in the hemolymph of uninfected larvae. As the NPV infection progressed, a decrease in concentration of the two major bands occurred and if the insects were heavily infected all of the protein fractions were decreased. This pattern was comparable to that of granulosis-infected larvae as will be noted under the next heading.

One rather distinctive study was that of Komano et al. (1966) in which the lipids in the hemolymph of normal and diseased silkworm larvae were assessed. The lipid content actually is higher than normal in jaundiced silkworms and the amount of phospholipid (2.5 fold increase) and cholesterol (2.6 fold) increase markedly after the release of polyhedra into the hemocoel. The authors speculate that viral infection may modify the lipid metabolism of the host *or* some of the substances could be released in the blood stream of the larvae through rupture of cytomembrane. The authors do not take a definite stand for either hypothesis.

Cytoplasmic Polyhedrosis Viruses. Cytoplasmic polyhedrosis viral infections (CPV) are probably less often reported than the nuclear polyhedroses, but even so they may exceed the nuclear polyhedrosis viruses in numbers; thus the list of insects that have been reported infected is quite impressive (Smith, 1967). The first site of multiplication is of course the cytoplasm of mid gut cells, extending to the foregut and hindgut as the disease progresses. Apparently the fat body is not affected. Kawash (1965) carried out one of the earlier investigations on constituents of hemolymph from *B. mori* infected with CPV. He compared the free amino acids and found that the concentration of most decreased markedly and was only one-half to one tenth of that in normal hemolymph. It was interesting however to note that glycine and lysine contents were higher than in normal hemolymph. The amount of ammonia in the hemolymph of the infected larvae was also higher. The author suggested this may have been due to the fact that the viral infection in some way activated deamination of amino acids.

Kawash and Hayashi (1965) carried out a subsequent investigation in which they investigated daily changes of nucleic acid and protein in the blood and mid gut of silkworm larvae infected with CPV. A distinct decrease of blood protein was found in diseased larvae as compared to the significant rise in healthy larvae. It is interesting however, that the trend to a slight rise was apparent even in the CPV-infected larvae. The RNA content in the hemolymph of diseased silkworm larvae tended to be lower. In contrast the DNA content in diseased hemolymph was similar to that in normal hemolymph.

Sikorowski et al. (1971) reported the presence of virus particles in the hemolymph of larvae and adults of the tobacco budworm, *Heliothis virescens* infected with CPV. They made no effort at this time to study blood components.

Granulosis Viruses. In a comparable study to that of NPV in sixth instar larvae of *Peridroma saucia* (Martigoni and Milstead, 1964), the same investigators determined total solids and total protein in hemolymph of cutworm larvae injected with granulosis virus (Martigoni and Milstead, 1966). There was a noticeable decrease in blood-plasma solids by the fourth day after inoculation. Hypoproteinemia was apparent as with the

NPV infection with the maximum decrease occurring slightly later in the granulosis infection. The fat body is the target organ here as in the case with nucleopolyhedrosis.

Tanada and Watanabe (1971) analysed by disc electrophoresis the hemolymph proteins of two groups of larvae of the armyworm, *P. unipuncta* infected with different granulosis virus strains. They observed a decrease in concentration of the two major bands in the same manner as a decrease occurred with NPV (see previous section). The authors contend that this is not too surprising in view of the fact that the fat body is the major target organ in both NPV and granulosis infections. They further suggest that a similar functional lesion impeding the synthesis of hemolymph proteins may occur in both the NPV and granulosis infections.

Infections Due to Agents Other Than Bacteria or Viruses, Fungi. Although entomogenous fungi receive less attention than bacteria or viruses, there is a considerable volume of references on these microorganisms. One comprehensive review is that of Müller-Kögler (1965). The mechanisms of virulence in fungi have been less well investigated and we can only deal here with direct references to hemolymph changes brought about by fungal infections or fungal toxemia. Many years ago it was reported that deaths occurred in fungal infection while the organisms were confined to the hemolymph and it was assumed that toxin production probably occurred in the hemolymph. There appear to be few references which lend weight to this theory. Burges and Hussey (1971) have suggested that fungal toxins might be responsible for digestion of hemolymph.

Roberts and Yendol (1971) also point out nine necessary areas of research on fungi with relation to their use as microbial control agents. They do not specifically state that an investigation of the mechanism of action is necessary though they list as number one - the elucidation of aspects of disease induction. Perhaps it is inherent in the nature of the organism and obvious from the existing literature that the main target tissue will be the cuticle and underlying tissue. For this reason perhaps the hemolymph can play only a minor role in the disease process and less investigation may be warranted here than for other microorganisms.

Protozoa. There are numerous pathogenic entomogenous protozoa and no effort will be made here to discuss the complex host-parasite relationships that may occur. Excellent biological studies such as that of Kellen and Lindegren (1968) have been made but without any reference to effects in hemolymph. It is perhaps important to state some obvious differences in the activity of protozoa as opposed to bacteria or viruses. They are not highly virulent and rather than rapidly upsetting vital host functions, they cause lessened fecundity, shorter life-span and lessened responses to stimuli, etc. The nature of this response has not encouraged investigation into changes in hemolymph during this type of host-parasite association. This oversight may not be warranted and it may well be profitable to study hemolymph changes concomitant with these more prolonged pathological effects.

Contrast and Comparison of Effects Induced by Various Agents. Since bacteria and viruses are such very different disease agents it may seem illogical to compare the changes they bring about in the hemolymph of insects during infection. It is interesting however, to contrast their effects on hemolymph and in some instances to compare them. Viral mechanisms of pathogenicity in insects have probably been studied with greater intensity than have infectious processes due to pathogenic bacteria. It is apparent there are hemolymph changes common to both bacterial and viral infections (Table I lists the most frequently reported changes in some bacterial infections and Table II shows changes in viral infections). The most commonly reported change in hemolymph is a depletion of total protein. Table II shows much more frequent reporting of this change in viral infections. In some instances where investigations have been more comprehensive it has been found that the depletion was mainly in major proteins occurring in normal insects (Young and Scott, 1970; Young and Lovell, 1971). Investigators of viral pathology have also sometimes extended their research to make detailed analyses of amino acid composition of normal and diseased hemolymph (van der Geest and Craig, 1967). There is little comparable information for bacterial pathogens. The amino acids in hemolymph are depleted not only in viral infections but also during parasitism of the tobacco budworm, *Heliothis virescens* by its habitual parasite *Cardiochiles nigriceps* (Barras et al., 1968). It is interesting that both parasitism by a microorganism and a macro-parasite can result in similar hemolymph changes. In the case of viral infection the amino acids show some qualitative changes whereas only quantitative differences occurred with the parasite *C. nigriceps*. Barras et al. found an increase of amino acids in this parasite suggesting its absorption and utilization of the amino acids. Similar processes have been postulated for viruses (van der Geest and Craig, 1967). The possibility of bacterial parasites acquiring protein or amino acids from their host also exists.

TABLE I

HEMOLYMPH CHANGES FOLLOWING SOME BACTERIAL INFECTIONS

Bacterium	Insect	Change
B. Popilliae	Japanese Beetle	Increase in total hemocytes. Lower oxygen concentration. Decrease in major proteins and in total protein.
B. Thuringiensis	Several Lepidoptera	Increased pH. Toxin interaction with hemolymph protein.
B. Cereus	Greater Wax Moth	Decrease in proteins.
P. Aeruginosa	Greater Wax Moth	Decrease in some proteins. Increased rate of melanization. Destruction of some hemocytes.

TABLE II

SOME COMMON HEMOLYMPH CHANGES DUE TO VIRAL INFECTIONS IN SOME REPRESENTATIVE INSECTS

Virus	Insect	Change
NPV	Silkworm	Lipids increased.
NPV	Greater Wax Moth	Total hemocyte count decreased.
NPV	Variegated Cutworm	Proteins decreased.
NPV	Bollworm	Proteins and amino acids decreased.
NPV	Cabbage Looper	Proteins decreased
NPV	Fall Webworm	Proteins decreased
NPV	Armyworm	Proteins decreased
CPV	Silkworm	Free amino acids decreased
CPV	Silkworm	Proteins decreased
Granulosis	Variegated Cutworm	Proteins decreased
Granulosis	Armyworm	Proteins decreased

Though there are still unanswered questions to some of the mechanisms of pathogenicity of *B. thuringiensis* and *P. aeruginosa*, these remain the most carefully studied of the bacteria. After surveying the existing literature the impression is gained that the effects on hemolymph of infection by many potential pathogens remain to be investigated.

CHANGES IN TICK HEMOLYMPH FOLLOWING INFECTION

Ticks are important vectors of rickettsial and some virus diseases of man. Řehàček (1965), in one of the few reviews dealing with the relationship of animal viruses and rickettsiae to ticks and mites, stated that there were relatively few papers which analysed in any detail the relationship between viruses, rickettsiae and these vectors. Since the review of Řehàček, the situation appears to have remained unchanged. In most instances these agents are not actually harmful to their arthropod vectors, (an exception is *R. prowazekii* which kills its vector, the louse), but it is not inconceivable that during even a non-pathogenic type of relationship some hemolymph changes might occur. Reports of such changes are obscure.

The interaction of a rickettsial species with a given host is very likely specific for that relationship. Hence it is unwise to generalize here at any length on the host-vector relationships or pathology. Chamberlain and Sudia (1961) believed that it was theoretically possible, if circumstances are favorable, for any blood-feeding arthropod to become a mechanical vector of organisms received from the vertebrate host. The arthropod host would always take the microorganism into its digestive tract but would only transmit the parasite (to another host) if genetically capable of doing so. Apparently when biological transmission does occur the organism penetrates through the digestive tract into the hemocoel. This implies that a minimal number of rickettsiac or viral particles are necessary for this to happen.

In the opinion of most authorities virus in a tick vector generally is transmitted to the host in secretions of the salivary glands. The occurrence of virus in the hemolymph or change in hemolymph brought about by virus have not been assigned an important role. Organisms undoubtedly must infect the hemolymph for a well known test for the detection of rickettsiae in ticks is based solely on contamination of the hemolymph by these organisms and their subsequent demonstration. The hemolymph test as described by Burgdorfer (1970) was designed to detect rickettsial pathogens in adult ticks. It was found that hemolymph provided a rapid means of detecting rickettsiae in ticks such as *Dermacentor andersoni* and *Dermacentor variabilis*. The rickettsiae multiply within hemocytes and within 24-48 hours after infection, are detectable by direct microscopic examination of hemolymph. A similar test was also described by Řehăcek et al. (1971).

A recent Russian investigation by Stepanchenok-Rudnik et al. (1969) is probably the first investigation on a mechanism of pathogenicity in ticks in any way comparable to the investigations of pathogenicity carried out in insects and other invertebrates. This is one of the few instances in which tick hemolymph has actually been analysed for shifting components due to infection. The authors analysed the hemolymph of normal ticks and that of ticks infected with rickettsiae. They showed shifts in the electrophoretic characteristics of proteins, both numbers, function and position on electropherograms. Because of the difficulty in translation, details of this work are not available but such investigations are necessary in order to be able to put pathogenic mechanisms on a comparative basis in many invertebrates. Such research is also promising because in the case of ticks which are vectors of medical importance, an immunochemical method potentially utilizable for the detection of infection could be invaluable.

CHANGES IN HEMOLYMPH FOLLOWING INDUCED IMMUNITY IN INSECTS

Recently there has been an increase both in interest and in the number of investigations on humoral mechanisms by which invertebrates, particularly insects, respond to and attempt to overcome parasites. There may be numerous reasons for increased interest in invertebrate immunity but some of the more important are: (1) a focus of attention on comparative immunology to elucidate the phylogeny of the immune response; (2) a need to understand both innate and acquired resistance in order to make wiser choices in the use of biological control agents and (3) quite probably often an academic interest in mechanisms which since the time of Pasteur and Metchnikoff have been superficially investigated, but for which a complete understanding still eludes most investigators. Several comprehensive reviews covering aspects of humoral immunity in considerable depth are available (Stephens, 1963a; Briggs, 1964; Chadwick, 1967; Whitcomb et al., 1974).

If the immune response in insects and ticks were analogous to that of mammals, the detection of antibodies would be the most obvious approach to an investigation of changes in the blood as a reflection of humoral immunity. Good and Papermaster (1964) denied the possibility of antibodies in invertebrates as have many other investigators. Gray (1969) summarizing the current position on the possibility of invertebrate antibodies stated that, from the available information on humoral substances, there was nothing to indicate components comparable to vertebrate antibodies. Gray did point out that the finding of any precursor type substance comparable to the subunit (or fraction thereof) of an immunoglobulin would be the most important and further studies in this area would be warranted. It has already been pointed out that Steinhaus (1969) stressed the value of starting at the bottom of the evolutionary tree to study disease processes. It would seem equally sensible to me that here too is the proper place to learn something about basic immune responses and possibly basic immune substances. One could say that efforts are now being focused in this area though many investigations still tend to be somewhat gross and diffuse, rather than precise and specific.

Because of the generality of some of the studies on insect immunity and the apparent lack of cohesiveness of many of the studies, there tends to be a great deal more information proving that insects can respond to foreign materials than on how they respond. Consequently the amount of information on hemolymph changes that occur during immunity is not vast. The existing information can be subdivided into two general areas: (1) hemolymph changes noted after induction of the immunity by bacteria or their products and (2) to a much lesser extent, hemolymph changes occurring as a result of immunity to viruses. Changes can be further subdivided into physico-chemical or serological.

Immunity Against Bacteria, Physico-chemical Changes. As previously mentioned much of the information dealing with the subject tends to be qualitative. Insofar as hemolymph changes are concerned one early observation that reflects a difference in the hemolymph of immunized insects was that of Stephens (1962a) on the absence of *in vitro* melanization of the hemolymph of insects immunized with bacterial vaccines. At the time I found that, although the hemolymph of vaccinated larvae did not darken on exposure to air, tyrosinase activity was present to a similar degree to that in normal hemolymph. Also I speculated on the possible relation of the absence of melanization to the immune mechanism. Since that report, several other investigators have made similar observations using different insects and occasionally different immunizing preparations. They too have attempted to link this observation to the immune mechanism.

Taylor (1969) went so far as to propose a system of non-specific bactericidal activity based on the polyphenoloxidase system. Ziprin and Hartman (1971), and Pye (1973) have used absence of melanization as a criterion of acquired resistance. Thus, some authors have looked to changes in melanization rates of hemolymph as an index of infection or toxicity, others as an index of immunity. Ziprin and Hartman (1971) investigated differing susceptibility in strains of *G. mellonella* to toxins present in bacterins of *P. aeruginosa*. They reported characteristics of the toxin and its typical pathological effects. They further investigated the effects of various treatments on the toxin with relation to the melanization times induced by the altered toxic products in previously susceptible and non-susceptible strains of wax moth. They concluded that if the strains of *Galleria* were susceptible to the toxin, treatments of toxin such as boiling would lower the melanization time and alter the ability of the material to induce an immune response. The meaning of this is presently not clear. Attempts to relate hemolymph changes in infection and immunity may well be useful. For example, if it is established that in an infectious process, melanization time is low and that it can be altered by immunization, experiments could be devised to learn something of the mechanism involved. Presently there are so many variables and so many unanswered questions that it is impossible to say whether this would be a useful approach.

Though I previously speculated (Stephens, 1962a) on the possible role of the polyphenoloxidase system in the immune system of *Galleria mellonella*, I now feel that some caution must be exerted in this assumption. It is well established that the endotoxin of *P. aeruginosa* serves as an excellent immunizing preparation against that organism in wax moth larvae (Chadwick, 1968). Endotoxin is even more efficient than heat-killed or formolized vaccine in the induction of immunity as evidenced by elevated LD_{50}'s for *P. aeruginosa* in insects thus immunized. However, the hemolymph from endotoxin-protected insects does not show as great a freedom from melanization as does the hemolymph from vaccine protected insects (Chadwick, unpublished). Such findings lead one to exert some caution in attempting to relate two phenomena which may be merely coincidental. Therefore, I feel inclined to concur with Whitcomb et al. (1974) that though the activities might be related it is unwise to make important conclusions in that assumption. Moreover, I would also agree that in this area we really need more evidence to be sure how, and if, polyphenols and the melanization reaction of hemolymph are implicated in insect defense mechanisms.

One gross change which is often a fore-runner of induced immunity is a change in total hemocyte count. Actually this is a cellular change but as it may be the first indication of change in the hemolymph and because it could trigger off humoral changes, it will be briefly mentioned herein. Wittig (1965) stated the total hemocyte counts were decreased in the armyworm, *Pseudaletia unipuncta*, and in wax moth larvae by injection of killed *B. thuringiensis*. Various effects occurred if the organisms were viable. Stephens (1963b) reported significantly decreased total hemocyte counts (THC) in *G. mellonella* larvae after immunization with formolized vaccine of *P. aeruginosa* whereas the total hemocyte counts were only slightly decreased after the injection of normal larvae with living *P. aeruginosa*. The results of Wittig and Stephens are in agreement suggesting that the total hemocyte number may fall during active immunity. Though

Wittig does not mention the THC in the armyworm after injection with virus capsules, she did report there was little, if any change in differential hemocyte counts.

The literature is almost devoid of references to the total protein content of hemolymph after induced immunity. One exception is the work of Schalwbe and Bouse (1971). They showed in several experiments an increase in total protein content of hemolymph obtained from wax moth larvae immunized with endotoxin. Because of variability in their results they question the validity of the finding. We have some evidence also that total protein may be increased during the height of active immunity (Chadwick, unpublished), but quantitative and confirmed information is still pending. Moreover our determinations were made electrophoretically. Gray (1969) has already pointed out that electrophoretic mobility may not be a good criterion for demonstrating the presence or absence of immunoglobulins so possibly it may not be the best technique to assess protein content, even initially.

Opposed to the findings of Chadwick (unpublished) were those of Kamp (1968). He was unable to demonstrate any differences between the hemolymph components of pre- and post-vaccinated larvae. On the other hand, Gingrich (1964) demonstrated that electrophoretically separated hemolymph from control and vaccinated milkweed bugs contained the same bands. However, one fraction from vaccinated hemolymph was much more intensely stained indicating a greater quantity of that protein.

From these somewhat contradictory results, one can only deduce that the technique of separation might contribute largely to discrepancy of findings. Until more research utilizing sophisticated techniques is reported, no final conclusions can be made. It is attractive to speculate that new or increased protein fractions are probably induced with the acquired response. I am inclined to feel the evidence does point in this direction.

Serological Changes. Serology implies the demonstration of antibodies. Insects and probably, most other invertebrates have no capacity to form immunolgobulins (Good and Papermaster, 1964; Chadwick, 1967; Weinheimer, 1970; Hildemann, 1974).

Therefore under the name of serology we really are not discussing antibody but can discuss any newly synthesized component or increased activity of blood occurring after the induction of immunity probably associated with the immune response and leading to the inactivation of the foreign inducer. In this case until it can be proved otherwise, we can consider that these substances, while not analogous to immunoglobulins in structure, may be somewhat analogous in function. Although many investigations on induced immunity in insects have demonstrated general features of the protective response such as eliciting agents, the rapidity of induction, short duration and relative non-specificity etc., a number of reports do refer to substances occurring in the hemolymph at the time of immunity. These immune substances have sometimes been partially characterized.

As early as 1931, investigators such as Zernoff were reporting the production of non-specific bacteriolysins in the hemolymph in response to external stimuli. At that time naturally no efforts were made to chemically characterize the bacteriolytic material. Frings et al. (1948) described in antibacterial principle in the hemolymph of the large milkweed bug, *Oncopeltus fasciatus* active against *Staphylococcus aureus* and a strain of *Bacillus Cohn*. They proceeded one step further than Zernoff in characterization by determining that the substance was heat stable. Briggs (1958) found that a heat stable antibacterial principle in the hemolymph of lepidopterous larvae increased upon vaccination and Stephens (1962b) demonstrated bactericidal activity against *P. aeruginosa* in the hemolymph of actively immunized wax moth larvae. This factor was later described by Stephens and Marshall (1962) as dialysable, heat stable, of relatively small molecular weight and probably non-protein. Hink and Briggs (1968) found two substances that were bactericidally active in the hemolymph of vaccinated wax moth larvae. Factor A was stable to heating for 5 min, present in unvaccinated larvae in low concentration, and had a molecular weight of 7000. The second factor, B, was present only in the blood of immunized insects and had a molecular weight of under 2000.

The above reports show that investigators are now finding substances in immune hemolymph that are appearing at the time of greatest immunity and many of them seem to have certain properties in common. There are many problems still underlying such findings however: (1) Characterizations are not complete. (2) Whether one or several substances are involved is really unknown. (3) The specificity of action of the substances is not known and more importantly, (4) the function of the isolated substance is sometimes not well demonstrated.

Since the preparation of some of the more exhaustive reviews on humoral immunity in insects, the question of the role of lysozyme in insect immunity has gained some prominence. Mohrig and Messner (1968) stated that lysozyme was the substance responsible for all immunity in insects. These authors published the results of some interesting experiments on the basis of which, such a conclusion was reached. They found increased concentrations of lysozyme in hemolymph after injection of insects with several non-specific substances. This increase was correlated with increased lysis and killing of bacteria. Chadwick (1970) however, although confirming the lysozyme increase upon injection, found no increase in immunity and indeed found high lysozyme concentrations after immunity had disappeared. Kawarabata (1971) on the other hand found similar lysozyme activities in the hemolymph of both vaccinated and non-vaccinated silkworm larvae.

Some of the conclusions made as to the importance of lysozyme (and also some of those drawn on the role of polyphenoloxidase) point to an underlying weakness in many of the reports on hemolymph substances associated with immunity. Perhaps because many questions concerning invertebrate immunity remain unanswered and the solutions to others have appeared very slowly, investigators tend to link an observed and partially described substance to a function which it may not have, or where it may only have a partial or minor role in an immune process.

Even though many unsolved questions remain after more intensive research in insect immunity during the last decade, it is still apparent that the most commonly occurring change in the hemolymph associated with immunity is development of a bactericidal or bacteriolytic factor or activity. Bactericidal activity in mammalian sera often is enhanced by complement (Müller-Eberhard, 1968). We know that insects in particular do not possess mammalian-type complement. Morgun (1950) stated that insect hemolymph probably contained something analogous to the C3 component of complement.

More recently evidence has been provided for the presence in invertebrates of a hemolytic complement-like system which can be activated by a protein isolated from cobra venom (Anderson et al., 1972; Day et al., 1970). These limited studies point to a system reminiscent of the alternate pathway of complement activation in mammalian serum. The cobra venom factor is known to form a complex with a mammalian serum protein, the C3-proactivator. The resulting complex is capable of activating the terminal complement sequence to achieve the lysis of erythrocytes and other cells in the absence of specific antibody (Osler and Sandberg, 1973). It is, therefore attractive to consider that invertebrate hemolymph contains a similar or rudimentary alternate pathway of complement activation implicating hemolymph factors and analogous to the C3-proactivator and the terminal complement components of vertebrate serum.

Such a finding in itself may not relate directly to a hemolymph change with immunity. However, the importance of such studies must not be minimized for they could give useful information on the possible presence in hemolymph of a substance analogous to a complement component or C3-proactivator. If present, the role of such component(s) could later be investigated in relation to an immune reaction in insect hemolymph (bacteriolysis). The *in vivo* effect of cobra venom factor on protective immunity, (particularly if it should interfere with immunity) would be another fascinating approach in the elucidation of the mechanism.

Interesting recent studies on antibacterial defence systems in insects are those of Boman and associates. Boman et al. (1972) described antibacterial activity that could be induced in *Drosophila melanogaster* males only with non-virulent, living bacteria. They stated that the inducing mechanism was inhibited by cycloheximide. Further to this, Boman et al. (1973) described antibacterial activity induced by living bacteria of *E. coli* K12 in hemolymph or body extracts of *Drosophila* or *Samia cecropia*. The material was characterized as heat stable, sensitive to trypsin, inhibited by lipopolysaccharide and sensitive to alkaline pH. It is unclear as to the range of its activity.

Rasmuson et al. (1973) made an interesting conclusion from work they reported on the *in vivo* induction of antibacterial factors in *Drosophila*. They reported that while *P. aeruginosa* kills Drosophila adults, *E. coli* strain D31 and *Enterobacter cloacae* B_1 did not. They concluded that, in the case of non-pathogens the insects were induced to release antibacterial substances presumably stimulated by the parasite itself. Both actinomycin D or cycloheximide were believed to stop the induction and the insects died with infections of *E. coli* or *Ent. cloacae*. This is a very interesting concept but careful controls and further experimentation are necessary to determine if the change in infective pattern of the two organisms is (a) actually due to substances affected by RNA or protein inhibitors and (b) if such a substance is truly responsible for protection.

It is well known that protection against *P. aeruginosa* in *G. mellonella* can be induced by vaccine or endotoxin (Chadwick, 1968) and that bactericidal activity occurs at the time of maximal immunity (Stephens, 1962b). We have since found that the hemolymph of *G. mellonella* normally supports *in vitro* the growth of *P. aeruginosa* but after immunization considerable inhibition of growth is evident (Chadwick, unpublished). A typical growth pattern of *P. aeruginosa* in hemolymph from variously treated insects is shown in Figure 1. Whether this inhibition is due to material similarly induced as described by Boman et al. (1973) is unknown. It would be most rewarding to isolate the supposed bactericidal factor by chromatography, gel electrophoresis or some other suitable method and then determine if it alone had similar inhibitory effects on the *P. aeruginosa* as shown in Figure 1. The results as they are shown in this figure do not indicate whether we are dealing with a factor or factors in hemolymph and whether *in vivo* and *in vitro* function are related. An attempt to relate the work of Hink and Briggs, Chadwick, and Boman and associates in such a fashion might be far-reaching.

Immunity Against Viruses. If the over-all question of insect immunity is intriguing, the possibility of immunity either natural or acquired, against viruses is even more questionable than that against bacterial pathogens. Kawarabata and Aizawa (1967) indicated the importance of acquired resistance to microbial insecticides in the field of applied microbiology. This is especially true with respect to viruses which in several instances have proved to be important agents of biological control. The authors also stated that immune responses to virus infection are generally more difficult to study. Certainly references to immunity against infections are few.

In 1953, Aizawa reported that it was possible to vaccinate against silkworm jaundice virus. Immunity lasted for about one week. In 1962, Aizawa used passive immunization against the nuclear polyhedrosis virus of the silkworm in *G. mellonella* larvae. Infection was inhibited if antiserum was injected at least 24 hours prior to challenge.

An actual report of hemolymph changes is that of Kawarabata and Aizawa (1967) who isolated a proteinaceous substance from the hemolymph of silkworm larvae infected with nuclear polyhedrosis virus. The substance apparently was a viral inhibitory factor. It was inhibitory against NPV *in vitro*, and could be destroyed by acid at pH 2.0. Interestingly enough, no similar substance could be isolated from the hemolymph of normal insects when the same separatory techniques were carried out. The substance then was assumed to play a large part in virus immunity. Unfortunately there are no reports of similar inhibitory substances active against other viruses.

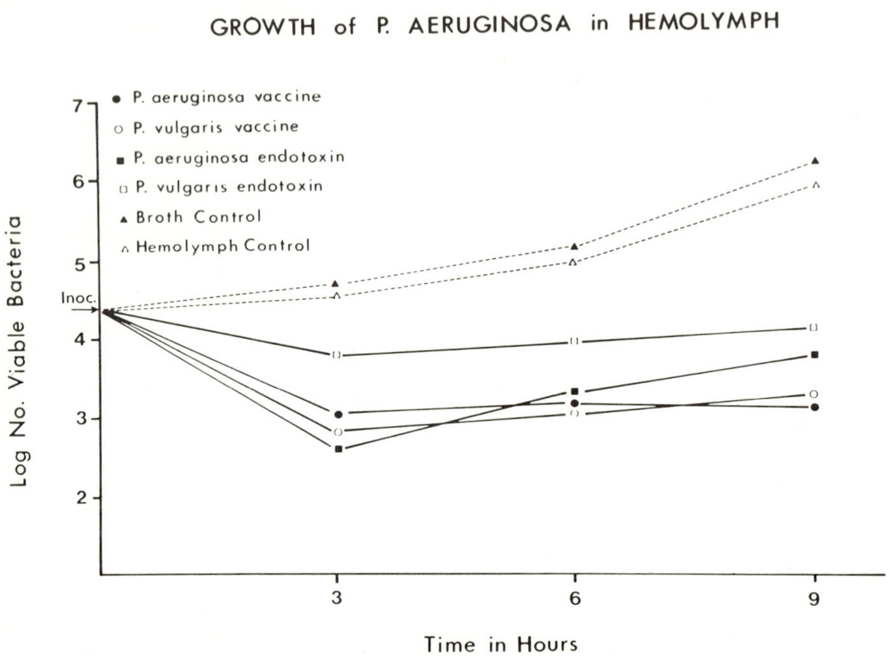

Figure 1. Growth curves of Pseudomonas aeruginosa in hemolymph of Galleria mellonella immunized with various preparations and in control hemolymph from normal larvae. Bacterial counts determined over a 9-hr period at 30°C.

●——● *denotes immunization with P. aeruginosa vaccine;*
○——○ *immunization with Proteus vulgaris vaccine;*
■——■ *immunization with P. aeruginosa endotoxin, and*
□——□ *with P. vulgaris endotoxin.*
Broth control denoted by ▲——▲ *and hemolymph control by* △—△

More recently, Raheja and Brooks (1971) reported the inability of the forest tent caterpillar *Malacosoma disstria* to acquire resistance to NPV. The survivors of a group of larvae previously fed about 7.53 x 10^5 polyhedra per larva were more susceptible than controls on subsequent feeding. Whether this represents inability of the larvae to acquire immunity or an improper immunizing preparation is difficult to say.

INDUCED IMMUNITY IN TICKS

The literature on immunity in ticks is virtually non-existent. The reasons for this are unclear. The association of many rickettsial species, and also viruses, with ticks has been known for a long time. Because in many instances rickettsial species often have no deleterious effect on the vector tick, it would seem that they could serve as a useful model for studying a host-parasite relationship, in invertebrates, where considerable resistance (probably natural) is evident.

COMPARISON AND CONTRAST OF HEMOLYMPH CHANGES IN INFECTION AND IMMUNITY

It is very difficult to compare or contrast the reactions seen in infection and immunity in insects and ticks because very frequently comparable investigations were not undertaken in the two situations. It would be simple to assume that, because immunity militates against infection, variation in an individual hemolymph feature during infection would be complemented by an opposite variation in the immune state. Such situations should not often be expected, nor do they occur.

Table III shows generally the concensus of findings in infection and immunity (mainly against bacteria and viruses) for several frequently investigated properties. From the limited amount of information available it is difficult to choose the salient hemolymph characteristics worthy of more investigation in either infection or immunity. It would seem that too few investigations have been carried out on many hemolymph properties. Most valuable information in either situation may well be derived from studying the chemical composition of hemolymph.

TABLE III

COMPARISON OF HEMOLYMPH PROPERTIES IN INSECTS DURING INFECTION AND INDUCED IMMUNITY

Property	Infection	Immunity
Total Hemocyte Count	Generally Unchanged	Decreased
Coagulation Time	No Change	Not Reported
Melanization Rate	Occasionally Decreased	Decreased
Hemolymph Volume	Not Established	Often Increased
Hemolymph pH	Variable Reports	Not Reported
Protein Content	Decreased	Increased
Free Amino Acids	Decreased	Not Reported
Lipid Content	Increased	Not Reported

PRESENT STATUS AND THOUGHTS ON THE FUTURE

In one sense, since Paillot's outstanding treatise on infection and immunity in insects (Paillot, 1933) we have made considerable inroads in the study of these areas;

in another, considering the lapse of four decades we are still faced with a deplorable lack of information and ignorance of mechanisms. With specific reference to changes in hemolymph during infection and immuity this paper does not contain a complete bibliography but only a short review with illustrative references of significant findings, so it may well be broader in omissions than inclusions. I will not attempt both to highlight the progress that has been made and to indicate a few of the gaps.

It is apparent that ticks have been rarely investigated from the viewpoint of infectious mechanisms and, almost never, on mechanisms of induced immunity. There are a few articles on the composition of normal tick hemolymph currently appearing in the literature and recently, the occasional attempt such as that of Stepanchenok-Rudnik et al. (1972) to compare components in hemolymph from normal and diseased ticks. Hopefully such investigations will pave the way for some insight into infectious processes in ticks. It would be a decided advantage to be able to compare hemolymph changes in ticks and insects.

From the literature reviewed herein, it would seem that in terms of attempting to investigate infection and consequent hemolymph changes, the greatest progress has been made with viral agents. The suggestion (van der Geest and Craig, 1967) that depleted hemolymph components such as protein actually contribute, in part, to viral protein is interesting and merits further investigation.

In contrast, in terms of changes occurring in induced immunity, the viruses lag behind the bacteria in investigation and understanding. This may be due to the fact that we have never conclusively shown that artifically induced immunity can occur against a virus. Research on acquired immunity has been the forerunner of all studies on insect immunity to bacteria. Efforts in the same direction attempting to use non-toxic viral products e.g. inactivated virions and polyhedral protein appear both desirable and necessary. Inclusion disease viruses have been investigated at length in terms of infectious processes and have shown considerable promise as biological control agents (Burges and Hussey, 1971). They well may present the best initial approach to studies of acquired immunity to viruses.

There is no reason to suppose that bacteria and viruses should be responsible for the same type of hemolymph changes either in infection or immunity. It would be most attractive to utilize as test animals an insect species which has both a known bacterial and viral pathogen. From Tables I and II it appears that there are some common changes with infection. We do not know however if the same type of protein depletion occurs with both a viral and bacterial pathogen. An extension of studies and some co-operative efforts in increasing both the number of hosts and number of parasites investigated would be useful.

In a contrast of changes in hemolymph (Table III) it is apparent that some opposites do occur. For example, there is scanty but potential evidence that proteins increase with immunity but decrease with infection. There have been some determinations of the total protein content of normal insect hemolymph, and a few determinations of total protein content in infected insects. Rather incongruously, quantitative determinations of the total protein content of actively immunized insects are non-existent. These simple determinations might well be the forerunner of more tedious and exacting determinations of individual protein fraction changes.

Although it is evident that there are still many questions concerning the pathogenicity of *P. aeruginosa*, Lysenko's work has broken much new ground. It is interesting to speculate whether the protein bands reported by Lysenko and Kŭcera (1968) as being destroyed might be identical to those which increase in the hemolymph upon active immunization of a susceptible insect (Chadwick, unpublished).

Fascinating and exciting are the suggestions of comparative immunologists that precursor units of immunoglobulins may exist in invertebrates. Research in this area is slight and an interesting field which may have much promise is waiting for investigation.

Pertinent to this is the recent work of Anderson and associates (1972). The thought that a complement-like component may exist even in insects is controversial, and if it can be proven will undoubtedly pave the way to much greater understanding of immune mechanisms.

Although at the beginning of this section I decried the lack of knowledge on this subject, it may be well to conclude with a note of optimism. Since Paillot's time there has been an obvious change in the depth and sophistication of investigations. An increased persistence and attempts to integrate and compare existing information may lead investigators using hemolymph changes as a major key to unlock doors which will reveal the relationship between infection and immunity in invertebrates.

REFERENCES

Aizawa, K. (1953). *Jap. J. Appl. Zool., 18*:141-142.

Aizawa, K. (1962). *J. Insect Pathol., 4*:122-127.

Aizawa, K. (1963). In "Insect Pathology. An Advanced Treatise." (E.A. Stanhaus, ed.), Vol. I, pp. 382-412. Academic Press, New York.

Anderson, R. S., Day, N. K. B., and Good, R. A. (1972). *Infect. Immun., 5*:55-59.

Angus, T. A. (1968). *World Rev. Pest Control., 7*:11-26.

Barras, D. J., Wiygul, G., and Vinson, S. B. (1969). *Comp. Biochem. Physiol., 31*:707-714.

Bennett, G. A., and Shotwell, O. L. (1970). *J. Invertebr. Pathol., 16*:157-164.

Bennett, G. A., Shotwell, O. L., Hall, H. H., and Hearn, W. R. (1968). *J. Invertebr. Pathol., 11*:112-118.

Bergold, G. H. (1963). In "Insect Pathology. An Advanced Treatise." (E.A. Steinhaus, ed.), Vol. I, pp. 413-456. Academic Press, New York.

Boman, H. G., Nilsson, I., and Paul, K. (1973). *Abstracts 5th Intern. Colloq. Insect Pathol. and Microbial Control*, Oxford, p. 99.

Boman, H. G., Nilsson, I., and Rasmuson, B. (1972). *Nature, 237*:232-235.

Briggs, J. D. (1958). *J. Exp. Zool., 138*:155-188.

Briggs, J. D. (1964). In "The Physiology of Insecta." (M. Rockstein, ed.), Vol. IV, pp. 259-283. Academic Press, New York.

Bucher, G. E. (1960). *J. Insect. Pathol., 2*:172-195.

Burgdorfer, W. (1970). *Amer. J. Trop. Med. Hyg., 19*:1011-1014.

Burges, H. D., and Hussey, N. W. (1971). "Microbial Control of Insects and Mites." Academic Press, London, New York.

Chadwick, J. S. (1967). *Fed. Proc. Fed. Amer. Soc. Exp. Biol., 26*:1675-1679.

Chadwick, J. S. (1968). *In Vitro, 3*:120-128.

Chadwick, J. S. (1970). *J. Invertebr. Pathol., 15*:455-456.

Chadwick, J. S. (1971). *J. Invertebr. Pathol., 18*:117-120.

Chamberlain, R. W. and Sudia, W. D. (1961). *Ann. Rev. Entomol., 6*:371-390.

Day, N. K. B., Gewarz, H., Johannsen, R., Finstad, J., and Good, R. A. (1970). *J. Exp. Med., 132*:941-950.

Dulmage, H. T. and Rhodes, R. A. (1971). In "Microbial Control of Insects and Mites." (H.D. Burges and N.W. Hussey, eds.), p. 57. Academic Press, New York, London.

Dutky, S. R. (1940). *J. Agr. Res., 61*:57-68.

Fast, P. G., and Angus, T. A. (1965). *J. Invertebr. Pathol., 7*:29-32.

Faust, R. M. (1968). *J. Invertebr. Pathol., 11*:465-475.

Faust, R. M., and Dougherty, E. M. (1969). *J. Invertebr. Pathol., 13*:155-157.

Frings, H., Goldberg, E., and Arentzen, J. C. (1948). *Science, 108*:689-690.

Gingrich, R. E. (1964). *J. Insect Physiol., 10*:179-194.

Good, R. A., and Papermaster, B. W. (1964). In "Advances in Immunology." (F.J. Dixon and J.H. Humphrey, eds.), Vol. IV, pp. 1-115. Academic Press, New York.

Grey, H. M. (1969). In "Advances in Immunology." (F.J. Dixon and H.G. Kunkel, eds.), Vol. X, pp. 51-104. Academic Press, New York.

Harvey, T. L., and Howell, D. E. (1965). *J. Invertebr. Pathol., 7*:92-100.

Heimpel, A. M. (1955). *Can. J. Zool., 33*:99-106.

Heimpel, A. M. (1967). *Ann. Rev. Entomol., 12*:287-322.

Heimpel, A. M., and Angus, T. A. (1959). *J. Insect. Pathol., 1*:152-170.

Heimpel, A. M., Angus, T. A. (1963). In "Insect Pathology. An Advanced Treatise." (E.A. Steinhaus, ed.), Vol. II, pp. 21-73. Academic Press, New York.

Heimpel, A. M., and Harshbarger, J. C. (1965). *Bacteriol. Rev., 29*:397-405.

Hildemann, W. H. (1974). *Life Sciences, 14*:605-614.

Hink, W. R., and Briggs, J. D. (1968). *J. Insect. Physiol., 14*:1025-1034.

Huger, A. (1963). In "Insect Pathology. An Advanced Treatise." (E.A. Steinhaus, ed.), Vol. I, pp. 531-575. Academic Press, New York.

Kamp, H. (1968). *Z. Vgl. Physiol., 58*:441.

Kawarabata, T., and Aizawa, A. (1967). *Proc. Joint U.S.-Japan Seminar on Microbial Control of Insect Pests*, Fukuoka, April, 1971, pp. 143-145.

Kawase, S. (1965). *J. Invertebr. Pathol., 7*:113-116.

Kawase, S., and Hayashi, Y. (1965). *J. Invertebr. Pathol., 7*:49-54.

Kellen, W. R., and Lindegren, J. E. (1968). *J. Invertebr. Pathol., 11*:104-111.

Komano, T., Himeno, M., Ohno, Y., and Onodera, K. (1966). *J. Invertebr. Pathol., 8*:67-74.

Kůcera, M., and Lysenko, O. (1968). *Folia Microbiol., 13*:288-294.

Liu, P. V., Abe, Y., and Bates, J. L. (1961). *J. Inf. Dis., 108*:218-228.

Lowry, O. H., Rosebrough, N. J., Farr, A. L., and Randall, R. J. (1951). *J. Biol. Chem., 193*:267-275.

Lysenko, O. (1963a). *J. Insect Pathol., 5*:78-82.

Lysenko, O. (1963b). *J. Insect Pathol., 5*:83-88.

Lysenko, O. (1963c). *J. Insect Pathol., 5*:89-93.

Lysenko, O. (1963d). *J. Insect Pathol., 5*:94-97.

Lysenko, O. (1964). *Proc. XII Intern. Congr. Ent., London*, 1964. (Section II, Insect Pathology, p. 717).

Lysenko, O. (1966). In "Insect Pathology and Microbial Control." (P.A. Van der Laan, ed.), pp. 219-237. North-Holland Publishing Co., Amsterdam.

Lysenko, O. (1972). *Folia Microbiol., 17*:228-231.

Lysenko, O., and Kůcera, M. (1968). *Folia Microbiol., 13*:295-299.

Madziara-Borusiewicz, K., and Lysenko, O. (1971). *J. Invertebr. Pathol., 17*:138-140.

Martigoni, M. E. (1964). *Ann. Rev. Entomol., 9*:179-206.

Martigoni, M. E., and Milstead, J. E. (1964). *J. Insect Pathol., 6*:517-531.

Martigoni, M. E., and Milstead, J. E. (1966). *J. Invertebr. Pathol., 8*:261-263.

Meinke, G., and Beck, R. S. (1970). *Proc. Soc. Exp. Biol. Med., 135*:360-363.

Mohrig, W., and Messner, B. (1968). *Sond. Biol. Zent., 87*:439-470.

Morgun, G. I. (1950). *Mikrobiol. Zh.* (Ukraine) *11*:43-50. (Abstract only seen *in Chem. Abstr., 46*:8776.)

Morris, O. N. (1962). *J. Insect Pathol., 4*:446-453.

Muller-Eberhard, H. J. (1968). In "Advances in Immunology." (F.J. Dixon and H.G. Kunkel, eds.), Vol. VIII, pp. 1-80. Academic Press, New York.

Muller Kogler, E. (1965). "Pilzkrankheiten bei Insekten." Parey, Berlin.

Osler, A. G., and Sandberg, A. L. (1973). *Progr. Allergy, 17*:51-92.

Paillot, A. (1933). "L'infection Chez les Insectes. Imprimerie de Trevoux," Paris.

Pye, A. K. (1973). "Current Topics in Comparative Pathobiology" *2*.

Raheja, A. K., and Brooks, M. A. (1971). *J. Invertebr. Pathol., 17*:136-137.

Rasmuson, B., Boman, H. G., and Gleerup, A. (1973). *Abstracts 5th Intern. Colloq. Insect Pathol. and Microbial Control*, Oxford, p. 98.

Rehacek, J. (1965). *Ann. Rev. Entomol., 10*:1-24.

Rehacek, J., Brezina, R., Kovacova, E., and Zupancicova, A. M. (1971). *Acta Virol.* (Prague) *15*:237-240.

Roberts, D. W. and Yendol, W. G. (1971). In "Microbial Control of Insects and Mites." (H.D. Burges and N.W. Hussey, eds.), pp. 125-149. Academic Press, New York, London.

Schwalbe, C. P., and Bouse, G. M. (1971). *J. Invertebr. Pathol., 18*:85-88.

Schwartz, P. H., and Townshend, B. G. (1968). *J. Invertebr. Pathol., 12*:288-293.

Shapiro, M. (1967). *J. Invertebr. Pathol., 9*:10-25.

Shapiro, M., and Ignoffo, C. M. (1971). *J. Invertebr. Pathol., 17*:327-332.

Sikorowski, P. P., Andrews, G. L., and Broome, J. R. (1971). *J. Invertebr. Pathol., 18*:167-168.

Smirnoff, W. A. (1971). *J. Invertebr. Pathol., 18*:296-297

Smith, K. M. (1963). In "Insect Pathology. An Advanced Treatise." (E.A. Steinhaus, ed.), Vol. I, pp. 457-497. Academic Press, New York.

Smith, K. M. (1967). "Insect Virology." Academic Press, New York.

Smith, K. M., and Xeros, X. (1954). *Parasitology, 44*:71-80.

Stairs, G. R. (1971) In "Microbial Control of Insects and Mites." (H.D. Burges and N.W. Hussey, eds.), pp. 97-124. Academic Press, New York, London.

Steinhaus, E. A. (1963). In "Insect Pathology. An Advanced Treatise." (E.A. Steinhaus, ed.), Vol. I, pp. 1-27. Academic Press, New York.

Steinhaus, E. A. (1969). *Fed. Proc. Fed. Amer. Soc. Exp. Biol., 28*:1810-1814.

Stepanchenok-Rudnik, G. I., Krasnobaeva, Z. N., Grokhovskaia, I. M., and Kryuchechnikov, O. N. (1972). *Med. Parazitol* (Mosk). *41*:279-281. (Abstract only seen in *Biol. Abstr.*, 1973-43623.)

Stephens, J. M. (1962a). *Can. J. Microbiol., 8*:597-602.

Stephens, June M. (1962b). *Can. J. Microbiol., 8*:491-499.

Stephens, June M. (1963a). In "Insect Pathology. An Advanced Treatise." (E.A. Steinhaus, ed.), Vol. I, pp. 273-297. Academic Press, New York.

Stephens, J. M. (1963b). *J. Insect Pathol., 5*:61-65.

Stephens, J. M., and Marshall, J. H. (1962). *Can. J. Microbiol., 8*:719-725.

Tanada, Y. (1971). In "Pest Control." (W.W. Kilgore and R.L. Doutt, eds.), pp. 31-88. Academic Press, New York.

Tanada, Y., and Watanabe, H. (1971). *J. Invertebr. Pathol., 17*:127-129.

Taylor, R. L. (1969). *J. Invertebr. Pathol., 14*:427-428.

van der Geest, L. P. S., and Craig, R. (1967). *J. Invertebr. Pathol., 9*:43-54.

van der Geest, L. P. S., and Wassink, H. J. M. (1969). *J. Invertebr. Pathol., 14*:419-420.

Watanabe, H. (1967). *J. Invertebr. Pathol., 9*:570-571.

Weiner, B. A., Kwolek, W. F., St. Julian, G., Hall, H. H., and Jackson, R. W. (1966). *J. Invertebr. Pathol., 8*:308-313.

Weinheimer, P. F. (1970). *Ala. J. Med. Sci., 7*:451-460.

Whitcomb, R., Shapiro, M., and Granados, R. R. (1974). In "The Physiology of Insecta." (M. Rockstein, ed.), Vol. 5. Academic Press, New York. (Still in Press)

Wittig, G. (1964). *J. Invertebr. Pathol., 7*:474-488.

Wyatt, G. R. (1974). Abstracts of Tropical Medicine and Parasitology. Study Section Workshop on Invertebrate Immunity.

Young, S. Y. (III), and Lovell, J. S. (1971). *J. Invertebr. Pathol., 14*:410-418.

Young, S. Y., and Scott, H. A. (1970). *J. Invertebr. Pathol., 16*:57-62.

Zernoff, V. (1931). *Ann. Inst. Past., 46*:565-571.

Ziprin, R., and Hartman, P. A. (1971). *J. Invertebr. Pathol., 17*:265-269.

THE HAEMOCOEL AS BARRIER TO PARASITE INFECTION IN INSECTS

A. Burns Weathersby

Entomology Department
University of Georgia
Athens, Georgia

It seems inappropriate to trace the history of this subject by way of an introduction since most of the published reports pertinent to the subject will be included in the various areas to be discussed later. It is appropriate though to give the parameters of barriers to parasite infection in insects and to indicate the limits of the subject to be considered.

The theme of this workshop on invertebrate immunity is concerned primarily with vector-parasite relations; therefore most examples of haemocoel barriers to infection will involve both the invertebrate and vertebrate hosts and relate to medical or verterinary diseases. Examples from other areas will be included only if they illustrate specific mechanisms.

There is very little information on haemocoelomic interactions between viruses, rickettsia, and bacteria and discussion of these agents of necessity will be limited.

An exhaustive literature review is not attempted for this paper, but a number of excellent reviews have been used extensively; Briggs (1974), Cantacuzene (1923), Chadwick (1967), Garnham (1964), Huff (1940), Jones (1962), Manson-Bahr (1921), Metchnikoff (1905), Nelson (1964), Paillot (1933), Poinar (1969), Steinhaus (1949), Stephens (1964), Taliaferro (1961), Tripp (1969), Weiser (1969), Wenyon (1926).

Since I did not compose the title but accepted it as one I could analyze and discuss, I should indicate my concept of "Barrier to Parasite Infection." A barrier is anything in the environment of the parasite that limits its progress or development. Such a barrier may be of many and varied forms and be all inclusive. The nature of barriers may be biological, biophysical, biochemical, and, in many instances, genetically related.

In this discussion reference will be made to susceptibility and immunity of certain animals to a specific parasite. There are varying degrees of susceptibility (Briggs, 1964; Huff, 1940; Steinhaus, 1949) but the lack of susceptibility should indicate immunity (freedom from disease). Since the literature has sanctioned varying degrees of immunity (Briggs, 1964), I propose the word *totimmune* to mean complete or absolute immunity.

Most parasites of vertebrates that have insects as biological vectors are directly or indirectly affected by the haemocoel of the vectors. The haemolymph supports every living cell in the body of the insect (Steinhaus, 1949). The factors that govern susceptibility (or immunity) may be located in the haemocoel or haemolymph whether or not the immunity is cellular or humoral.

Except for mechanical, transovarian, and transstadial transmission, most pathogens are introduced orally into the potential vector by means of blood or other body fluids. If the parasite survives the bloodclot, digestive juices, or peritrophic membrane (Garnham, 1964; Manson-Bahr, 1921), its first exposure to haemocoelomic influence is in contact with the cells of the midgut wall (Ball, 1965; Garnham, 1964; Weathersby, 1952). This tissue may serve as a physical or mechanical barrier to the parasite (Huff, 1940; Weathersby, 1952) and cause it to die or seek development in the anterior or posterior alimentary tract (Manson-Bahr, 1921).

In discussing the haemocoel as barrier to parasite infection we must differentiate between acquired and innate (or natural) immunity. Included under both headings is individual immunity which frequently is considered independently.

ACQUIRED IMMUNITY

The immune response in insects of an acquired nature pertains to individuals and has not received extensive attention (Briggs, 1964; Gingrich, 1964; Huff, 1940; Paillot, 1933; Stephens, 1964; Weiser, 1969). It most likely originates in the haemocoel of most, if not all, insects. This type of barrier is difficult to study in natural infections and most workers attempted to produce acquired immunity through artificial technique. Early attempts to produce such immunity were unsuccessful and frequently the techniques employed were those in use for mammals (Paillot, 1933). Evaluations of results also were based on mammalian immunity (Briggs, 1964; Huff, 1940; Stephens, 1964).

Several reviews that included acquired immune investigation have been thorough (Briggs, 1964; Huff, 1940; Stephens, 1964; Tripp, 1969; Wagner, 1961; Weiser, 1969) and there is little reason to cover that same material here since most of the reports concerned nonvector insects and pathogens that were of little relation to medical or verterinary parasites. Frequently the information on immunity was derived from investigations that were not immunologically oriented.

Acquired immunity in mammals and insects is different in many respects and early interpretations of the phenomonen probably were misleading (Stephens, 1964). In some instances the pathogens employed to produce immune responses were not highly infective for the insects.

Although insects are known to produce specific substances in response to foreign substances and antigenic stimulation (Brown, 1963; Chadwick, 1967; Huff, 1940; Weiser, 1969) insects appear to have poor powers of overcoming many parasites (Huff, 1940) and some authorities believe acquired immunity is completely absent in the mosquito (Garnham, 1964).

There may be some evidence of acquired immunity in *Triatoma infestans* to *Trypanosoma rangelli* since parasites reach a peak of 15 million/ml of haemolymph in 15 days, with a sharp decline to 1/2 million at 20 days (Hanson, 1974).

Many vertebrate parasites that might be expected to produce acquired immunity in insects either do not penetrate to the haemocoel or they produce very high mortality in the vectors (Weijers, 1958; Weijers and Willett, 1960; Zeledon, 1966). Frequently the vector (many mosquitoes) lives only sufficient time to transmit the disease one time and there is little opportunity to determine immunity.

Individual Acquired Immunity. The phenomenon of acquired immunity is of an individual nature and of little practical use to vectors. Some insects may have acquired the ability to transmit a parasite and pass it on to offspring but it is not likely that an antigenic stimulus as we think of it will affect the organism genetically.

INNATE IMMUNITY

The barrier to infection in the haemocoels of insect vectors that may be considered to be naturally immune are characteristic of the species (Huff, 1940, Stephens, 1964). In most instances innate or natural immunity is species immunity. Totimmune species do not exhibit individual variance but it has been shown that, with susceptible species, many individuals may escape infection (Garnham, 1964; Huff, 1927, 1929, 1930, 1934, 1940; Steinhaus, 1949) and they are categorized under individual immunity.

Individual Innate Immunity. Individual immunity in insects has received little attention (Garnham, 1964; Stephens, 1964) for it is difficult and time consuming to investigate the individual insects that escape infection. It has been possible to utilize individual immunity to an advantage in analyzing susceptiblity. Through selective breeding and genetic manipulation changes in susceptibility are possible (Huff, 1931; Kartman, 1953; Kilama and Craig, 1969; Rockefeller Foundation, 1950; Trager, 1942; Ward, 1963). Hereditary factors relate directly to haemocoelomic barriers. Although susceptibility may be altered through genetic technology, a genetically manipulated insusceptible strain is little different from a naturally immune strain unless it is shown that specific barriers or factors have been altered which render the strain different from its parents.

Species Immunity. Extrinsic factors that prevent an insect from exposure to a parasite are simple (Garnham, 1964) but frequently relegate numerous species to insusceptibility. There are numerous species of insects that have opportunities for infection yet the species is totimmune because of intrinsic barriers. These barriers are complex (Garnham, 1964) and we have just begun to understand the simplier relations. These immune species do not provide an environment that will permit development of the parasite. In most instances we know only that infectious stages are not produced and are not aware of where or how the barrier is functioning.

It is possible that mechanical factors in the mid-gut such as the peritrophic membrane in certain mosquitoes (Garnham, 1964) which inhibit the penetration of the haemocoel by infective microfilariae of *Brugia pahangi* (Esslinger, 1962), *Trypanosoma gambiense* in *Glossina palpalis* (Wijers, 1958), and *Onchocerca volvulus* in *Simulium damnosum* (Lewis, 1953) determine susceptibility of a species. It is possible that digestive enzymes (Janoff and Hawrylko, 1964; Kartman, 1953; Nelson, 1964) that are produced by cells that are haemocoel oriented may restrict the parasite development in certain species of insects. There may be other local or tissue factors that are responsible for immunity and which are produced by haemocoelomic tissues. Such conditions are logically referred to as local or *tissue immunity* (Taliaferro, 1961) if the factors are restricted to specific areas or tissue.

It was shown that ookinetes of *Plasmodium cathemerium* were produced in the insusceptible as well as susceptible mosquito but the ookinetes degenerated and died in the intestinal wall of the insusceptiblele *Culex pipiens* (Huff, 1934, 1940; Weathersby, 1954). It was considered that "in this natural immunity the barrier seems to be the intestinal wall" (Huff, 1940). In natural infections this is true. "The oocysts of *Plasmodium* probably develop in the wall of the mosquito stomach because that is the part of the body which it encounters first" (Ball, 1965; Weathersby, 1952). This also is where the ookinete first meets the adverse environment of the totimmune mosquito (Weathersby, 1952).

Tissue Immunity. It is now known that the factors that govern susceptibility of mosquitoes to the *Plasmodium* are not restricted to the stomach wall tissues but they are present throughout the haemocoel and in the haemolymph (Weathersby, 1952, 1954, 1960, 1963; Weathersby and McCall, 1968; Weathersby et al., 1971). All the stages of the *Plasmodium*, endogenous and exogenous, were capable of development to infectious stage when the stomach wall was bypassed by means of injections directly into the haemocoels of susceptible mosquitoes. No further development of these stages was evident in the haemocoles of refractory species (Weathersby, 1952, 1954, 1960, 1963; Weathersby and McCall, 1968). It was concluded that susceptibility of mosquitoes to malarial parasites was not of the tissue immune type.

Atreptic Immunity. Atreptic immunity is one of two other types of immunity that may be defined for insects. It has been called "atrepsins" (Ehrlich, 1910), "atrepsie" (Poinar, 1969; Zinsser, 1916; Zinseer et al., 1939), or "athreptic" (Taliaferro, 1929,

1961). I have chosen to use atreptic because of priority. There are few, if any, reports of atreptic immunity in insects.

Antiblastic Immunity. The type of immunity encountered when the parasites are killed as a result of antagonistic agents (toxins, antitoxins, antibodies, agglutinins, phagocytosis, encapsulation, melanization, lysis, etc.) is defined antiblastic immunity.

Parabiotic twinning of mosquitoes was utilized to determine the type of immunity displayed by mosquitoes to malaria (Weathersby, 1965). Infected *Aedes aegypti*, highly susceptible to *Plasmodium gallinaceum* were joined to immune *Culex pipiens*, thorax to thorax, by means of small glass capillaries so that they shared common haemolymph. If immunity was due to the lack of required substance in the *C. pipiens* these substances could be supplied by the susceptible *A. aegypti* and the parasites would develop in both species. On the other hand, if toxic substances were responsible for the immunity of *C. pipiens* they would be transported to the susceptible *A. aegypti* and kill the parasites there.

Sixteen *A. aegypti*, whose cage mates were above 90% infected, failed to become infected with *P. gallinaceum* when joined parabiotically to refractory *C. pipiens* (Weathersby, 1965, 1966). Additional results such as these (unpublished) have led to the conclusion that immunity of mosquitoes to malaria is *antiblastic*. This is a haemocoelomic barrier, it is systemic, and probably is the result of cellular activity.

Antiblastic immunity may be predominantly defensive in nature but at times it may be highly active. Most defensive measures against intrahaemocoelomic parasites may fall into humoral or cellular activities. Originally most investigators placed greater importance on cellular immunity factors (Stephens, 1964) but more recently humoral responses have increased in importance.

Humoral factors are of great importance in vertebrate immunity and in some instance are of greatest importance such as with malaria (Garnham, 1964). Any immune response in insects that is not attributable to the cell is considered humoral and usually implies the production of antibodies (Huff, 1940; Steinhaus, 1949; Stephens, 1964). Some authorities have indicated doubts about conventional antibodies in insects (Huff, 1940; Stephens, 1964).

Humoral immunity involving antibody factors such as complement, precipitins (Briggs, 1964), lyses, agglutinins, antitoxins (Chorine, 1929), opsonins (Paillot, 1920), have been indicated in very few insects and all examples have been from nonvector species (Huff, 1940; Stephens, 1964; Tripp, 1969; and Wesier, 1969). Most of the reports have not shown evidence of agglutinins in insect blood but a few investigators have reported agglutination (Feir and Walz, 1963; Gary et al., 1949; and Glasser, 1918).

Cellular factors are responsible for several types of haemocoelomic barriers to parasitic infections. Phagocytosis was considered by many invertebrate immunologists to be the chief cellular defense against foreign bodies in the haemocoel (Toumanoff, 1940; Weiser, 1969). It is not now considered to be the sole manifestation of insect immunity (Stephens, 1963).

Phagocytosis is exhibited in a variety of forms and, where considered in general, are powerful defenses against objects that are introduced into the haemocoel either naturally through the digestive tract wall or by artificial means. Phagocytosis begins immediately following the introduction of foreign bodies (Tripp, 1969). There are numerous types of haemocytes (Jones, 1962) which have been shown to engulf colloidal substances, india ink, carmine, mycetomes, bacteria, fungi, protozoa, vertebrate erythrocytes and gas bubbles (Jones, 1962). Several factors, including external ones, such as temperature (Ball, 1965; Chao and Ball, 1962; Hu 1939c; Ragab, 1949) and humidity (Garnham, 1964; Hu, 1939b; Ragab, 1949) may have adverse effects on phagocytosis. They vary also with species, stages, and age (Duxbury, et al., 1961; Huff, 1940; Terizan, et al., 1956; Wijers, 1958, 1960).

Ameboid cells or haemocytes (Jones, 1962; Toumanoff, 1940), either fixed (microphages) or wandering (macrophages) (Paillot, 1933) engulf foreign bodies (Chorine, 1929; Metchnikoff, 1905; Tobie, 1968) and, in general, phagocytize small particles such as bacteria and encapsulate objects that are too large to engulf.

Many of the reports of phagocytosis by single haemocytes have been on bacteria (Stephens, 1964; Tripp, 1969; Weiser, 1969) or other small organisms and resulted from experimental or artificial means of inoculation. These, for the most part, have not included biological vectors of vertebrate diseases. Recently several flagellates were shown to be phagocytised in *Rodnius prolixus* (Tobie, 1968) and phagocytic activity was demonstrated with *Plasmodium gallinaceum* in *Culex pipiens* (McCall, 1973).

The sporozoites of *Plasmodium* seem not to be subject to phagocytosis unless by chance this is the means by which they gain entrance into the cells of the salivary glands. Haemocytes will engulf vertebrate erythrocytes that have been injected into the haemocoels of mosquitoes (McCall, 1973). When the erythrocytes were infected with a *Plasmodium*, malarial pigment was evident in haemocytes for long periods of time (McCall, 1973). Malarial pigment and avian erythrocyte nuclei have been seen in a variety of haemocytes such as giant cells, plasmocytes, pericardial cells and phagocytes in fat body tissues. Of course the erythrocytic stages are not found in the haemocoel in natural infections. There is no evidence that oocysts and sporozoites are molested.

The presence of foreign bodies in the haemocoel seems to stimulate the wandering phagocytes and this may trigger reproduction of these cells. Phagocytes ingest numerous organisms and digestion may vary with the type of phagocyte. Many organisms may remain alive for long periods of time following phagocytosis. Several flagellates of *Crithidia arili* have been observed in phagocytes of *Drosophila virilis* with the flagellum protruding from the phagocyte (Schmittner and McGhee, 1970).

It seems unlikely that phagocytosis by single cells is of much importance in the immunity of insect vectors to vertebrate diseases except possibly with viruses and bacteria.

Encapsulation of foreign bodies in the haemocoel by means of mass attack of the haemocytes may be important in haemocoelomic resistance to disease organisms. This phenomenon has been described for numerous invertebrates including many vectors of diseases (Briggs, 1932; Burton, 1963; Esslinger, 1962; Highby, 1943; Hu, 1935, 1938, 1939a, 1939b; Huff, 1940; Kartman, 1953, 1956; Manson-Bahr, 1921; Metchnikoff, 1905; Phillips, 1966; Poinar, 1969; Salt, 1956, 1957a, 1957b, 1959, 1961, 1967; Schmittner and McGhee, 1970; Stephens, 1964; Yen, 1938).

Encapsulation of Filariodea has been reported in various species of mosquitoes for over 70 years (Manson-Bahr, 1921, Noe, 1901). The early reports were on the encapsulation of *Dirofilariae immitis* and *Wuchereria bancrofti* but later reports included *Dipetalonema, Thelazia, Habronema* and others. Although there have been questions regarding the significance of encapsulation it seems now to be accepted as a true defensive measure. It probably is not as effective in protecting the host as some other measures of defense.

Frequently the encapsulated agent may be viable for a long period of time and may continue its development if freed from its capsule.

Chitinization has been considered by some researchers separately from encapsulation and questioned as a means of defense against living parasites (Brug, 1932). Actually Chitinin may enclose a parasite but is different from the encapsulation in that the encapsulating material is a product of the cells and not composed of the cells themselves (Brug, 1932; Esslinger, 1962; Huff, 1940; Kartman, 1953, 1956; Manson-Bahr, 1921; Yoeli, et al., 1958).

Melanization sometimes is considered apart from strict encapsulation and chitinization. It also would be a product of cells and indeed may form a capsule about a parasite (Brug, 1932; Chorine, 1929; Esslinger, 1962; Hu, 1938; Kartman, 1956; Nappi, 1973; Salt, 1956, 1961, 1963).

Black objects associated with parasites in the haemocoels of adult mosquitoes have captured the attention of many investigators since Ross noticed that some oocysts of *Plasmodium* ceased development and changed into "black spores" (Garnham, 1964). Various interpretations have been given regarding the nature of Ross' black spores. Usually only a small percent of oocysts on the stomach deteriorate into black spores and these probably occur when a host is resistant or when temperatures are too high (Garnham, 1964). The change may be the result of melanin deposition rather than chitinization (Toumanoff, 1940) or attack by phagocytic cells.

Biochemical factors play a significant role as barriers to parasite development in the haemocoel of insects. Eventually they may be the most important factors of immunity to pathogens. This phase of the subject has received very little attention and has been thought to be one of the major areas for basic malaria research.

Many factors that permit parasites to develop in susceptible hosts are biochemical in nature. The metabolic constitution of the host must be adequate for the survival of the parasites. Certainly there are biochemical factors that are barriers to parasite development. The free amino acid constitution has been shown to relate to susceptibility of mosquitoes to malaria (Clark and Ball, 1952). The susceptibility of mosquitoes has been affected by various metabolites, carbohydrates, antibiotics, vitamins, hormones, bases and acids, (Ball, 1965; Terzian, 1950; Terzian, et al., 1952, 1953; Terzian, 1955; Terzian and Stahler, 1959).

Undetermined antiblastic factors (Weathersby, 1963, 1965, 1967; Weathersby and McCall, 1968) have been shown to be factors in the innate immunity of mosquitoes to malaria. An effect "which not only inhibits development but destroys the parasite" has been called a "poison" (Garnham, 1964). Lethal environments have been suggested for *Plasmodium brasilianum* in *Anopheles aztecus* and *Plasmodium gonderi* in *Anopheles maculipennis* since sporozoites are produced in the oocysts but are destroyed immediately in the haemolymph. There was no evidence of phagocytic activity (Garnham, 1964).

As mentioned previously, the factors that govern innate immunity of mosquitoes to *Plasmodium* are systemic and can be transferred to other mosquitoes by injection or by parabiosis.

These also may be effective through oral administration to the mosquitoes. Atrepsins (the lack of a required factor) could not be transferred in this manner. Various concentrations of whole body extracts of refractory mosquitoes, administered *per os*, reduced oocyst development in susceptible mosquitoes in direct proportion to the concentration (Weathersby, et al., 1971). It was surprising to note that whole body extracts of the susceptible mosquitoes also had an effect on the development of the oocysts. With 75% susceptible mosquito extract, oocyst development was decreased 45%, which was close to the 43% depression obtained with 25% refractory mosquito extract. It seems that both the susceptible and refractory mosquitoes have toxins that affect parasite development but in different concentrations.

Whole body extracts were not as concentrated as the pure haemolymph that was transferred from mosquito to mosquito in parabiosis; therefore, complete inhibition of parasite development was not observed as was seen in the parabiotic twinning experiments.

The factor(s) of susceptibility of mosquitoes to malaria is systemic, in the haemolymph, and is a chemical. It can be transferred directly or absorbed through the digestive system.

Sporozoites of *Plasmodium gallinaceum*, injected into the haemocoels of refractory *Culex pipiens*, may remain infective for 41 1/2 hours (Weathersby, 1967). Infectivity decreases sharply after 20 hours but prior to this time the sporozoites may produce high parasitemias in the vertebrate host and with shorter prepatent periods than sporozoites from the susceptible host. Evidence of increased infectivity for short term exposure of sporozoites to resistant haemolymph was seen in parabiotic experiments when long and narrow capillary tubes were used. Haemolymph exchange was inhibited and oocyst development proceeded in the susceptible host. Sporozoites passed to the resistant species and were capable of infecting the vertebrate host.

Sporozoites in capillary tubes next to the resistant host were observed to be very active. Salivary glands that contained sporozoites of *P. gallinaceum* were dissected in susceptible *A. aegypti* haemolymph and then surrounded by refractory *C. pipiens* haemolymph. Sporozoites expressed from the glands so that they traversed both types of haemolymph were irritated or agitated to great activity in the refractory area. This stimulation for a short time may cause the sporozoites to be highly infective.

This unknown barrier in the haemolymph of mosquitoes is toxic, in high concentrations, to the *Plasmodium*. At low concentrations or exposure for short periods of time it is stimulating to the parasites. It is systemic and chemical in nature. Such a factor(s) or barrier may be effective in other invertebrate parasite relations.

CONCLUSIONS

There are numerous barriers to parasite infection in the haemocoels of insects. One type of defense may be effective against certain types of organisms but may be ineffective against others. Immune factors so efficient in the vertebrate host may not be important in invertebrates. Many of the barriers that have been exposed seem not to be responsible for totimmunity of certain species to given parasites. There are many biophysical approaches that need to be made for exposing the interactions between the parasite and its host. The biochemical basis for immunity has tremendous potential but the field has scarcely been touched. There are great opportunities in this area and this may be the most important direction for research in barriers to parasite infection in insects.

REFERENCES

Ball, G. H. (1965). Adaptation of the malarial parasite, particularly to its insect host. In "Host-parasite Relationships." (McCauley, J. E., ed.). Corvallis, Oregon State University Press, pp. 73-96.

Briggs, J. D. (1964). Immunological responses. In "The Physiology of Insecta." (Rockstein, M., ed.). New York, Academic Press, Inc., *3*:259-283.

Brown, K. N. (1963). The antigenic character of the "*Brucei*" trypanosomes. In "Immunity to Protozoa." (Garnham, P. C. C., A. E. Pierce, and I. Roitt, eds.). Oxford, Blackwell Scientific Publications, pp. 204-212.

Brug, S. L. (1932). Chitinization of parasites in mosquitoes. *Bull. Ent. Res., 23*:229-231.

Burton, G. J. (1963). Encapsulation of *Wuchereria bancrofti* in seven species of mosquitoes in British Guiana. *Amer. J. Trop. Med., 12*:870-876.

Cantacuzene, J. (1923). Le problem de l'immunite ches les invertebres. *Comptes Rendus Societe de Biologie*. Paris (Celebr. 75e Ann) 48-119.

Chadwick, J. S. (1967). Serological responses of insects. *Federation Proceedings, 26:*126-127.

Chorine, V. (1929). Immunité antitoxique ches les chenilles de *Gallirea mellonella. Ann. Inst. Pasteur, 43*:955-958.

Chao, J. and G. H. Ball. (1962). The effect of low temperature on *Plasmodium relictum* in *Culex tarsalis. Ann. Parasit., 48*:252-254.

Clark, E. W. and G. H. Ball. (1952). The free amino acids in the whole bodies of culicid mosquitoes. *Exper. Parasit., 1*:339-346.

Duxbury, R. E., A. P. Moon, and E. H. Sadun. (1961). Susceptibility and resistance of *Anopheles quadrimaculatus* to *Dirofilaria uniformis. J. Parasit., 47*:687-691.

Ehrlich, P. (1910). "Studies in Immunity." John Wiley and Sons, N. Y. pp. 685.

Esslinger, J. H. (1962). Behavior of microfilariae of *Brugia pahangi* in *Anopheles quadrimaculatus. Amer. J. Trop. Med., 11*:749-758.

Feir, D. and M. A. Walz. (1963). An agglutinating factor in insect haemolymph. *Ann. Ent. Soc. Amer., 57*:388.

Gary, N. D., C. I. Nelson, and J. A. Munro. (1949). Serological evidence of resistance of larvae and workers to *Bacillus* larvae. *J. Econ. Entomol., 41*:661-662.

Garnham, P. C. C. (1964). Factors influencing the development of protozoa in their arthropodan hosts. In "Host-Parasite Relationships in Invertebrate Hosts." (Taylor, A. E. R., ed.). Oxford, Blackwell Scientific Publications, pp. 33-50.

Gingrich, R. E. (1964). Acquired humoral immune response of the large milk-weed bug, *Oncopeltus fasciatus* (Dallas), to injected materials. *J. Insect. Physiol., 10*:179-194.

Glaser, R. W. (1918). On the existence of immunity principles in insects. *Psyche, 25*:39-46.

Hanson, W. L. (1974). Personal Communication.

Highby, P. R. (1943). Mosquito vectors and larval development of *Dipetalonema arbuta* Highby (Nematoda) from the porcupine, *Erethizon dorsatum. J. Parasit., 29*:243-252.

Hu, S. M. K. (1935). Studies on the susceptibility of Shanghai mosquitoes to experimental infection with *Wuchereria bancrofti* Cobbold. III. *Culex tritaeniorhynchus* Giles. *Peking Natural History Bulletin., 10*:39-43.

Hu, S. M. K. (1938). Studies on the susceptibility of Shanghai mosquitoes to experimental infection with *Wuchereria bancrofti* Cobbold. V. *Culex fuscanus* Wiedemann. *Peking Natural History Bulletin, 13*:39-52.

Hu, S. M. K. (1939a). Studies on the susceptibility of Shanghai mosquitoes to experimental infection with *Wuchereria bancrofti* Cobbold. VII. *Culex vorax* Edwards. *Peking Natural History Bulletin, 13*:287-292.

Hu, S. M. K. (1939b). Studies on the susceptibility of Shanghai mosquitoes to experimental infection with *Wuchereria bancrofti* Cobbold. VIII. *Culex bitaeniorhynchus* Giles. *Peking Natural History Bulletin, 14*:15-22.

Hu, S. M. K. (1939c). Observations of the development of filarial larvae during the winter season in Shanghai region. *Amer. J. Hyg., 29*:67-74.

Huff, C. G. (1927). Studies on the infectivity of plasmodia of birds for mosquitoes, with special reference to the problem of immunity in the mosquito. *Amer. J. Hyg., 7*:706-734.

Huff, C. G. (1929). The effects of selection upon susceptibility to bird malaria in *Culex pipiens* Linn. *Ann. Trop. Med. Parasit., 23*:427-443.

Huff, C. G. (1930). Individual immunity and susceptibility of *Culex pipiens* to various species of bird malaria as studied by means of double infectious feedings. *Amer. J. Hyg., 12*:424-441.

Huff, C. G. (1931). The inheritance of natural immunity to *Plasmodium cathemerium* in two species of *Culex. J. Prevent. Med., 5*:249-259.

Huff, C. G. (1934). Comparative studies on susceptible and insusceptible *Culex pipiens* in relation to infection with *Plasmodium cathemerium* and *P. relictum. Amer. J. Hyg., 19*:123-147.

Huff, C. G. (1940). Immunity in invertebrates. *Physiol. Rev., 20*:68-88.

Jafri, R. H. (1961). Interaction of protozoan infections, toxins and radiation in insects. In "Progress in Protozoology." (Ludvik, J., et al., eds.). Prague, Acad. Publishers, pp. 510-515.

Janoff, A. and E. Hawrylko. (1964). Lysosomal enzymes in invertebrate leukocytes. *J. Cell. Physiol., 63*:267-271.

Jones, J. C. (1962). Current concepts concerning insect hemocytes. *Amer. Zool., 2*:209-246.

Kartman, L. (1953). Factors influencing infection of the mosquito with *Dirofilaria immitis* (Leidy, 1856). *Exp. Parasit., 2*:27-78.

Kartman, L. (1956). Notes on the encapsulation of *Dirofilaria immitis* in the mosquito *Aedes aegypti. Am. J. Vet. Res., 17*:810-812.

Kartman, L. (1957). The vector of canine filariasis; A review with special reference to factors influencing susceptibility. *Rev. brasil. malariol., 5*:41.

Kilama, W. L. and G. B. Craig, Jr. (1969). Monofactorial inheritance of susceptibility to *Plasmodium gallinaceum* in *Aedes aegypti. Ann. Trop. Med.* and *Parasit., 63*:419-432.

Lewis, D. J. (1953). *Simulium damnosum* and its relation to onchocerciasis in the Anglo-Egyptian Sudan. *Bull. Ent. Res., 43*:597-644.

Manson-Bahr, P. H. (1921). "Manson's Tropical Disease." 7th Ed. London.

Metchnikoff, E. (1905). "Immunity in Infective Diseases." (F. G. Binnie Transl). 591 pp. Cambridge Univ. Press. London and New York.

McCall, J. W. (1973). Personal Communication.

Nappi, A. J. (1973). The role of melanization in the immune reaction of larvae of *Drosophila algonguin* against *Pseudeucoila brochei. Parasit., 66*:23-32.

Nelson, G. S. (1964). Factors influencing the development and behavior of filarial nematodes in their arthropodan hosts. In "Host-Parasite Relationships in Invertebrate Hosts." (Taylor, A. E. R., ed.). Oxford, Blackwell Scientific Publications, pp. 75-119.

Noe, G. (1901). Sul Ciclo evolutivo della *Filaria bancrofti* (Cobbold) e della *Filaria immitis* (Leidy). *Ric. Lab. Anat. norm. Univ. Roma., 8*:275-353.

Paillot, A. (1920). L'immunite acquise chex les insectes. *Compt. rend. soc. biol.*, *83*:278-280.

Paillot, A. (1933). "L'infection chex les insectes." 535 pp. G. Patissier, Imprimerle de Trévoux. Paris.

Phillips, J. H. (1966). Immunological processes and recognition of foreignness in the invertebrates. In "Phylogeny of Immunity." (Smith, R. T., P. A. Miescher, and R. A. Good, eds.). Gainesville, University of Florida Press, pp. 133-140.

Poinar, G. O., Jr., (1969). Arthropod immunity to worms. In "Immunity to Parasitic Animals." Vol. I. (G. J. Jackson, R. Herman, and I. Singer.). Appleton-Century-Croft. N.Y.

Ragab, H. A. (1949). Effect of humidities and temperatures on the size and number of oocysts of *Plasmodium gallinaceum* transmitted by a mosquito. *Nature*, *163*:643-643.

Rockefeller Foundation. (1950). Annual Report, p. 29.

Salt, G. (1956). Experimental studies in insect parasitism. VIII. Host reactions following artificial parasitization. *Proc. Roy. Soc. Biol.*, *144*:380-398.

Salt, G. (1957a). Experimental studies in insect parasitism. IX. The reactions of a stick insect to an alien parasite. *Proc. Roy. Soc. Biol.*, *146*:93-108.

Salt, G. (1957b). Experimental studies in insect parasitism. X. The reactions of some endopterygote insects to an alien parasite. *Proc. Roy. Soc. Biol.*, *147*:167-184.

Salt, G. (1960). Experimental studies in insect parasitism. XI. The haemocytic reaction of a caterpillar under varied conditions. *Proc. Roy. Soc. Biol.*, *151*:446-467.

Salt, G. (1961). The haemocytic reaction of insects to foreign bodies. In "The Cell and the Organism." (Ramsey, J. A., and V. B. Wigglesworth, eds.). Cambridge, Cambridge University Press, pp. 175-192.

Salt, G. (1963). The defence reactions of insects to metazoan parasites. *Parasitology*, *53*:527-642.

Salt, G. (1967). Cellular defense mechanisms in insects. *Fed. Proc.*, *26*:1671-1674.

Sande, M. van and D. Karcher. (1960). Species differentiation of insects by haemolymph electrophoresis. *Science N.Y.*, *131*:1103-1104.

Schmittner, S. M. and R. B. McGhee. (1970). Host specificity of various species of

Crithidia Leger. *J. Parasit.*, *56*:684-693.

Steinhaus, F. A. (1949). "Principles of Insect Pathology." 1st Ed. pp. 199-217. McGraw-Hill Book Co., Inc., New York.

Stephens, J. M. (1963). Bactericidal activity of haemolymph of some normal insects. *J. Invertebr. Path.*, *5*:61-65.

Stephens, J. M. (1964). Immunity in insects. In "Insect Pathology: An Advanced Treatise." (Steinhaus, E. A., ed.). New York, Academic Press Inc. Vol. 2, pp. 273-297.

Taliaferro, W. H. (1929). "The Immunology of Parasitic Infections." New York, The Century Co.

Taliaferro, W. H. (1961). Cellular and humoral factors in immunity to Protozoa. In "Immunity to Protozoa." (P. C. C. Garnham, ed.). Blackwell Scientific Publication, Oxford.

Terzian, L. A. (1950). The sulfonamides as factors in increasing susceptibility to parasitic invasion. *J. Infect. Dis.*, *87*:285-290.

Terzian, L. A., N. Stahler, and P. A. Ward. (1952). The effect of antibiotics and metabolites on the immunity of mosquitoes to malaria infection. *J. Infect. Dis.*, *90*:116-130.

Terzian, L. A., N. Stahler, and H. Miller. (1953). A study of the relation of antibiotics, vitamins and hormones to immunity to infection. *J. Immun.*, *70*:115-123.

Terzian, L. A. (1955). The comparative morphological and physiological effects of various drugs on the sporogonous cycle of *Plasmodium gallinaceum* in *Aedes aegypti. J. Cell. Physiol.*, *46*:279-300.

Terzian, L. A. and N. Stahler. (1959). Some inorganic acids, bases and salts as determinants of innate immunity in the mosquito. *J. Infect. Dis.*, *106*:45-52.

Terzian, L. A., N. Stahler, and F. Irrevere. (1956). The effects of aging, and the modifications of these effects, on the immunity of mosquitoes to malarial infection. *J. Immun.*, *76*:308-313.

Tobie, E. J. (1968). Fate of some culture flagellates in the haemocoel of *Rodnius prolyxus. J. Parasit.*, *54*:1040-1046.

Toumanoff, C. (1940). Les "Black spores" de Ross et la "dégénérsecence brune" des microfilaires chex les moustiques; conception nouvelle sur leur nature et origine possibles. Considérations sur la "chitinisation" défensive chez les insectes. *Rev. Med. Franc. d'Extrême-Orient, 4-5*:173.

Trager, W. (1942). A strain of the mosquito *Aedes aegypti* selected for susceptibility to the avian malaria parasite *Plasmodium lophurae. J. Parasit., 28*:457-465.

Tripp, M. R. (1969). General mechanisms and principles of invertebrate immunity. In "Immunity to Parasitic Animals" (G. J. Jackson, R. Herman, and I. Singer, eds.). pp. 111-128, Vol. I. Appleton-Century-Crofts.

Wagner, R. R. (1961). Acquired resistance to bacterial infection in insects. *Bact. Rev., 25*:100-110.

Ward, R. A. (1963). Genetic aspects of the susceptibility of mosquitoes to malarial infection. *Exper. Parasit., 13*:328-341.

Weathersby, A. B. (1952). The role of the stomach wall in the exogenous development of *Plasmodium gallinaceum* as studied by means of haemocoel injections of susceptible and refractory mosquitoes. *J. Infect. Dis., 91*:198-205.

Weathersby, A. B. (1954). The ectopic development of malarial oocysts. *Exper. Parasit., 3*:538-543.

Weathersby, A. B. (1960). Further studies on the exogenous development of malaria in the haemocoels of mosquitoes. *Exper. Parasit., 10*:211-213.

Weathersby, A. B. (1963). Innate immunity of mosquitoes to malaria. *Proceedings of 7th International Congress of Tropical Medicine and Malaria, 5*:Div. B., 30-31.

Weathersby, A. B. (1965). Parabiotic twinning of mosquitoes. *Mosquito News, 25*:44-45.

Weathersby, A. B. (1966). Mosquito transmission of malaria. *Military Medicine, 131*:929-931.

Weathersby, A. B. (1967). The survival time for sporozoites of *Plasmodium gallinaceum* Brumpt in the haemocoels of refractory *Culex pipiens pipiens* L. (Diptera: Culicidae). *Journal Georgia Entomological Society, 2*:31-35.

Weathersby, A. B. and J. W. McCall. (1968). The development of *Plasmodium gallinaceum* Brumpt in the haemocoels of refractory *Culex pipiens pipiens* Linn. and susceptible *Aedes aegypti* (Linn.). *J. Parasit., 54*:1017-1022.

Weathersby, A. B., J. W. McCall, Hyong-Sun Ah, and M. W. Nelms. (1971). *Culex pipiens pipiens* and *Aedes aegypti*: Whole body extracts and development of *Plasmodium gallinaceum* in *Aedes aegypti. Exper. Parasit., 29*:42-46.

Weiser, Jaroslav. (1969). Immunity of insects to protozoa. In "Immunity to Parasitic Animals" (G. J. Jackson, R. Herman and I. Singer, eds.). pp. 129-147, Vol. I. Appleton-Century-Crofts.

Wenyon, C. M. (1926). "Protozoology" Vol. II. Baillere, Tindall and Cox, London.

Wijers, D. J. B. (1958). Factors that may influence the infection rate of *Glossina palpalis* with *Trypanosoma gambiense*. I. The age of the fly at the time of the infective feed. *Ann. trop. Med. Parasit.*, *52*:385-390.

Wijers, D. J. B. and K. C. Willett. (1960). Factors that may influence the infection rate of *Glossina palpalis* with *Trypanosoma gambiense*. II. The number and the morphology of the trypanosomes present in the blood of the host at the time of the infective feed. *Ann. trop. Med. Parasit.*, *54*:341.

Yen, C. H. (1938). Studies on *Dirofilaria immitis* Leidy, with special reference to the susceptibility of some Minnesota species of mosquitoes to the infection. *J. Parasit.*, *24*:189-205.

Yoeli, M., N. Alger and H. Most. (1958). Studies in Filariasis. I. The behavior of microfilariae of *Dirofilaria immitis* in the wax moth larvae (*Galleria mellonella*). *Amer. J. Trop. Med.*, 7:249.

Zeledon, R. and E. de Monge. (1966). Natural immunity of the bug *Triatoma infestans* to the protozoan *Trypanosoma rangeli*. *J. Invert. Path.*, *8*:420-424.

Zinsser, H. (1916). "Infection and Resistance." The MacMillan Co., N.Y.

Zinsser, H., J. F. Inders, and L. D. Fathergill. (1939). "Immunity, Principles and Application in Medicine and Public Health." The MacMillan Co., N.Y.

HEMOLYMPH COMPONENTS IN INVERTEBRATE IMMUNITY

M. A. Stirewalt
Biomedical Research Institute
American Foundation for Biological Research
Rockville, Maryland

Several conditions indicate that today's research environment is favorable to productive study of the roles of the components of hemolymphs of invertebrates in their immune reactions to invading organisms. For example, there is presently a predominant interest in immunity, whether protective or not, in the research community. Trained and interested immuno-zoologists are active in the field. The double line of evolutionary animal development may provide unusual variability among invertebrates with respect to their immune reactions and this applies to the role of hemolymph. The unanswered questions are legion.

Findings presented this afternoon suggest that the state of the art requires that each host type must be investigated individually as a reactor to its invading viral, bacterial, fungal, protozoan and metazoan parasites. Each invertebrate group, perhaps even closely-related species, may have developed characteristic immune reactions. Several basic questions have been especially emphasized by the speakers who have addressed themselves to the topic of the role of hemolymph in invertebrate immunity. What are the composition, variability and functions of normal hemolymphs? What physical, biochemical and serological changes occur in hemolymph with infection? What are the effects of physical and chemical factors on these changes? Are the antibody-like humoral factors of invertebrates similar to the antibodies of vertebrates? Where are the antibody-like components of hemolymph synthesized? What are the interreactions of the various cell types of the hemolymph with each other and with invading organisms? How do the antibody-like humoral factors influence the cellular interactions?

Answers to some of these questions have been given this afternoon for a few ticks, insects and molluscs. It is certain, however, that the synthesis of principles of invertebrate immunity will be possible only after intensive examination of the immune reactions of many species of invertebrate hosts under myriad conditions of infection.

IV. Vector Destruction

PARASITE ENCAPSULATION IN INSECTS

A. J. Nappi

Biological Sciences
State University of New York
Oswego, New York

The defense reactions of multicellular organisms against various infectious and oncogenic agents are adaptations of great complexity. Of fundamental importance in the evolution of effective immunologic mechanisms is the development of a coordinated system of cellular communications which involves the ability of host cells to differentiate between 'self' and 'not-self' and to react appropriately to eliminate foreignness.

Vertebrates generally have evolved with adaptive immune mechanisms characterized by both humoral and cell-mediated responses. Humoral responses involve the proliferation and differentiation of certain lymphoid cells which synthesize and release antibodies (immunoglobulins). Cell-mediated immunity is caused by lymphocytes which apparently do not secrete antibody, but which are responsible for the rejection of most tissue grafts and for delayed hypersensitivity responses. Immune competence is presently attributed in large part to the presence of specific immunoglobulin antigen-binding receptors on the cell surface of lymphocytes, and to various collaborative responses of lymphocytes ('helper effect', 'carrier effect'). Unfortunately, much still remains to be learned of the physiology of the antigen receptor sites, cellular cooperation, and the mechanism of lymphocyte proliferation and differentiation in primary lymphoid organs (Ada, 1970; Eldelman and Gall, 1969; Metzger, 1970; Nossal and Ada, 1971; Feldmann and Nossal, 1972).

At present, invertebrate immune systems do not appear to be homologous to vertebrate mechanisms for distinguishing foreignness. In fact, if the immune reactions of vertebrates were used as evaluative criteria, insects and many other invertebrates would not be considered immunocompetent (Good and Papermaster, 1964, Saunders, 1970). Two major features of the immune responses of vertebrates lacking in insect immunity are specificity (antigen-antibody complementarity), and increased responsiveness to foreign substances as a result of prior anitgenic exposure (immunologic memory). Nevertheless, insects are capable of effectively discerning and combating foreignness and abnormal or effete host tissues.

Against various invading microbial organisms some insects produce relatively non specific substances (antimicrobial, antibacterial, lytic) that afford the host some defense. Unfortunately, virtually nothing is known of the origin and chemical nature of these humoral immune substances (Briggs, 1958, 1964; Stephens, 1959, 1962a,b, 1963a-d; Stephens and Marshall, 1962; Gingrich, 1964; Chadwick, 1967, 1970; Mohrig and Messner, 1968; Hink and Briggs, 1968, 1969; Chadwick and Vilk, 1969; Bakula, 1970, 1971; Boman et. al., 1972; Gotz, 1973; Whitcomb et al., 1974).

The cellular immune mechanisms of insects are mediated by blood cells, or hemocytes, which eliminate invading foreign organisms by phagocytosis and/or encapsulation. Encapsulation is a coordinated response involving the aggregation, adhesion, and flattening of hemocytes over foreign surfaces too large to be engulfed by individual cells. Generally, encapsulation reactions in insects are accompanied by the intra- and extracellular deposition of the pigment melanin (Salt, 1963, 1970; Poinar, 1969; Shapiro, 1969; Whitcomb et al., 1974).

Although there have been many important contributions to the study of cellular immunity in insects, fundamental questions concerning the mechanism of activation of the immune response, the mobility of hemocytes, and the specificity of cell interactions remain unanswered. This paper attempts to focus attention on these and other problems concerned with encapsulation reactions in insects.

CAPSULE FORMATION

Much of the available evidence indicates that capsules in insects are formed principally, if not exclusively, by hemocytes (Salt, 1970). However, other cells and tissues of the host may be incidentally incorporated into hemocytic capsules. In a few cases parasite encapsulation may result from the action of hypodermal cells (Schell, 1952; Burns, 1961; Courturier, 1963; Matz et al., 1971). However, it is not certain whether the participation of cells other than blood cells in the formation of capsules is a primary or secondary response.

The manner in which insect hemocytes form capsules around endoparasites has been carefully described by several workers, and has been reviewed by Salt (1963, 1970). The brief account which follows describes only some of the salient morphological and behavioral changes that occur in the hemocytes during capsule formation.

Encapsulation and phagocytic processes are fundamentally alike. The spreading or flattening of hemocytes during encapsulation may represent unsuccessful attempts of the 'dwarf' cells to engulf large objects. Presumably, the increase in the surface area of flattened hemocytes contributes to their ability to adhere to foreign surfaces and cohere with one another to form capsules. The hemocytes comprising a capsule may be arranged in distinct layers and regions forming a more or less fixed association of cells, many of which do not flatten, but retain their original shape and show no pathological changes. Some investigators have interpreted the structure of capsules as being syncytial, and formed by the fusion of hemocytes. In other cases hemocytes form capsules without losing their structural integrity.

Poinar et al. (1968) studied by means of electron microscopy the formation of melanotic capsules in beetle larvae of *Diabrotica* spp. in response to parasitization by the nematode *Filipjevivermis leipsandra*. The first hemocytes to make contact with the parasite lysed and released their cytoplasm over the cuticle of the parasite. The liberated cytoplasm entered the cuticular folds and was transformed into an electron dense layer, identified as melanin, which surrounded the nematode. The initial reaction was considered an autolysis resulting from contact of the hemocytes with the surface of the parasite and not brought about by any mechanical disruption due to movement of the nematode. However, the majority of cells that formed the capsule did not lyse, but exhibited various degrees of necrosis and formed many membrane-bound electron opaque inclusions. The latter were considered intracellular stages in the production of melanin. The frequency of the inclusions decreased toward the periphery of the capsule.

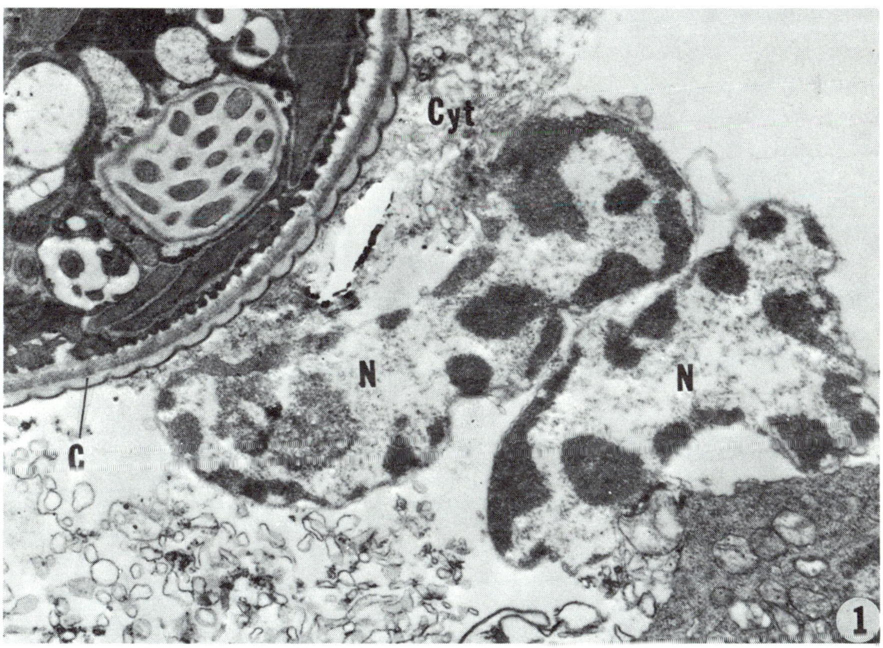

Figure 1. Isolated nuclei (N) and fragments of cytoplasm (Cyt) from lysed hemocytes of Diabrotica which had made initial contact with the cuticle (C) of the nematode parasite Filipjevivermis leipsandra 1-2 hours after infection. x 15,000. Courtesy of Dr. George O. Poinar, Jr.

Definitive capsules examined 72 hours after infection were comprised of 4 distinct regions. (1) An inner noncellular region composed of melanin. (2) A layer of entirely necrotic hemocytes, some of which were completely pigmented. (3) A third region consisting of extremely flattened hemocytes. (4) An outer region of loosely attached cells that closely resembled free hemocytes (Poinar, et al., 1968). Except for the cells attached to the layer of melanin, most of the hemocytes forming the capsule retained their integrity and a syncytium was not formed (Figures 1-6).

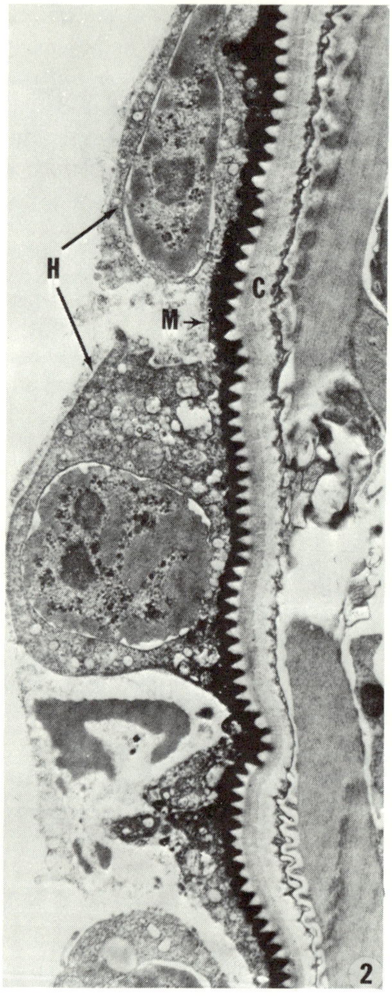

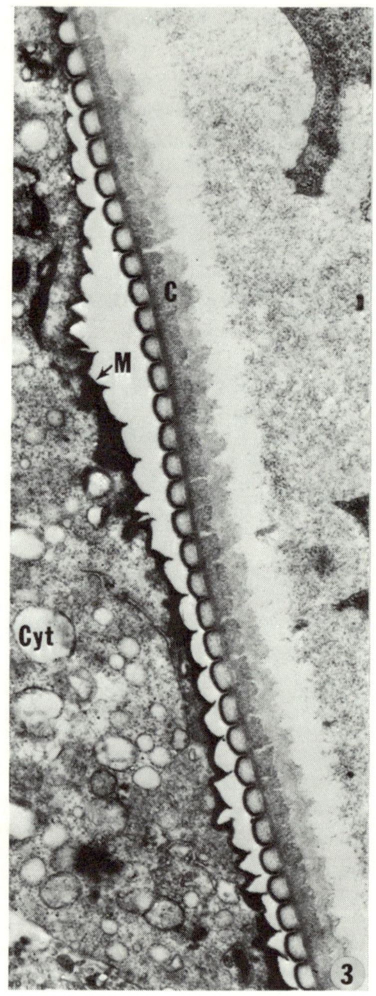

Figure 2. Necrotic hemocytes (H) from a third-instar parasitized Diabrotica larva attached to a layer of melanin (M) surrounding the cuticle (C) of Filipjevivermis leipsandra 6-8 hours after infection. x 12,000. Courtesy of Dr. George O. Poinar, Jr.

Figure 3. Layer of melanin (M) and fragments of highly vacuolated cytoplasm (Cyt) of host hemocytes of Diabrotica separated from the cuticle (C) of Filipjevivermis. x 24,000. Courtesy of Dr. George O. Poinar, Jr.

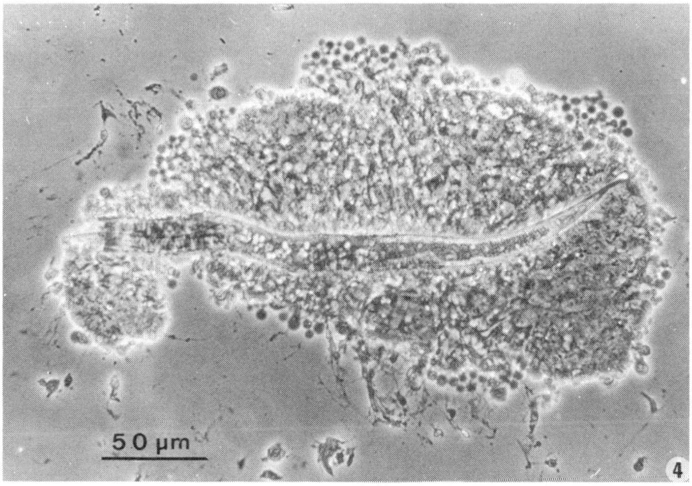

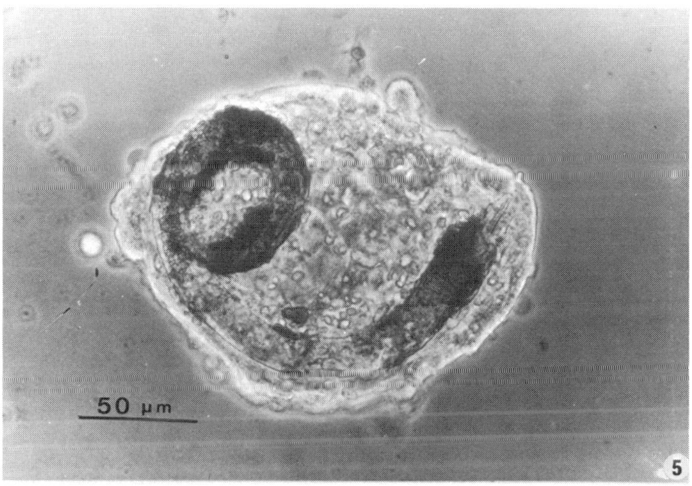

Figures 4 and 5. Stages in the encapsulation and melanization of the nematode Filipjevimermis by hemocytes of larvae of Diabrotica. Figure 4, 12-24 hours after infection. Figure 5, 48 hours after infection. Courtesy of Dr. Peter Götz.

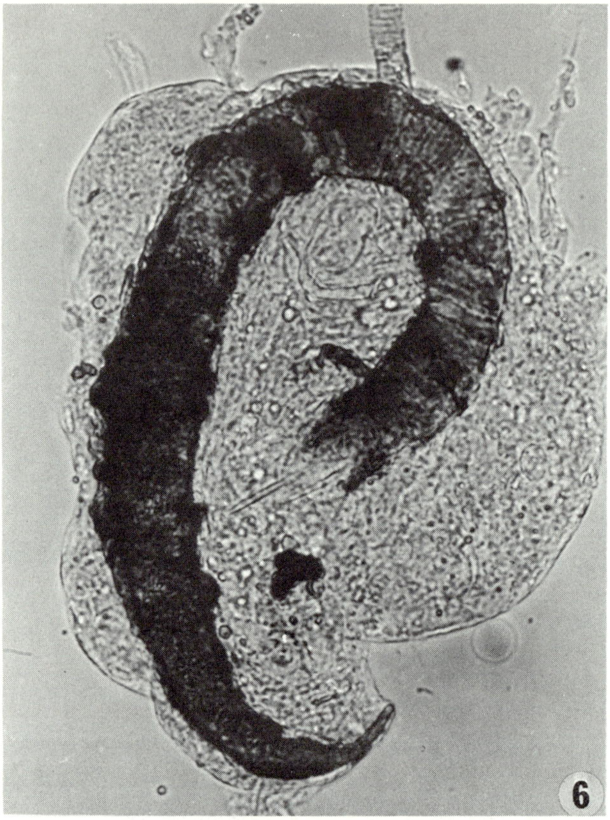

Figure 6. Encapsulated and melanized juvenile of the nematode Filipjevivermis 72 hours after penetration into the body cavity of a third-instar host larva of Diabrotica. x 190. Courtesy of Dr. George O. Poinar, Jr.

In a very detailed study of encapsulation in the cockroach, *Periplaneta americana* Misko (1972) found the hemocytes at the surface of the nematode, *Caenorhabditis briggsae,* to be more electron dense and flattened than those in other regions. In addition, the cells contained numerous membrane-bound cytoplasmic inclusions varying in size, shape and electron density. During the early stages of capsule formation the extracellular spaces at the surface of the parasite contained an amorphous material in which were found free nuclei and cytoplasmic fragments of disrupted hemocytes. As encapsulation progressed the membrane-bound vesicles became pigmented and coalesced extracellularly to form extensive masses of pigment that were deposited on the parasite. The cells comprising the capsule around *Caenorhabditis* did not fuse to form a syncytium (Figures 7-12).

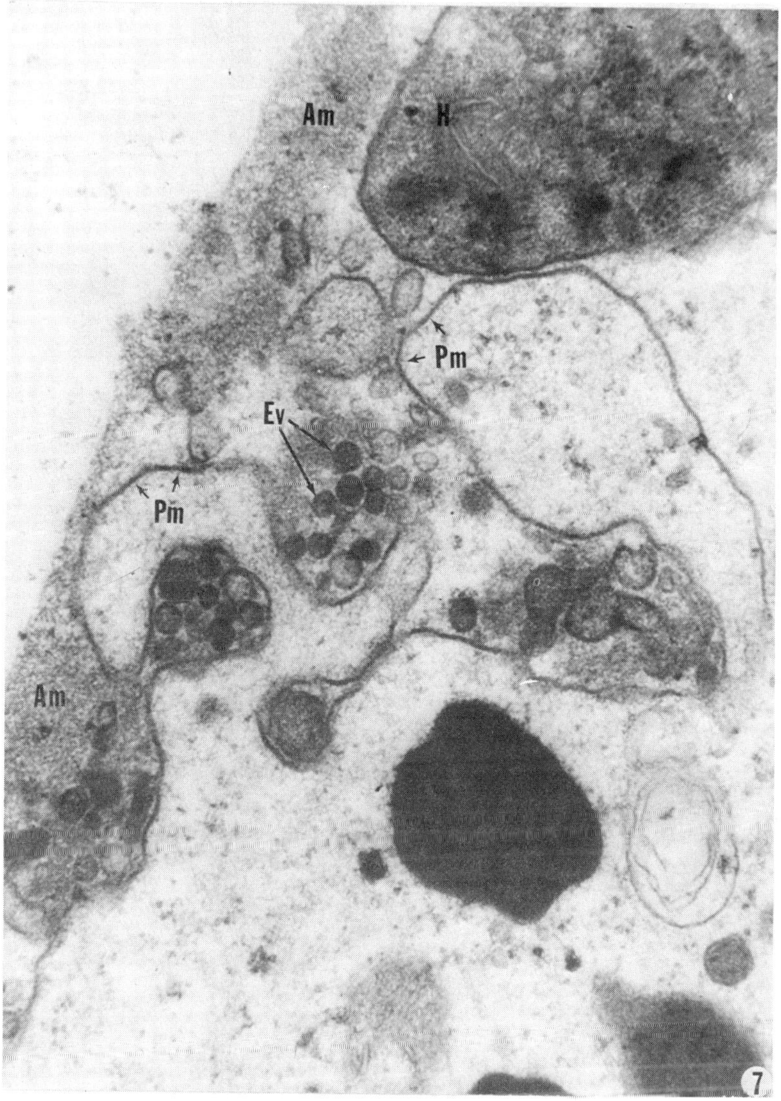

Figure 7. The innermost part of a developing capsule formed by Periplaneta americana around the nematode Caenorhabditis 1 hour after infection. The nematode is just out of the field of view at top left. Distinct extracellular vesicles (Ev) of varying size and density are present in an amorphous extracellular matrix (Am). Some vesicle appear to have discharged their contents into the extracellular space, while others may have coalesced with one another. Plasma membrane (Pm) of one hemocyte is incomplete. H, electron dense hemocyte. Courtesy of Dr. I. S. Misko.

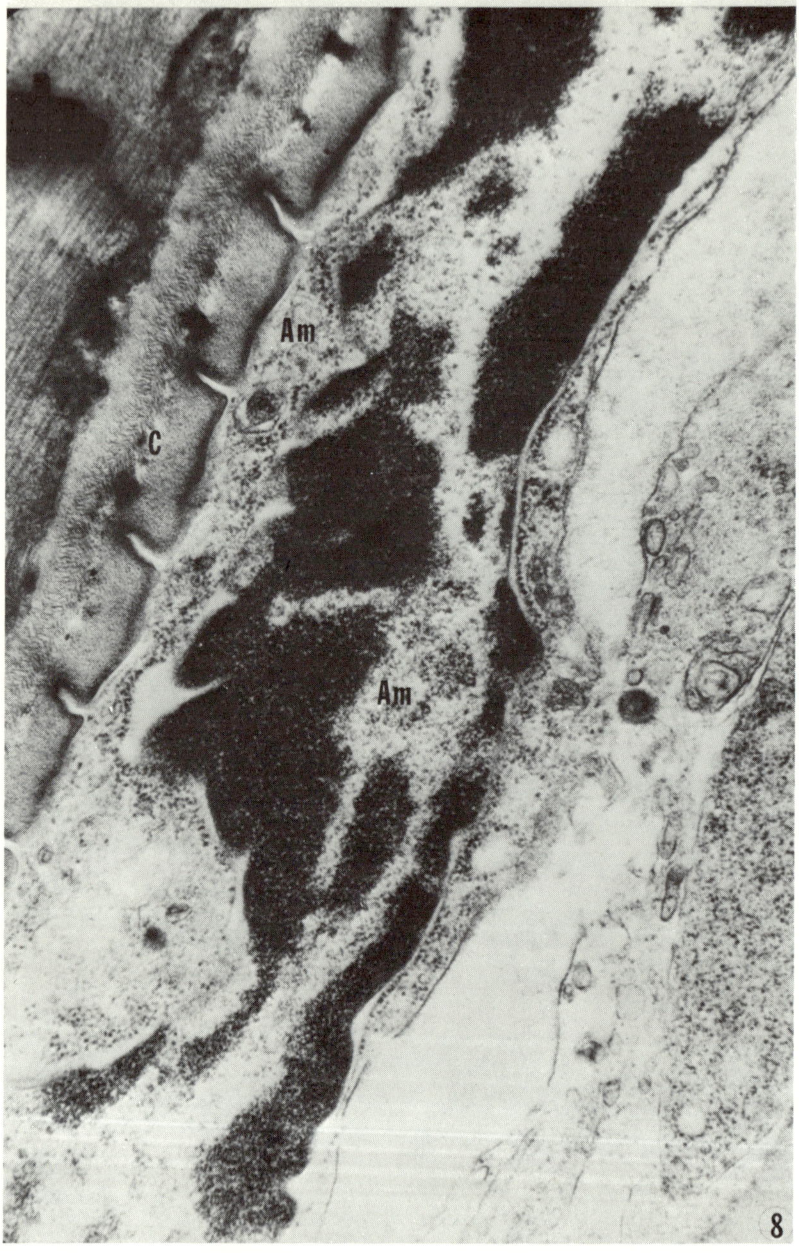

Figure 8. Transverse section of an early capsule formed by Periplaneta around Caenorhabditis 1 hour after infection. Conspicuous extracellular amorphous material (Am) is present near the cuticle (C) of the nematode. Courtesy of Dr. I. S. Misko

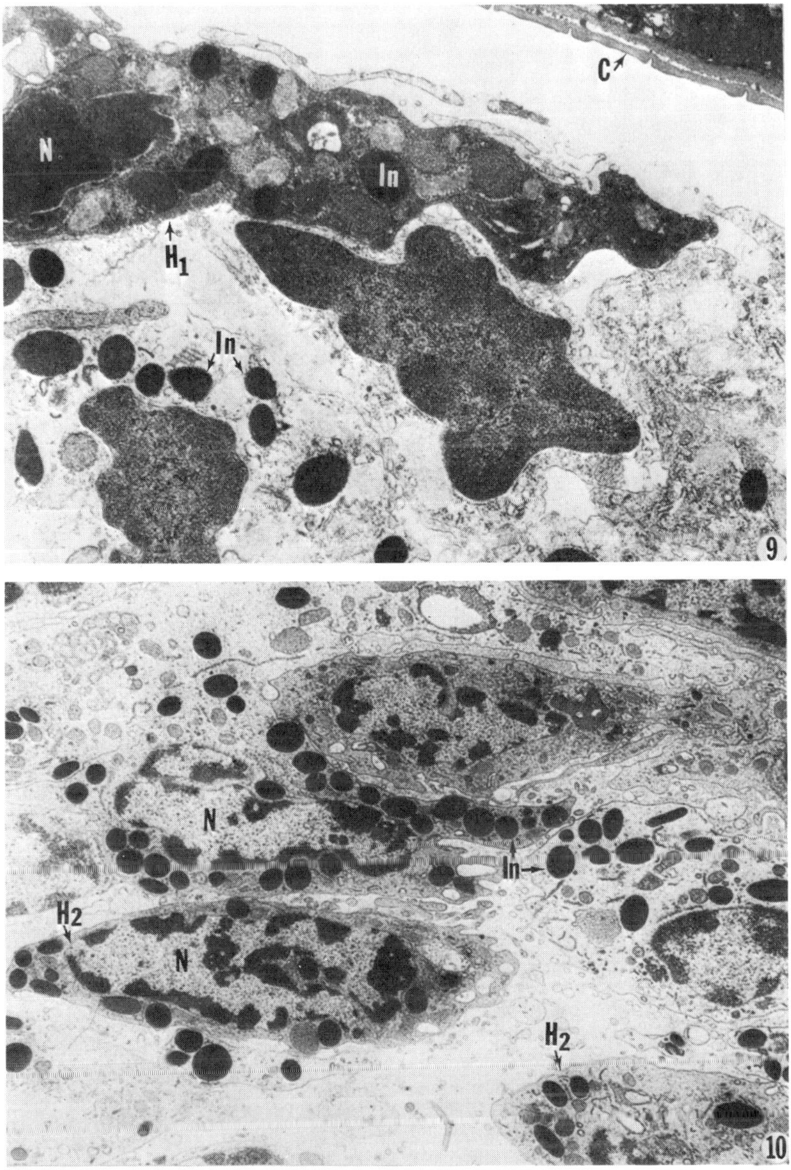

Figures 9 and 10. *Transverse sections through portions of a capsule formed by Periplaneta around Caenorhabditis 1 hour after infection. The innermost hemocytes (H_1) appear to be more electron dense and flattened than the outer hemocytes (H_2). The hemocytes comprising the capsule at this stage typically contain numerous electron dense inclusions (In). N, nucleus. Courtesy of Dr. I. S. Misko.*

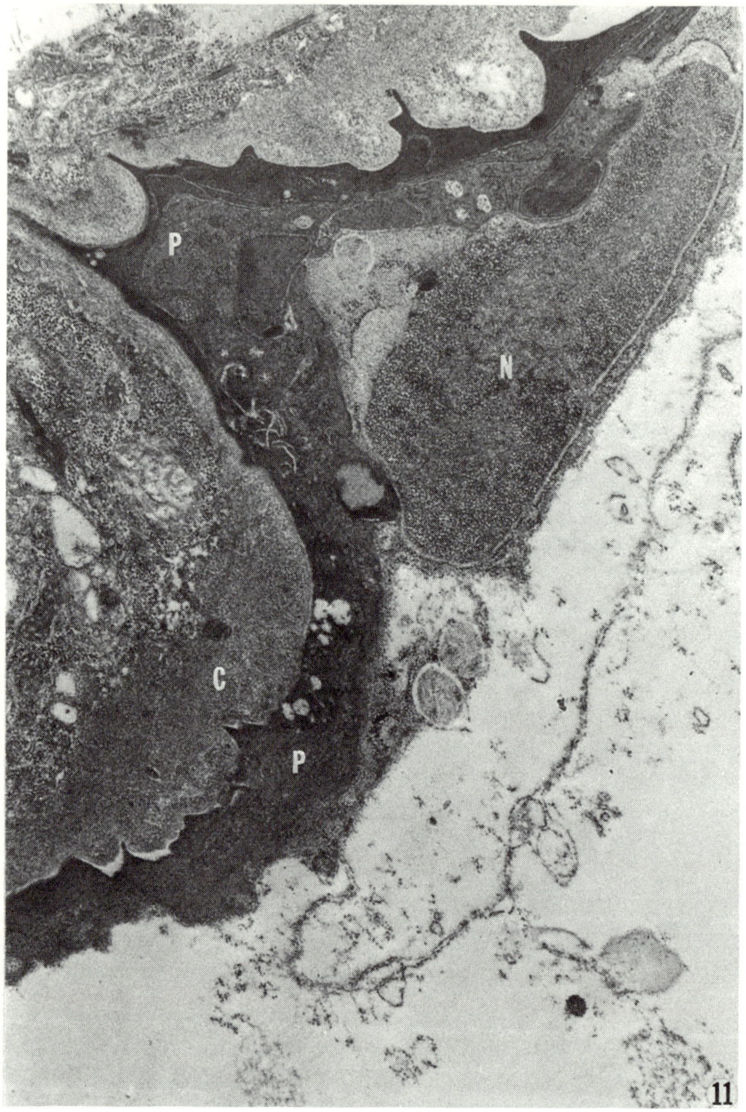

Figure 11. *Layer of pigment (P) closely applied to the cuticle of the nematode Caenorhabditis 12 hours after infection. N. nucleus of disrupted hemocytes. Courtesy of Dr. I. S. Misko.*

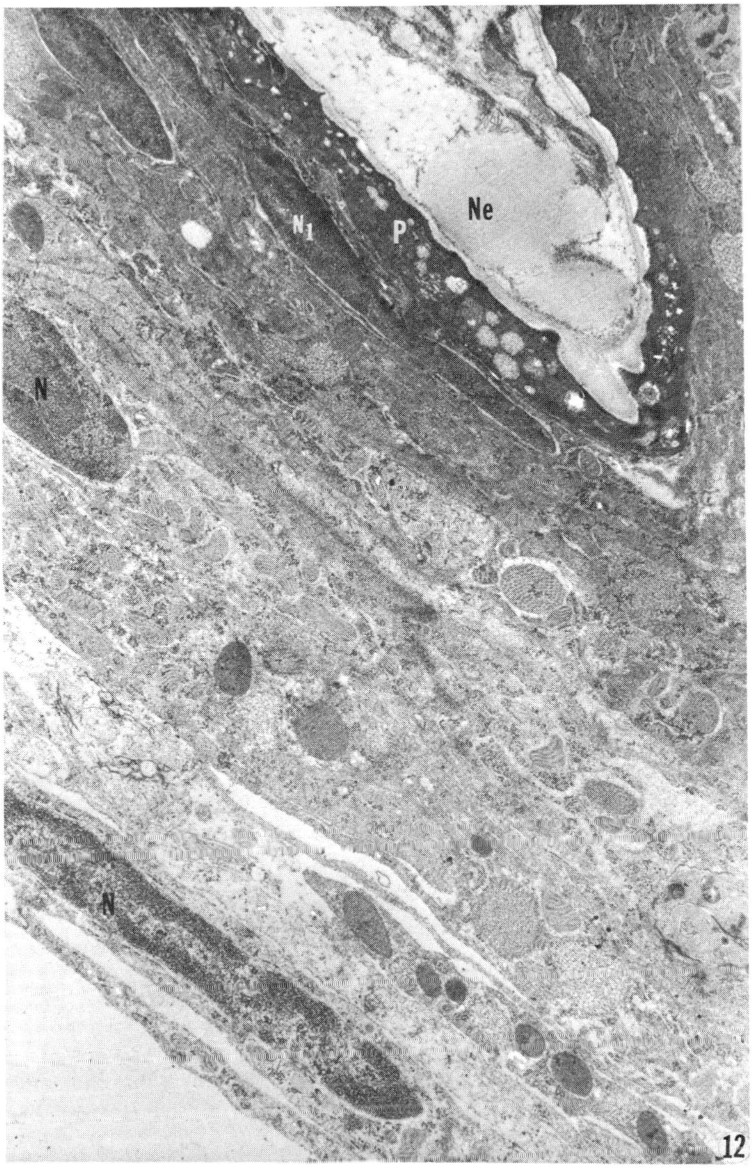

Figure 12. A transverse section of a fully pigmented capsule formed by Periplaneta around Caenorhabditis (Ne) 1 week after infection. The surface of the nematode is covered by a heterogeneous, electron dense layer of pigment (P). The nucleus (N_1) of the innermost hemocyte is exceptionally elongated and flattened. There is an obvious lack of electron dense inclusions in the hemocytes comprising the capsule at this stage. Courtesy of Dr. I. S. Misko.

When styrene particles were implanted into *Periplaneta* and later examined they were found encapsulated by hemocytes (Misko, 1972). The cells at the surface of the implant contained numerous electron dense microvesicles in their cytoplasm. It was proposed that these small vesicles coalesced to form an amorphous, non-pigmented layer at the surface of the implant which functioned in cellular adhesion. The innermost region of the capsules contained various cell structures but no discernable cell boundaries, indicating the formation of a syncytium (Misko, 1972).

Grimstone et al. (1968) studied the formation and fine structure of hemocytic capsules around small pieces of Araldite which were implanted into the hemocoel of the Mediterranean flour moth, *Ephestia kuehniella.* Aside from the absence of evidence of cell lysis and melanization, the capsule structure was similar to that found in nematode-infected insects (Poinar et al., 1968; Misko, 1972). The hemocytes retained their integrity throughout the thickness of the capsule and did not appear to form a syncytium. However, the authors noted the possibility that some fusion of cells might have occurred.

Completed capsules were comprised of 3 regions. Both inner and outer regions contained relatively isodiametric cells that were not markedly different from free blood cells. The cells of the innermost layer adhered to the implant and covered it completely. Only the cells in contact with the implant appeared necrotic. The cytoplasm contained many membrane-lined, electron dense inclusions termed cytolysomes. In addition, the majority of mitochondria were enlarged several times their normal diameter, contained only a few, irregular cristae, and lacked the dense matrix material characteristic of mitochondria in the free blood cells of *Ephestia.*

The electron microscope studies of capsule formation in *Ephestia* and *Diabrotica* indicate that hemocytes were added to the capsules as morphologically unaltered cells and only subsequent to their contact and adhesion with foreign surfaces did the cells undergo flattening and exhibit pathological changes. However, in larvae of *Drosophila* parasitized by the wasps *Pseudeucoila bochei* or *P. mellipes* marked morphological and behavioral changes occurred in the free hemocytes prior to capsule formation (Walker, 1959; Nappi and Streams, 1969, Nappi, 1973a). Plasmatocytes, the most numerous type of hemocyte, rapidly enlarged and differentiated into extremely flat, disc-shaped cells termed lamellocytes. The lamellocytes aggregated around the egg of the parasite and adhered to its surface forming a capsule. The deposition of melanin was believed to result from the lysis of oenocytoids or crystal cell (Nappi and Streams, 1969) which contain substrate(s) for phenoloxidase activity (Rizki and Rizki, 1959). The large number of lamellocytes found in the hemolymph during the early stages of parasitization before any visible evidence of encapsulation and melanization suggests the cells were stimulated to react before they made contact with the parasite.

One question that has never been completely answered is what causes hemocytes to cohere to form a capsule of several layers around a parasite when only the innermost layer of cells is in contact with the foreign surface? It may be that as the cells aggregate and are applied to the developing capsule their surface properties change so they become 'foreign' to other cells (Grimstone et al., 1967). However, this proposal can only account for the continued growth of the capsule, and not for the termination of capsule formation.

The idea that a chemical stimulus is conveyed from the foreign object through the inner layers of cells to circulating hemocytes was rejected by Salt (1970). He pointed out the diffusion of a chemical substance was conceivable when the object being encapsulated was a parasite, but the fact that inert objects were also encapsulated by many layers of hemoçytes argued against the proposal.

Perhaps concomitant with the establishment of initial contact with foreign surfaces intact or disrupted hemocytes release substances which promote cellular cohesion. In the case of inert objects substances released from cells in contact with the object would wane as additional cell layers are added. In the case of parasite encapsulation substances released from the parasite may augment those released from disrupted host hemocytes and cause cellular cohesion.

MELANIZATION AS A MECHANISM OF DEFENSE

Melanins constitute an important group of pigments formed principally by the oxidation and polymerization of phenols, such as tyrosine and dopa by enzymes collectively referred to as phenoloxidases (Figure 13).

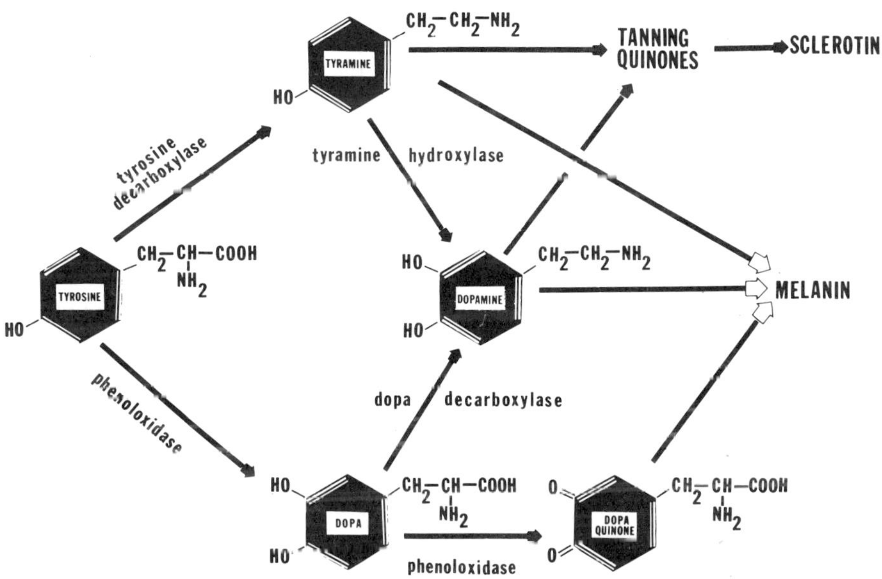

Figure 13. General scheme of tyrosine metabolism in insects.

Unfortunately, very little is known of the role of melanin in the immune reactions of insects. In some hosts melanotic capsules around parasites may serve merely as impermeable barriers and prevent the exchange of metabolites (Muldrew, 1953; Bronskill, 1960; Salt, 1963). However, several investigators have assigned a more active role for certain products or components of tyrosine metabolism in the immune reactions of insects (Eckstein, 1931; Ermin, 1939; Schneider, 1950; Schell, 1952; Stephens, 1962a,b; Heimpel and Harshbarger, 1965; Poinar et al., 1968; Taylor, 1969a; Götz, 1969, 1973, 1974; Nappi, 1970, 1973b; Brewer and Vinson, 1971; Stoffolano and Streams, 1971; Maier, 1973). Indeed, since quinones and other intermediary metabolites are potentially extremely reactive substances capable of inactivating various enzyme systems, biochemical reactions associated with melanin formation may serve not only in immune mechanisms, but also in certain pathophysiological disorders such as wound healing processes and the formation of melanotic lesions (pseudotumors or tumors). The merits of an inactive tyrosinase system in the development of insects and during injury or infection were considered by Cottrell (1964). Such a system could, upon activation, synthesize metabolites which would deactivate biologically active substances, isolate foreign material and repair wounds.

A few investigators have attempted to demonstrate the importance of melanization in the encapsulation reactions of insects. In *Carausius morosus* Salt (1956) blocked the deposition of melanin on the eggs of *Nemeritis canescens* by injecting phenylthiourea (PTU) into the hosts. When the hosts were later dissected and examined more live parasites were found in insects injected with PTU than in the controls. Although PTU prevented melanization, live parasites were found encapsulated by hemocytes. Brewer and Vinson (1971) found that injecting certain melanin inhibitors reduced the encapsulation rate in host larvae of *Heliothis zea* parasitized by *Cardiochiles nigriceps.* Recent attempts to block the melanization of the eggs of the hymenopterous parasite *Pseudeucoila bochei* were made by feeding PTU to host larvae of *Drosophila algonquin.* (Nappi, 1973b). In all hosts fed PTU, the immune reaction rates were less than those in the controls. Moreover, not only was the deposition of melanin prevented, but also hemocytic encapsulation.

According to Misko (1972) there is a drop in the endogenous levels of tyrosine in the hemolymph of starved cockroaches which causes reduced pigment synthesis. If the starved insects are injected with nematodes, the parasites are encapsulated but remain alive in non-pigmented capsules.

Among the more important and still unanswered questions about insect immune responses are those concerned with the origin and formation of capsule pigment. Although the deposition of melanin in some insect hosts may occur without visible evidence of the intervention of hemocytes (Bronskill, 1962; Esslinger, 1962; Götz, 1969; Poinar and Leutenegger, 1971), generally the melanization of parasites is associated with hemocytic responses.

From the available evidence it would appear that certain insect hemocytes normally synthesize and/or concentrate various amounts of phenols and enzymes associated with tyrosine metabolism at different times during development for possible use in cuticular sclerotization (Dennell, 1947; Jones, 1956, 1962, 1964; Vercauteren and Aerts, 1958; Rizki and Rizki, 1959; Decleir and Vercauteren, 1961; Crossley, 1964; Chadwick, 1966; Mills et al., 1968; Maier, 1969; Preston and Taylor, 1970; Whitehead, 1969, 1970; Mills and Whitehead, 1970; Whitehead and Mills, 1970; Pye and Yendol, 1972; Misko, 1972). Presumably melanization is normally controlled in intact insect cells because of a structural isolation of either enzyme and substrate, or the inactive proenzyme and activator (Ohnishi, 1953, 1954, 1958, 1959; Rizki and Rizki, 1959; Mitchell and Weber, 1965; Lai-Fook, 1966; Ashida and Ohnishi, 1967; Hackman and Goldberg, 1967; Evans, 1967, 1968a,b; Hackman, 1971). Although no organelles resembling vertebrate melanosomes have been identified in insect hemocytes, certain electron dense intracellular inclusions have been implicated as possible sites of synthesis and storage of pigment and/or phenolic precursors (Poinar et al., 1968; Taylor, 1969b; Grimstone et al., 1967; Misko, 1972). Poinar et al. (1968) suggested that the electron dense inclusions in *Diabrotica* hemocytes were intracellular stages in the production of melanin. The authors were unable to identify acid phosphatase activity within the inclusions, but they found enzyme activity on adjacent lysosomes, vesicles, and on the cell membrane. Acid phosphatase activity was found within the electron dense inclusions of *Ephestia* hemocytes which formed the inner layer of a capsule around implanted pieces of Araldite (Grimstone et al., 1967). Phenoloxidase, peroxidase, and acid phosphatase have been histochemically localized within certain hemocytes in *Periplaneta*, and the activity of each of these enzymes has been demonstrated in pigmented capsules formed around nematodes (Misko, 1972).

Some investigators have proposed that disrupted hemocytes release substances which initiate the localized synthesis of melanin at the surface of parasites (Poinar et al., 1968; Nappi and Streams, 1969). Unfortunately, the nature of the stimulus responsible for the lysis of free hemocytes or those in contact with foreign surfaces is not known.

Normally in insects undergoing a larval molt or puparium formation the metabolism of tyrosine is switched from the synthesis of various proteins to the production of N-acetyldopamine, a precursor of sclerotization (Karlson and Sekeris, 1962). The N-acetyldopamine is then oxidized to an O-quinone which tans cuticular proteins (Burnet, 1967; Hackman, 1971). Apparently two hormones influence tyrosine metabolism during sclerotization. Bursicon is believed to mediate cuticular tanning in some adult insects (Mills, 1967; Fraenkel and Hsiao, 1965; Fraenkel et al., 1966), possibly by increasing the permeability of the cell membrane of the hemocytes to tyrosine so that N-acetyldopamine is synthesized more rapidly (Mills and Whitehead, 1970). Bursicon may also inhibit the formation of melanin by shifting the metabolism of dopa towards the synthesis of N-acetyldopamine (Mills and Fox, 1972). A second hormone, ecdysone, is known to promote the formation of dopa decarboxylase (Hackman, 1971). This enzyme converts dopa to dopamine which may then be acetylated into N-acetyldopamine. Thus, ecdysone and bursicon may have a complementary role in stimulating the production and use, respectively, of tanning quinones (Wyatt, 1968).

Perhaps parasitization temporarily shifts the host metabolism of tyrosine and dopa from the synthesis of N-acetyldopamine towards the production of melanin. In larvae of *D. melanogaster* infected with *P. mellipes* the percentage of crystal cells, hemocytes known to contain phenoloxidases and tyrosine (Rizki and Rizki, 1959), decreases when the eggs of the parasites are melanized and encapsulated (Nappi and Streams, 1969). Also, there is considerable variation in the degree and rate of melanization of the hemolymph of parasitized larvae of *D. euronotus* (Nappi, 1970).

In larvae of *Galleria mellonella* Stephens (1962a,b) found bactericidal activity to be associated with the lack of visible melanization of the hemolymph. Although tyrosinase was present in immune blood, it was either less active or in lower concentration than normal. More recently, Pye and Yendol (1972) showed that when larvae of *Galleria* were injected with bacteria the percentage of hemocytes exhibiting phenoloxidase activity increased.

In adult cockroaches injected with the nematode *Caenorhabditis*, Misko (1972) found both mono- and diphenolic activities. In non-infected individuals only diphenoloxidase activity was present. Presumably, infected insects could use monophenoloxidase to convert available tyrosine in the hemolymph to precursors of capsule pigment or sclerotin. Of importance was the fact that capsule pigment was found to be comprised of both melanin and sclerotin.

Recently, Götz and Vey (1974) found that fungal spores injected into *Chironomus* larvae were rapidly enclosed in a 'capsule substance' that was composed of a polyphenol-protein complex (Figures 14-17). The formation of capsule substance (Humoral encapsulation) was attributed to the activity of phenoloxidases. Material for the formation of capsule substance was found not only in the cuticle, hypodermal cells, and hemolymph, but also appeared to be released from hemocytes when they lysed.

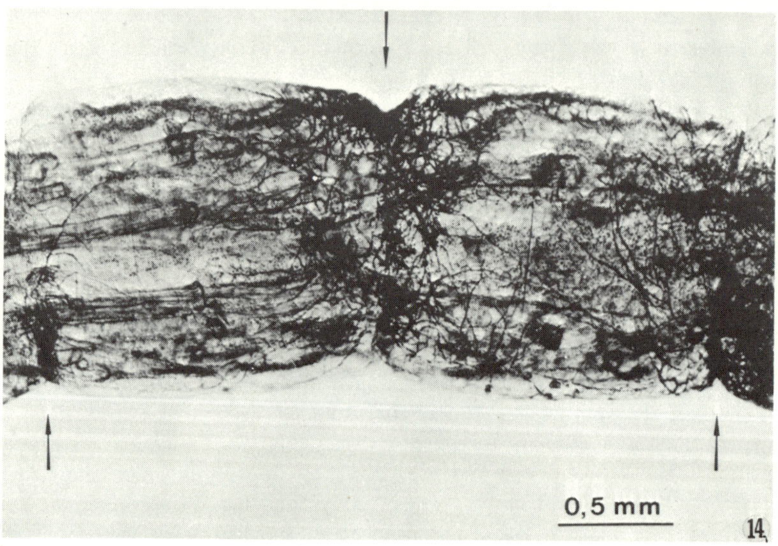

Figure 14. Humoral encapsulation (melanization) of the fungus Beaveria bassiana in a host larva of Chironomus luridus. Courtesy of Dr. Peter Götz.

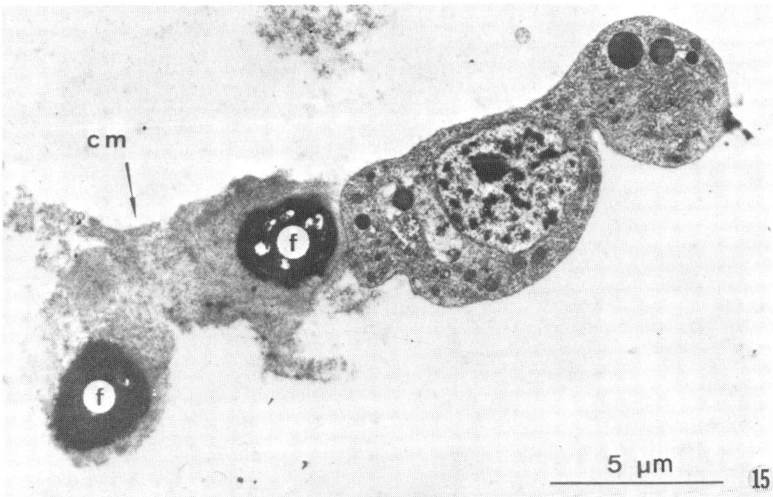

Figure 15. *Formation of capsule substance (Cm) by host larva of Chironomus (5 minutes after injection) around spores of Beaveria. Courtesy of Dr. Peter Götz.*

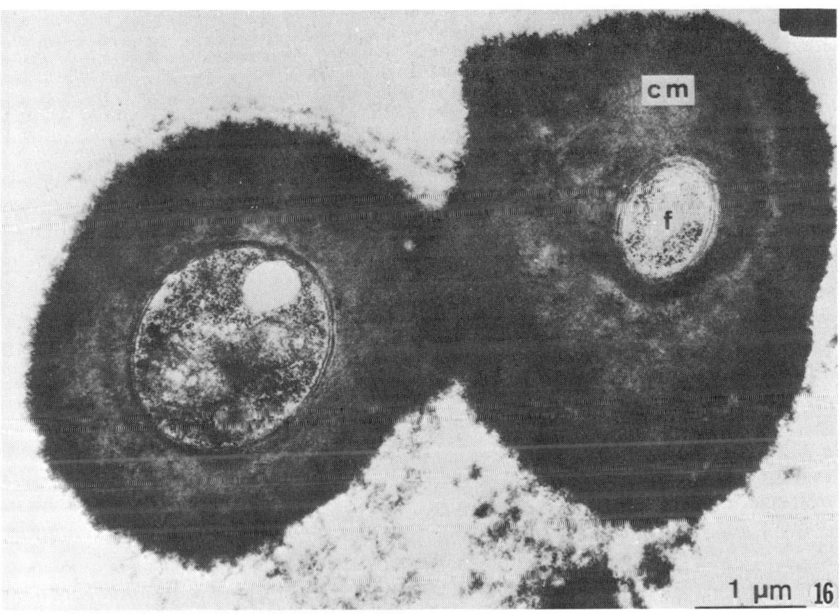

Figure 16. *Final stage (3 days after injection) in the deposition of capsule substance (Cm) by Chironomus larva around spores of Beaveria. Courtesy of Dr. Peter Götz.*

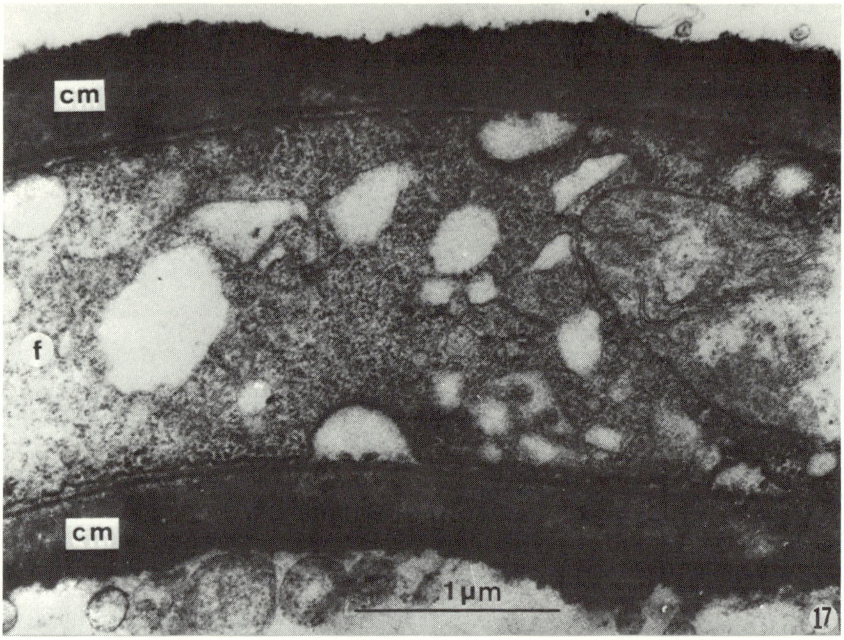

Figure 17. Fungal filament (f) of Beaveria encased in a solid crust of capsule substance (Cm) in the host C. luridus. Courtesy of Dr. Peter Götz.

Since phenoloxidases are involved in the repair of cuticular wounds (Lai-Fook, 1966; Evans, 1967), the localized deposition of melanin at the site of wounded tissues may also result in part from a temporary alteration in the metabolic pathway of tyrosine. It is also of interest to note that in newly-molted adult cockroaches encapsulation and melanization reactions against parasites (Misko, 1972) and interspecific ovary implants (Peterson, 1968) are either weakened or delayed until sclerotization of the cuticle is completed.

ACTIVATION OF CELL-MEDIATED RESPONSES

Numerous studies have shown that insect hemocytes recognize cellular kindships and respond discriminately, but just how is still obscure. Each hypothesis that has been put forward to elucidate the mechanism of activiation of the cellular immune reactions in insects is in need of supportive experimental evidence.

One theory states that cell-mediated responses are initiated only after fortuitous contacts of the host hemocytes with the parasite, rather than the cells being attracted to the parasite from a distance (Salt, 1970). So far it is not known whether hemocytes possess receptor sites on the cell membrane for distinguishing foreign objects with which they make contact. The fact that virtually any inert object of appropriate size implanted into insects elicits an encapsulation reaction does not necessarily prove that recognition is achieved solely by contact stimuli as has been suggested. For it is questioned just how such objects introduced into the body of an insect can avoid being even partially contaminated with substances released from injured tissue and, thus, attract hemocytes (Nappi, 1974). In some insects substances ('injury factors') released from injured tissues activate hemocytes causing them to increase in motility and adhesiveness, and to aggregate at the site of injury (Harvey and Williams, 1961; Cherbas, 1973). Recently, Cherbas (1973) identified an injury factor, called hemokinin, in both epidermal cells and hemolymph of saturniid pupae. She suggested, however, that hemokinin may also be localized within other cells including hemocytes. Of interest were the findings that extracts containing hemokinin activity accelerated melanization, and that sensitivity of hemocytes to hemokinin changed during diapause.

A second theory is that cell-mediated immune reactions of insects against internal metazoan parasites are activated by changes in the normal titer of host hormones (Nappi, 1973b,c). The hormonal imbalance caused by parasitization is believed to change the cell membrane permeability to certain metabolites, perhaps tyrosine and/or tyrosinase, and to bring about the premature differentiation and migration of hemocytes. Such altered cell permeability may cause the lysis of hemocytes at the surface of the parasite and the release of substances that result in melanization, aggregation, and adhesion of additional hemocytes to form pigmented capsules around foreign organisms.

The above proposal was based in part on the observation that hemocytic changes occurring during encapsulation reactions in some insect hosts were precocious, resembling those changes taking place in normal insects at pupation (Walker, 1959; Nappi and Streams, 1969; Nappi and Stoffolano, 1971, 1972). Recently it was shown that when host larvae of *D. algonquin* were ligated to prevent hormones from the anterior ring gland from reaching the hemocytes and parasites located posterior to the ligature, the percentage of melanized and encapsulated parasites was reduced from an average of 52% in non-ligated hosts, to less than 3% (0-8%) when hosts were ligated within 6 hrs after parasitization. Moreover, the total immune reaction rates, including unsuccessful pigment reactions, were reduced from 90-95% to less than 17% (7-25%) (Figure 18).

Consistant with the hypothesis that encapsulation and melanization of parasites results from a hormonal imbalance is the fact that the incidence of tumorigenesis in *Drosophila*, a process very similar to the formation of melanized capsules around parasites, is also affected by neck ligation which deprives posterior regions of the body of hormones from the anterior ring gland (Burdette, 1954a; Rizki, 1960). In the tu^{W} tumor stock of *D. melanogaster* Rizki (1957) found a precocious formation of lamellocytes followed by the encapsulation and melanization of the cells of the caudal fat body. He suggested the stimulus for the hemocyte transformation came from the caudal fat body cells which apparently developed asynchronously with respect to other larval tissues. Rizki (1960) found that neck ligation of tumorous larvae reduced the percentage of tumors. He showed the earlier during development larvae were ligated, the more tumor formation was suppressed (Figure 19).

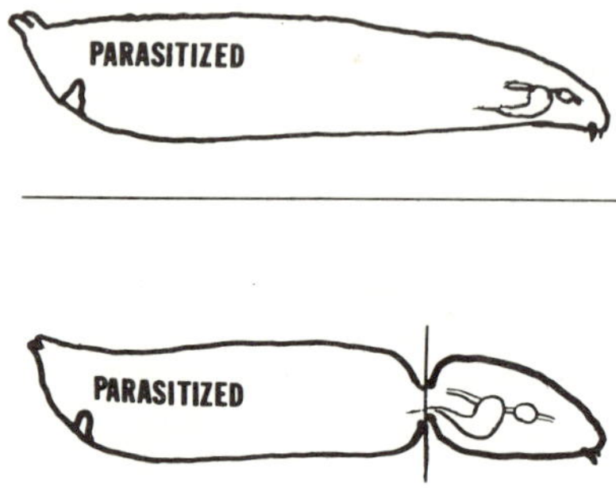

	HR AFTER PARASITIZATION WHEN LIGATED	% MELANIZED & ENCAPSULATED PARASITES	% TOTAL IMMUNE REACTIONS
Parasitized		52	97
Parasitized, ligated	6	2.7	16.6
	12	15.2	67.3
	18	30.5	75

Figure 18. *Effects of neck ligation on the ability of host larvae of Drosophila algonquin to make an immune response against eggs of the hymenopterous parasite Pseudeucoila bochei. Host larvae were parasitized during the second-instar, and dissected 26 hours after ligation.*

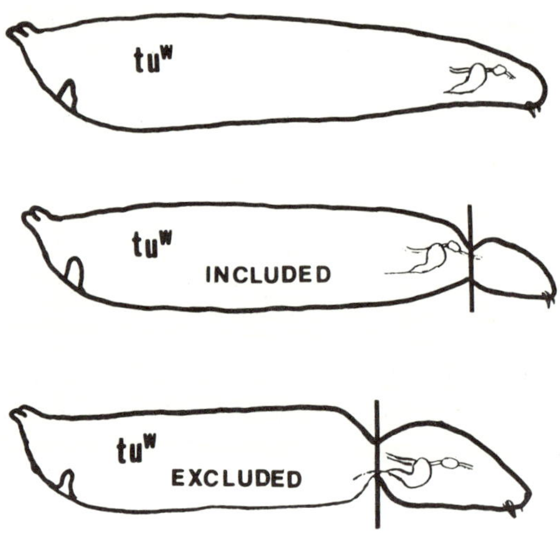

	AGE (HR) WHEN LIGATED	% CAUDAL TUMORS
Control		90-95
Included	60	70
	69	75
	72	85
	73	85
Excluded	60	0
	69	9
	72	10
	73	30

Figure 19. *Percentage of caudal tumors in control and ligated tu^W larvae 60-65 hours after ligation. When larvae were ligated so as to prevent hormones from the anterior ring gland from influencing cells in the posterior region, the number of melanotic tumors in the caudal fat body decreased until complete supression occurred at 60 hours. (From Rizki, 1960).*

On the other hand, Burdette (1954a) reported an increase in tumor frequency when larvae of the tu^{wps} strain of *D. melanogaster* were ligated, and when genes which damage the ring gland were introduced into the genotype of other tumor stocks (Burdette, 1954b). The disparity in the results of studies by Rizki and Burdette may be due to the use of different stains, to methods used in scoring melanotic tumors, or to the age of the larvae at the time of ligation.

Bryant and Sang (1969) found that various juvenile hormone mimics were active tumorigenic agents when fed to certain tumor strains. Recently, Madhavan (1972) was able to induce the formation of melanotic tumors in non-tumerous stocks by topical application of juvenile hormone. He concluded that metabolic processes responsible for tumorigenesis are sensitive to juvenile hormone. From these and other studies (Oster, 1954; Burdette, 1959, 1964; Sang and Burnet, 1963; Sang, 1969) it seems evident that changes in the distribution and titer of hormones alters the incidence of formation of melanotic tumors in *Drosophila.*

The hemocytic transformations during parasitization and in response to asynchronous growth of host tissues in *Drosophila* serve to localize and destroy foreign and abnormal cells. The evidence suggests the cell-mediated responses are controlled by host hormones. This does not negate the role of contact stimuli in initiating the immune reactions. The proposal suggests, however, that chemotactic stimuli are involved in the rapid differentiation of numerous hemocytes and in the directional movement of these cells.

A seriously neglected aspect of insect as well as invertebrate immunity concerns the possible relationship between neuroendocrine activities and the activation of cellular mechanisms of defense. However, it is unlikely that significant advances will be made to our understanding of how hormonal regulatory systems influence atypical growths and capsule formation in insects until we first learn how hormones interact with cells at the molecular level to initiate normal physiological and biochemical changes during development

Since the vast majority of cellular interactions appear to be mediated by contact, mechanisms of cellular recognition and communication among homologous cells may be attributable to cell surface molecular compatability. Accordingly, the loss of surface coadaption by abnormal host cells, or the presence of non-conforming molecular surfaces of parasites, would initate hemocytic responses. According to Holley (1972), a fundamental change in a malignant or transformed cell is an alteration in the transport systems at the cell surface which is regulated by hormones or growth factors. Unfortunately, there is no unifying concept explaining the functional significance of membrane changes in transformed cells. One reason for this is that it is not always possible to determine whether a specific difference (i.e. intracellular concentrations of various nutrients, or membrane changes in enzyme-substrate complexes, glycoproteins, glycolipids, etc.), is a cause or effect of transformed cells.

That the molecular specificity of surfaces which is considered essential for cellular recognition, communication, and adhesion, may result from enzyme-substrate complexes is of interest, especially when consideration is given to the acquisition of conforming molecular surfaces in host-parasite interfaces. Hemocyte transformations during encapsulation may be associated with altered cell membrane permeability allowing certain enzyme-substrate complexes to interact more readily. On the other hand, successful parasites may be able to alter certain biochemical processes and to localize enzyme-substrate complexes on their surfaces so as to attain steric conformance with the host.

RESISTANCE OF PARASITES TO ENCAPSULATION

One of the more interesting aspects of host-parasite interactions concerns the ability of some parasites to successfully develop within potentially immunologically responsive hosts. Although a wide variety of foreign objects encounter hemocytic reactions in insects, certain parasites ('resistant', 'tolerant', or 'habitual') and most intraspecific organ transplants develop unmolested in otherwise immunocompetent hosts. These states of apparent immunological inactivity may represent specific failures of the immune system to recognize certain parasites as aliens, or they may be due to the ability of some parasites to weaken or totally abrogate the defense reactions of their hosts.

One proposal that has been advanced to explain how parasites avoid immune reactions states they are not recognized as foreign because their surface properties mimic the antigenic pattern(s) of host tissue. This host-parasite interface phenomenon has been termed molecular mimicry (Damian, 1964). Studies by Salt (1965, 1966, 1968), Lewis and Vinson (1968), and Vinson (1971) suggest the wasp parasites *Nemeritis canescens* and *Cardiochiles nigriceps* avoid recognition as aliens within their insect hosts.

A second theory suggests that certain parasites produce substances that actively suppress the immune responses of the host. Such an encapsulation-inhibiting capacity has been proposed for the parasitoids *Limnerium validum* (Timberlake, 1912), *Diplazon fissorius* (Schneider, 1950), *Bathyplectes curculionis* (Van den Bosch, 1964), *Pseudeucoila bochei* (Walker, 1959; Nappi and Streams, 1969; Streams and Greenberg, 1969), and *Pieris rapae crucivora* (Kitano, 1974). In *Tribolium confusum* Ubelaker et al. (1970) suggested the cysticercoid of *Hymenolepis diminuta* secreted substances which repelled host hemocytes.

Recently, Salt (1973) concluded that the hypothesis of molecular mimicry previously proposed as a mechanism of resistance by *Nemeritis* was untenable. Instead, eggs of *Nemeritis* were considered to be resistant to the hemocytic reactions of *Ephestia* by virtue of a coating on their surface, derived from the reproductive tract of the female parasite, which directly inhibited adhesion of host blood cells. Evidence for this conclusion was that capsules formed by *Ephestia* around implanted pieces of Nylon thread gradually dissociated when they were incubated in saline with sliced calyces and other debris of the genital tracts of the female parasites. However, these preliminary studies were viewed as inconclusive (Salt, 1973). Apart from the fact that 25% of the capsules incubated as controls with ganglia and nerve cords also dissociated, substances that may dissociate cells do not necessarily also function in suppressing cellular adhesion.

While it is beyond the scope of this review to discuss the merits of the experimental evidence in support of theories of host specificity at the molecular level in different host-parasite systems, the following points should be noted. (1) Theories of the origin of 'host-like' surface properties frequently fail to consider these as developments resulting from the integrating adaptations of *both* parasites and hosts. (2) The hemostatic adaptations which attend host-parasite relations may be the consequence of a *changing* cell surface allotypy. Thus, various degrees of compatibility may be manifested at different times during the evolution of a given host-parasite system, with parasites occasionally mounting substantial defense reactions against their alien hosts. The possibility that hosts and their parasites continually direct or alter certain biochemical processes so that the surface of the parasite acquires (at least temporarily) host-like antigenic determinants (Sprent, 1959; Damian, 1964, 1967; Clegg et al., 1971, Clegg, 1972, Smyth, 1973) offers challenging opportunities for studying the sequence of cytological events responsible for the development of immunologic competence and tolerance in various heterospecific relationships. (3) The idea that insect blood cells fail to react against connective tissues of intraspecific organ implants because of similar or compatible surface properties may be creditable, but that surface properties of all successful entomophilic parasites are so similar to those of their hosts that they do not provoke a hemocytic response fails to generate the same confidence. The mere observation of the successful development of a parasite with no discernable encapsulation reaction is singularly insufficient evidence to suggest the host failed to recognize the parasite as foreign.

In parasitized larvae of the tobacco budworm, *Heliothus virescens*, Vinson (1971) found no significant changes in total and differential hemocyte counts during the first few days of parasitization and he suggested the parasite, *C. nigriceps*, avoided recognition as a foreign object and thereby failed to elicit a host reaction. On the other hand, Kitano (1974) found marked changes in the differential hemocyte counts of larvae of the cabbage butterfly, *Pieris rapae crucivora* parasitized by the braconid wasp, *Apanteles glomeratus*. He suggested that some factor released from the egg of the parasite suppressed normal hemocyte transformation.

Both cytological and genetic aspects of capsule formation were considered by Walker (1959) in her investigations of the immune reactions of larvae of *D. melanogaster* against the wasp *P. bochei*. When larvae of a highly immune strain were parasitized by a susceptible strain of *P. bochei* (E), there was a marked increase in the total number of hemocytes and a precocious transformation of plasmatocytes to lamellocytes. The lamellocytes quickly aggregated around the eggs of *P. bochei* and encapsulated them. When infected by a resistant parasite strain (Br), marked hemocyte transformations did not readily occur (Figure 20). The encapsulation rate, which normally averaged 95% in E-infected hosts and 5% in Br-infected hosts, was reduced to less than 33% when host larvae were infected with both parasite strains (i.e. superparasitized). From these data Walker concluded that highly resistant parasite strains inhibited the formation of lamellocytes. The resistance was considered to have a biochemical basis and could also protect eggs of susceptible strains of *P. bochei* from encapsulation in superparasitized hosts.

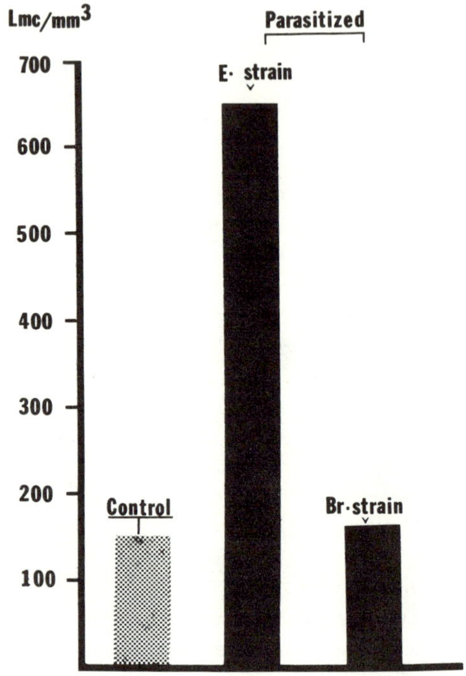

Figure 20. *Lamellocyte density in host larvae of D. melanogaster (Hindelbank stock) parasitized by a susceptible (E-strain) and a highly resistant (Br-strain) stock of the parasite Pseudeucoila bochei. The Br-strain parasites are capable of suppressing the encapsulation reactions of the host by inhibiting the formation of lamellocytes. (From Walker, 1959).*

In other experiments Walker (1959) examined the effects of parasitization on tumor formation in the *tu* B_3 stock of *D. melanogaster.* Of interest was the fact that the incidence of tumors was reduced in larvae parasitized by *P. bochei* that were normally highly resistant to encapsulation (Br-strain). On the other hand, the incidence of tumors was greater in *tu* B_3 larvae infected with susceptible *P. bochei* (E-strain) (Figure 21). Thus, it would appear that the ability of *P. bochei* to suppress the encapsulation reaction of its host served also to inhibit similar hemocytic transformations associated with the formation of melanotic tumors (Rizki, 1957).

Walker's conclusion regarding the ability of *P. bochei* to suppress the immune reactions of *D. melanogaster* was supported by comparative studies of encapsulation reactions against *P. bochei* and *P. mellipes*. In larvae of *D. melanogaster* the eggs of *P. mellipes* elicited an intense hemocytic reaction resulting in the rapid encapsulation and melanization of parasites (Figure 22). Differential hemocyte counts showed the premature mass transformations of large number of plasmatocytes to lamellocytes and the aggregation of the cells to form capsules around the eggs of the parasite. The melanization of the capsule was attributed to the lysis of crystal cells and the release of phenolic substances. However, differential hemocyte counts from larvae parasitized by *P. bochei* showed abnormally low percentages of lamellocytes, indicating the immune reactions of the host were suppressed by the parasite (Nappi and Streams, 1969).

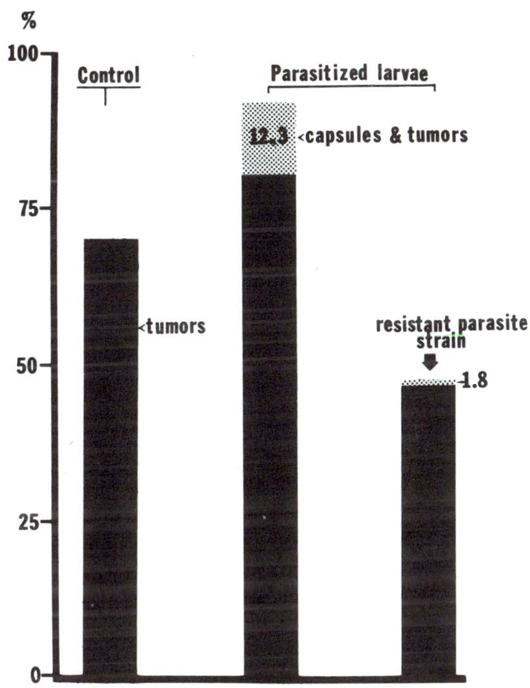

Figure 21. Effects of parasitization by P. bochei on tumorigenesis in D. melanogaster. Highly resistant P. bochei are capable of reducing the incidence of formation of melanotic tumors in certain mutant stocks of D. melanogaster, while susceptible E-strain parasites (middle bar) are unable to block tumor formation. (From Walker, 1959).

In another host, *D. algonquin,* differential hemocyte counts from larvae parasitized by *P. bochei* showed marked transformations of plasmatocytes to lamellocytes not only in immune larvae in which capsules were formed, but also in those hosts exhibiting unsuccessful 'pigment' reactions (Figure 23) (Nappi, 1973a). The presence of transformed lamellocytes in susceptible larvae suggests the hosts recognized the alien parasites but were either unable to react further or were prevented by the parasites from doing so.

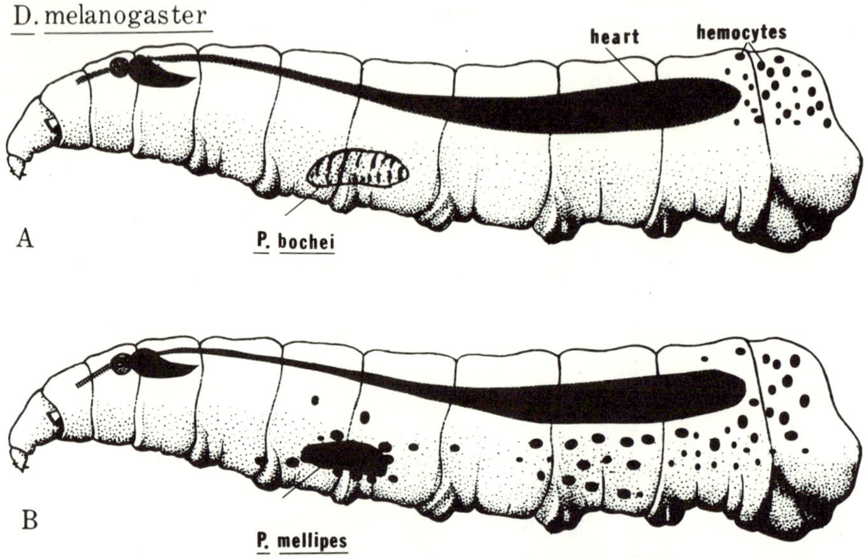

Figure 22. Diagrams illustrating the fates of two closely related parasite species in host larvae of D. melanogaster. (A) Eggs of P. bochei strains may be highly resistant to the encapsulation reactions of the host whereas eggs of P. mellipes elicit an intense host reaction which is characterized in part by the premature differentiation and migration of hemocytes.

The factor which affords resistance to *P. bochei* can also protect *P. mellipes* from encapsulation when both parasites are within the same host. The immune reaction rate against *P. mellipes* depends upon the period of time between the first infection with *P. mellipes* and the second infection with *P. bochei* (Figure 24). When the interval between multiple parasitization is 24 hours or less, 80% of the eggs of *P. mellipes* are protected by *P. bochei* from encapsulation. If the time elapsed between double infection is more than 36 hours, less than 8% of the *P. mellipes* were protected (Streams and Greenberg, 1969).

The factor which protects *P. bochei* in *D. melanogaster* may originate within the egg, or it may be a substance released into the host by the ovipositing female parasite. Regardless of its origin the ability of *P. bochei* to suppress hemocyte transformation decreases during the oviposition period of the female. According to Streams (1968) the increase in the susceptibility of the eggs of *P. bochei* to encapsulation was related to the amount of previous oviposition by the parasite. Females that lay a large number of eggs (partially spent) are more likely to produce eggs that will be encapsulated than are females of the same age but with no prior oviposition experience. Moreover, eggs from partially spent female *P. bochei*, since they are susceptible to encapsulation, fail to protect eggs of *P. mellipes* from encapsulation in multiparasitized hosts.

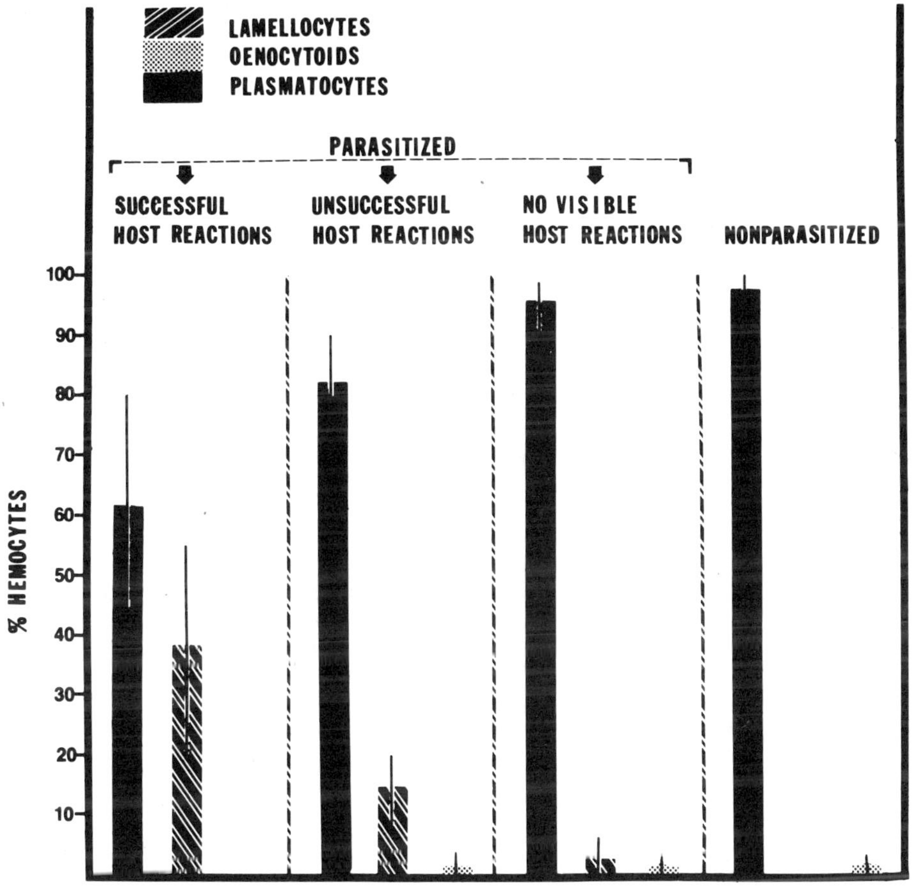

Figure 23. Differential counts of hemocytes of D. algonquin larvae 27 hours after parasitization by P. bochei. Successful immune responses involve the premature transformation of large numbers of plasmatocytes to lamellocytes which then encapsulate the parasite. Lamellocytes are also formed prematurely in hosts exhibiting unsuccessful 'pigment' reactions and in those hosts with no visible defense reactions. It would appear that susceptible hosts recognize P. bochei as foreign but are prevented from completing the reaction.

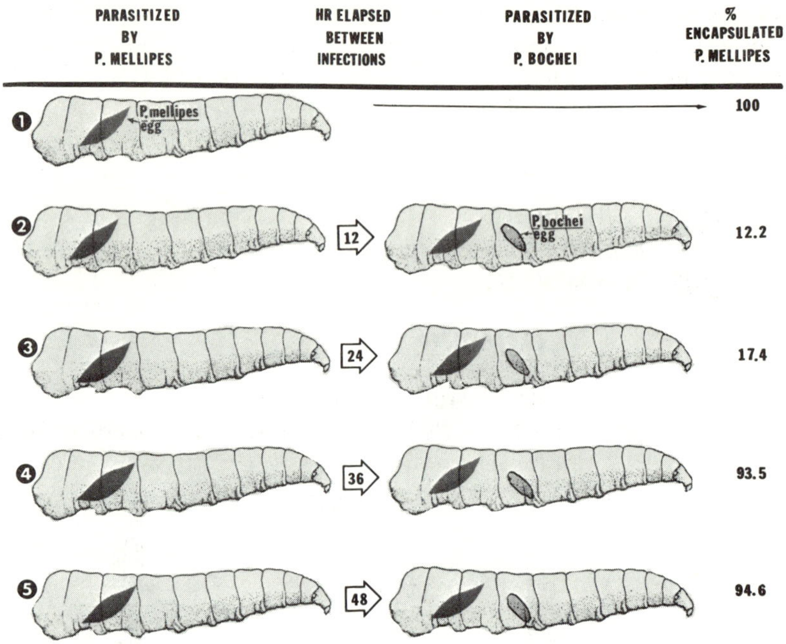

Figure 24. Effects of double infection, first by P. mellipes and then at different times by P. bochei, on the encapsulation and melanization reactions of host larvae of D. melanogaster. (From Streams and Greenberg, 1969).

CONCLUSION

The cellular immune reactions of insects and the mechansims of resistance of their parasites are homeostatic adaptations of great complexity. Our present lack of understanding of the mechanisms by which insects distinguish between 'self' and 'not-self' is complicated by our lack of understanding of the functions of specific blood cells and the mechanisms which control the activities of the cells during normal development. It would appear to this investigator that capsule formation in insects provides opportunities to study several basic biological phenomena, including cellular communication and recognition, aggregation, adhesion, degeneration and cell death. For with a homogeneous population of cells whose activities are predictable, attempts can be made to investigate the factors responsible for such activities, and to experimentally interrupt, reverse, or alter the processes.

ACKNOWLEDGMENTS

I thank Dr. George O. Poinar, Jr., of the Division of Entomology, University of California at Berkeley, and Dr. Peter Götz of the Zoologisches Institut, University of Freiburg, Freiburg, West Germany, for providing me with several photomicrographs. I especially thank Dr. I. S. Misko of the Department of Immunology, The John Curtin School of Medical Research, The Australian National University, Canberra, for allowing me to cite some results from his Ph.D. Dissertation, and for the use of several electron micrographs. I thank Richard Kampus of the Learning Resources Center and Cathie L. Hughes of the Department of Biology at Oswego for their technical assistance. Some of the investigations by the author cited in this paper were supported by grants from the National Science Foundation and from the Research Foundation of the State University of New York.

REFERENCES

Ada, G. L. (1970). *Transplant. Rev. 5,* 105-129.

Ashida, M., and Ohnishi, E. (1967). *Arch. Biochem. Biophys. 122,* 411-416.

Bakula, M. (1970). *J. Insect Physiol. 16,* 185-197.

Bakula, M. (1971). *J. Insect Physiol. 17,* 313-319.

Boman, H. G., Nilsson, I., and Rasumuson, B. (1972). *Nature 237,* 232-235.

Brewer, F. D., and Vinson, S. B. (1971). *J. Invertebrate Pathol. 18,* 287-290.

Briggs, J. D. (1958) *J. Exper. Zool. 138,* 155-188.

Briggs, J. D. (1964). *In* "The Physiology of Insecta" (M. Rockstein, ed.), Vol. 3, 259-283. Academic Press, New York.

Bronskill, J. G. (1960). *Can. J. Zool. 38,* 769-775.

Bronskill, J. G. (1962). *Can. J. Zool.* 40, 1269-1275.

Brunet, P. C. J. (1967). *Endeavour 26,* 68-74.

Bryant, P. J., and Sang, J. H. (1969). *Genetics 62,* 321-336.

Burdette, W. J. (1954a). *Cancer Res. 14,* 780-782.

Burdette, W. J. (1954b). *J. Nat. Cancer Inst. 15,* 367-376.

Burdette, W. J. (1959). Publ. Univ. Texas, No. 5914, pp. 57-68.

Burdette, W. J. (1964). *Cancer Res. 24.* 521-536.

Burns, W. C. (1961). *J. Parasit. 47*, 927-

Chadwick, J. S. (1966). *J. Invertebrate Pathol. 8*, 126-127.

Chadwick, J. S. (1967). *Fed. Proc. 26*, 1675-1679.

Chadwick, J. S. (1970). *J. Invertebrate Pathol. 15*, 455-456.

Chadwick, J. S., and Vilk, E. (1969). *J. Invertebrate Pathol. 13*, 410-415.

Cherbas, L. (1973). *J. Insect Physiol. 19*, 2011-2023.

Clegg, J. A. (1972). *Symp. Br. Soc. Parasitol. 10*, 23-40.

Clegg, J. A., Smithers, S. R., and Terry, R. J. (1971). *Int. J. Parasitol. 1*, 43-49.

Cottrell, C. B. (1964). *In* "Advances in Insect Physiology." (J. W. L. Beament, J. E. Treherne, and V. B. Wigglesworth, eds.), Vol. 2, pp. 175-218. Academic Press, New York.

Couturier, A. (1963). *Ann. Epiphyt. 14*, 203-267.

Crossley, A. C. S. (1964). *J. Exper. Zool. 157*, 375-398.

Damian, R. T. (1964). *Amer. Nat. 98*, 129-149.

Damian, R. T. (1967). *J. Parasit.* 53, 60-

Decleir, W., and Vercauteren, R. E. (1961). *11th Inter. Cong. Ent. Vienna*, 589-591.

Dennell, R. (1947). *Proc. R. Soc. Lond. B. 134*, 79-110.

Eckstein, F. (1931). *Anz. Schadlingskunde 7*, 49-55.

Eldelman, G., and Gall, W. (1969). *Ann. Rev. Biochem. 38*, 415-466.

Ermin, R. (1939). *Zeit. Zellforsch. Mikro. Anat., 28*, 613-669.

Esslinger, J. H. (1962). *Amer. J. Trop. Med. Hyg. 11*, 749-758.

Evans, J. J. T. (1967). *J. Insect Physiol. 13*, 1699-1711.

Evans, J. J. T. (1968a). *J. Insect Physiol. 14*, 107-119.

Evans, J. J. T. (1968b). *J. Insect Physiol. 14*, 277-291.

Feldman, M., and Nossal, G. J. V. (1972). *Quart. Rev. Biol. 47*, 269-302.

Fraenkel, G. and Hsiao, C. (1965). *J. Insect Physiol. 11*, 513-556.

Fraenkel, G., Hsiao, C., and Seligman, M. (1966). *Science 151*, 91-93.

Gingrich, R. E. (1964). *J. Insect Physiol. 10*, 179-194.

Good, R. A., and Papermaster, B. W. (1964). *Adv. Immunology 4*, 1-115.

Götz, P. (1969). *Zool. Anz. Suppl. 33*, 610-617.

Götz, P. (1973). *Naturw. Rdsch. 26,* 367-375.

Götz, P. (1974). *Parasitology 68,* (in press).

Grimstone, A. V., Rotheram, S., and Salt, G. (1967). *J. Cell Sci. 2,* 281-292.

Hackman, R. H. (1971). *J. Insect Physiol. 17,* 1065-1071.

Hackman, R. H., and Goldberg, M. (1967). *J. Insect Physiol. 13,* 531-544.

Harvey, W. R. and Williams, C. M. (1961). *J. Insect Physiol. 7,* 81-99.

Heimpel, A. M., and Harshbarger, J. C. (1965). *Bacteriol. Rev. 29,* 397-405.

Hink, W. F., and Briggs, J. D. (1968). *J. Insect Physiol. 14,* 1025-1034.

Hink, W. F., and Briggs, J. D. (1969). *J. Invertebr. Pathol. 13,* 308-309.

Holley, R. W. (1972). *Proc. Nat. Acad. Sci. 69,* 2840-2841.

Jones, J. C. (1956). *J. Morph. 99,* 233-257.

Jones, J. C. (1962). *Am. Zool. 2,* 209-246.

Jones, J. C. (1964). *In* "The Physiology of Insecta." (M. Rockstein, ed.), Vol. 3, pp. 1-106. Academic Press, New York.

Karlson, P., and Sekeris, C. E. (1962). *Nature 195,* 183-184.

Kitano, H. (1974). *J. Insect Physiol. 20,* 315-327.

Lai-Fook, J. (1966). *J. Insect Physiol. 12,* 195-226.

Lewis, W. J., and Vinson, S. B. (1968). *J. Insect Physiol. 14,* 613-626.

Madhavan, K. (1972). *Wilhelm Roux' Archiv. 169,* 345-349.

Maier, W. (1969). *Z. Zellforsch. 99,* 54-63.

Maier, W. A. (1973). *J. Insect Physiol. 19,* 85-95.

Matz, G., Monier, Y., and Vago, C. (1971). *Bull. Soc. Zool. France 96,* 209-215.

Metzger, H. (1970). *Ann. Rev. Biochem. 39,* 889-928.

Mills, R. R. (1967). *J. Insect Physiol. 13,* 815-820.

Mills, R. R., and Fox, F. R. (1972). *Insect Biochem. 2,* 23-28.

Mills, R. R., and Whitehead, D. L. (1970). *J. Insect Physiol. 16,* 331-340.

Mills, R. R., Androuny, S., and Fox, F. R. (1968). *J. Insect Physiol. 14,* 603-611.

Misko, I. S. (1972). The Cellular Defense Mechanisms of *Periplaneta Americana* (L.). Ph.D. Thesis. Australian National University.

Mitchell, H. K., and Wever, U. M. (1965). *Science 148,* 964-965.

Mohrig, W., and Messner, B. (1968). *Biol. Zbl. 87,* 439-470.

Muldrew, J. A. (1953). *Can. J. Zool. 31,* 313-332.

Nappi, A. J. (1970). *J. Invertebrate Pathol. 16,* 408-418.

Nappi, A. J. (1973a). *Amer. Zool. 13,* 1295.

Nappi, A. J. (1973b). *Parasitology 66,* 23-32.

Nappi, A. J. (1973c). *Exper. Parasit. 33,* 285-302.

Nappi, A. J. (1974). *In* "Contemporary Topics in Immunobiology." (E. Cooper, ed.) Vol. 4 (in press). Plenum Publishing Co., New York.

Nappi, A. J., and Stoffolano, J. G., Jr. (1971). *Exper. Parasit. 29,* 116-125.

Nappi, A. J., and Stoffolano, J. G., Jr. (1972). *Parasitology 65,* 295-302.

Nappi, A. J., and Streams, F. A. (1969). *J. Insect Physiol. 15,* 1551-1566.

Nossal, G. J. V., and Ada, G. L. (1972). "Antigens, Lymphoid Cells, and the Immune Response." Academic Press, New York.

Ohnishi, E. (1953). *Jap. J. Zool. 11,* 69-74.

Ohnishi, E. (1954). *Ann. Zool. Japan 27,* 188-193.

Ohnishi, E. (1958). *Japan J. Zool. 12,* 179-188.

Ohnishi, E. (1959). *J. Insect Physiol. 3,* 219-229.

Oster, I. I. (1954). *Cancer Res. 14,* 478-481.

Peterson, L. G. (1968). "Cellular Immune Responses of Insects to Foreign Tissue Implants." Ph.D. Thesis. University of Illinois.

Poinar, G. O., Jr. (1969). *In* "Immunity to Parasitic Animals." (G. J. Jackson, R. A. Herman, and I. Singer, eds.), Vol. 1, pp. 173-210. Appleton-Century-Crofts, New York.

Poinar, G. O. Jr., Leutenegger, R., and Götz, P. (1968). *J. Ultrastruct. Res. 25,* 293-306.

Poinar, G. O., Jr., aNd Leutenegger, R. (1971). *J. Ultrastruct. Res. 36,* 149-158.

Preston, J. W., and Taylor, R. L. (1970). *J. Biophys. Biochem. Cytol. 5,* 235-240.

Pye, A. E., and Yendol, W. G. (1972). *J. Invertebrate Pathol. 19,* 166-170.

Rizki, M. T. (1957). *J. Morph. 100,* 437-458.

Rizki, M. T. (1960). *J. Morph. 106,* 147-157.

Rizki, M. T., and Rizki, R. M. (1959). *J. Biophys. Cytol. 5,* 235-240.

Salt, G. (1956). *Proc. R. Soc. Lond. B. 146,* 93-108.

Salt, G. (1963). *Parasitology 53,* 527-642.

Salt, G. (1965). *Proc. R. Soc. Lond. B. 162,* 303-318.

Salt, G. (1966). *Proc. R. Soc. Lond B. 165,* 155-178.

Salt, G. (1968). *Biol. Rev. 42,* 200-232.

Salt, G. (1970). "The Cellular Defense Reactions of Insects." Cambridge University Press.

Salt, G. (1973). *Proc. R. Soc. Lond. B. 183,* 337-350.

Sang, J. H. (1969). *Nat. Cancer Inst. Monograph 31,* 291-301.

Sang, J. H., and Burnet, B. (1963). *Genetics 48,* 235-253.

Saunders, G. C. (1970). *In* "Biology of the Immune Response." P. Abramoft and M. LaVia, eds.) pp. 93-135. McGraw-Hill, New York.

Schell, S. C. (1952). *Trans. Amer. Micr. Soc. 71,* 293-302.

Schneider, F. (1950). *Vijschr. Naturf. Ges. Zurich. 95,* 22-44.

Schapiro, M. (1969). *In* "Immunity to Parastic Animals." (G. J. Jackson, R. A. Herman, and I. Singer, eds.), Vol. 1, pp. 173-210. Appleton-Century-Crofts, New York.

Smyth, J. D. (1973). *Can. J. Zool. 51,* 367-377.

Sprent, J. F. A. (1959). *In* "The Evolution of Living Organisms." (G. W. Leeper, ed). Melbourne University Press, Melbourne, Australia.

Stephens, J. M. (1959). *Can. J. Microbiol. 5,* 203-228.

Stephens, J. M. (1962a). *Can J. Microbiol. 8,* 491-499.

Stephens, J. M. (1962b). *Can. J. Microbiol. 8,* 597-602.

Stephens, J M (1963a). *In* "Insect Pathology" (E. A. Steinhous, ed.). Vol. 1, pp. 232-297. Academic Press, New York.

Stephens, J. M. (1963b). *J. Insect Pathol. 5,* 61-65.

Stephens, J. M. (1963c). *J. Insect Pathol. 5,* 152-156.

Stephens, J. M. (1963d). *J. Insect Pathol. 5,* 129-130.

Stephens, J. M., and Marschall, J. H. (1962). *Can. J. Microbiol. 8,* 719-725.

Stoffolano, J. G., Jr., and Streams, F. A. (1971). *Parasitology 63,* 195-221.

Streams, F. A. (1968). *J. Invertebrate Pathol. 12,* 379-387.

Streams, F. A., and Greenberg, L. (1969). *J. Invertebrate Pathol. 13,* 371-377.

Taylor, R. L. (1969a). *J. Invertebrate Pathol. 14,* 427-428.

Taylor, R. L. (1969b). *J. Invertebrate Pathol. 13,* 167-187.

Timberlake, P. H. (1912). *Tech. Ser. U. S. Bur. Ent. 19,* 71-92.

Ubelaker, J. E., Cooper, N. B., and Allison, V. F. (1970). *J. Invertebrate Pathol. 16,* 310, 312.

Van den Bosch, R. (1964). *J. Insect Pathol. 6,* 343-367.

Vercauteren, R. E., and Aerts, F. (1958). *Enzymologia 20,* 167-172.

Vinson, S. B. (1971). *J. Invertebrate Pathol. 18,* 94-100.

Walker, I. (1959). *Rev. Suisse Zool. 68,* 569-632.

Whitcomb, R. F., Shapiro, M., and Granados, R. R. (1974). *In* "The Physiology of Insects."(M. Rockstein, ed.), (in press) Academic Press, New York.

Whitehead, D. L. (1969). *Nature 224,* 721-723.

Whitehead, D. L. (1970). *FEBS Letters 7,* 263-266.

Whitehead, D. L., and Mills, R. R. (1970). *J. Insect Physiol. 16,* 331-340.

Wyatt, G. R. (1968). *In* "Metamorphosis: A Problem in Developmental Biology." (W. Etkin, and L. I. Gilbert, eds.). pp. 143-184. Appleton-Century-Crofts, New York.

MELANIN IN HOST-PARASITE INTERACTION

H. Lipke
Department of Biology
University of Massachusetts
Boston, Massachusetts

This symposium signals renewed intentions using contemporary technology to investigate the management of infectious and parasitic organisms by invertebrates. Aside from a common interest in the cellular components of body fluids, the diversity of the pathology, cytology and biochemistry is of such magnitude that the contribution of one laboratory often proves to be of minimal value to groups studying other families or genera (Acton *et al.*, 1972). The resolution of the many inconsistencies which speak against current attempts to devise useful generalizations require the careful definition of innumerable experimental situations. For what it is worth, the discipline currently under discussion, stands in need of extensive data-gathering, tedious as it may appear to the uninitiated. The present contribution will identify some of the problems that require solution in the general area of melanin production by beleaguered invertebrates. Particular emphasis will center on aspects of chemical specificity since the former consideration is rarely considered by parasitologists. More extensive treatments of melanin structure and biochemistry are presented by Nicolaus (1968), and Riley and Hutchinson (1972).

THE STRUCTURE OF MELANIN

Melanin is a copolymer comprised of five or more aromatic substituents joined by carbon-carbon bonds of great stability. Information on structure is restricted almost entirely to blacks available in generous supply, such as those from black hair, eyes, melanomas, puparial cases, banana skin, octopus ink and the terminal oxidation products of *ortho*-dihydric phenols polymerized *in vitro*. The backbone of the polymer bears additional substituents in the form of polypeptides, and complete characterization of melanin must include the proteinaceous component as well as the pigment. Central to the parasite-melanin relationship is the specificity of the melanogenic response. Among the questions posed are: a) Does the host elaborate a melanin of a given structure for each species of invader, i.e., do melanin-evoking parasites bear determinants that cause a unique melanoprotein to be produced? b) Does a given species of parasite cause chemically identical melanins to be produced regardless of host species? c) Is polypeptide an obligatory component of the defense reaction or a fortuitous occlusion captured during polyindole assembly? These questions can best be answered by a detailed analysis of the protective secretion, *per se*. Although each property of the polymer can be observed by appropriate methods, operationally, the amount of sample required for description of each parameter can differ by a factor of 10^4. This discussion of melanin analysis will include an assessment of sample limits.

Isolation of the melanoprotein complex. In the case of encapsulated parasites, serious consideration should be given to those methods which do not require extensive purification of the sample, such as radioautography, enzyme histochemistry or fluorescent antibody microscopy. The usual criticism directed at these techniques is based on the superficiality of the particular investigation, rather than the technique as such. For radioautography, all of the amino acids and most of the catecholamines and hydrozyindoles are available commercially as tritiated derivatives, and more obscure precursors are obtainable by custom synthesis. In the light of these opportunities, restriction of a study of melanogenesis to only one or two postulated intermediates does not constitute a rigorous examination of the metabolic pathways leading to pigment formation. There is also a widespread failure to examine the purity of labelled reagents, which in the case of phenols, may be a function of the storage history (Hempel and Männl, 1966).

Routine qualitative tests for melanin and its depolymerization products have been adapted to samples in the 0.3-3.0 mg range by Hackman and Goldberg (1971a). These weight limits do not refer to unpurified blacks as they occur in host tissues, extensive removal of adhering tissue from host and parasite by both surgical and chemical techniques is a necessary preliminary to standards methods of melanin analysis. Particularly lacking at this time is information on the contribution of preformed material of host and parasite origin to the melanoprotein complex. Since the major impurity of the inclusion is usually protein, most contaminants are removed with 1-6 N HCL at 100°. Unfortunately, yields of melanin as low as 5% have been reported for this procedure, and acid-labile components are lost (Fogal and Fraenkel, 1968; Swan and Waggott, 1970). The losses in chemical fine structure of the pigment are acceptable for many purposes, however, the growing interest in the protein component of the pigment-polypeptide complex requires gentler and more specific reagents. Serial extraction with neutral buffers, detergents, urea, and cold 0.2 M alkali removes the more soluble visceral proteins, leaving a pigmented residue with polypeptides tightly bound to the polyindole component. Current methods do not permit unequivocal identification of the bonding of the peptide fraction to the pigment, however, the above solvents are selected to retain the bulk of the stable covalent links between pigment and protein. The material, furthermore, is suitable for fingerprinting, since controlled release of peptides can be attempted with specific proteases and cleaving reagents cush as N-bromosuccinimide and chilled sodamide (Benisek *et al*., 1967; Witkop, 1968; Lipke and Geoghegan, 1971). A major increase in pigment purity can be affected by solution of the acid-stable residue in boiling 1 M NaOH followed by reprecipitation with acid. An unknown portion of the quinoid function is sacrificed by this approach since detailed studies of reflux temperature, duration, hydroxide and oxygen concentration are lacking.

Alkali-treatment and precipitation can be repeated to constant N, values of 3% to 10% have been reported. The range is indicative of the diversity of melanin structure within the animal kingdom. With respect to arthropods and related phyla, contaminants such as chitin, sclerotized protein, and mineralized components are predictable survivors of the procedures outlined above. Inadequate information exists at this time to consider "native" and "derived" melanoproteins, but the implications of these terms should be appreciated. If mild conditions are required for characterization of the native copolymer, new approaches to chain cleavage are in order. Unlike other biopolymers, the bonds between indolic, pyrrolic and phenolic residues in the melanin chain exhibit great diversity in energy content, no single set of conditions can release a majority of the chain components. A special need exists, then, for specific melanases. The slow but constant removal of melanized tissues from the biosphere by microorganisms suggest that soil enrichment and microbial adaptation may provide mild hydrolytic reagents of great value to melanin analysts. (Veldkamp, 1970).

Characterization of the pigment fraction. Given the poverty of the sample available from melanin inclusions and the losses incurred during the initial purification steps, little opportunity exists for thorough delineation of polymer morphology. The information required includes:

1. Characterization of the monomers
2. Monomer sequence
3. Molecular weight
4. Degree of branching
5. Oxidation state (ratio of quinonoid to dihydroxy groups)
6. Structure of the pigment-protein bridge
7. Three-dimensional orientation of all substituents

Current status of melanin analysis is akin to the situation with respect lignin chemistry, with the provision that samples of plant origin are rarely in such short supply. Reasonable agreement has been reached by the two most active laboratories in Newcastle and Naples concerning the more stable residues encountered in the pigment (Fig. 1). As used by Hackman and Goldberg (1970a), alkali fusion can reveal 5,6-dihydroxyindole and *o*-diphenols, if severe losses are countenanced. The cleavage products have been detected on thin layer and paper chromatograms with reagents of low specificity and sensitivity and by radiometric methods. Gas chromatography and mass spectrography should provide useful tools in this endeavor, providing standards become available. A semi-quantitative degradation is also possible with hydrogen peroxide or potassium permanganate in alkali. This procedure is indicative of the abundance of 2,3- and 2,3,5-

pyrrole carboxylic acids (Swan and Waggott, 1970). Ring-bound carboxyls have been assessed by heating to 150° in a vacuum system, the results generally supporting procedures based on the incorporation of carboxyl-labelled precursors (Hempel, 1966). Phenolic hydroxyls have been revealed by methylation (Piatelli *et al.*, 1962). The determination of molecular weight is not possible for most of the melanins as they are obtained from host tissues, almost all are insoluble in aqueous and nonpolar solvents. Varying degrees of solubility are observed with partly degraded material following alkali treatment, in which case gel filtration can provide useful distinctions between degraded melanins of identical preparative histories. Sorption to the gel matrix can be reduced with borate-NaOH solvent systems or sodium benzoate, in either case, care is indicated in selection of the gel from the many now available. Standards of comparable aromaticity and molecular weight to melanin can be obtained from producers of synthetic resins (Screaton *et al.*, 1973). Physical probes for Stokes radii and molecular weight, such as sedimentation and light scattering, encounter serious inconsistencies from a) non-covalent association- aggregation b) branching within the polyindole chain and c) branching from appended polypeptides. Aggregates may be detected by sedimentation in alkali with concentration as a variable (Roark and Yphantis, 1971). Branching is more difficult to discern, detection requiring end group analysis and isolation of a triply-substituted chain member (King *et al.*, 1970). These criteria have not been realized for the polyindole component, although the covalent nature of the pigment-protein bond has been established (Hackman and Goldberg, 1971b; Mills and Fox, 1971).

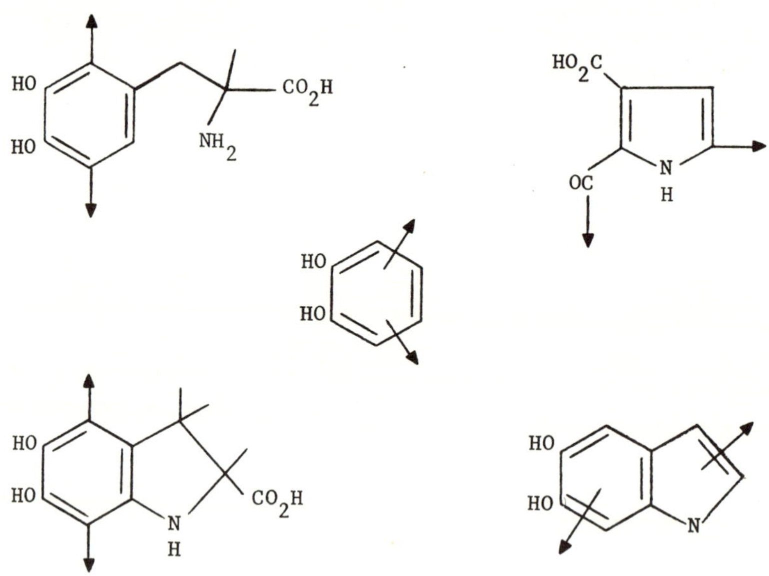

Figure 1. Constituents of polymerized DOPA Melanin. (After King et al., 1970).

The difficulties outlined are formidable to the biochemist and discouraging to the biologist seeking new tools for identification and quantitation of pigment. Many of the problems may be avoided if the investigator has access to specifically labelled precursors of melanin. These compounds have been used to estimate the ratio of the aryl residues illustrated in Figure 1 with minimal clean-up and reasonable accomodation to small sample size (Kirby and Ogunkoya, 1965; Hempel, 1966; Swann and Waggott, 1970). As outlined in Table I, the abundance of each substituent in the polymer chain can be evaluated by measuring retention of ^{3}H and α-^{14}C melanin following administration of two labelled precursors. Well-defined syntheses of the desired precursors are reported by the three laboratories mentioned above. Necessary precautions include a) simultaneous administration of two labelled precursors to simplify quantitation and b) measurement of incorporation into tissues other than the melanin capsule. In this respect, invertebrates frequently metabolize and sequester aryl derivatives for purposes distinct from melanin formation such as sclerotization. These compartments cannot only contribute intermediates to the melanin inclusion, but also indicate the extent to which contaminating tissues can confound measurement of radioactivity. Isotope analysis of cuticle and viscera will aid in the evaluation of these factors. Of particular relevance would be comparison of melanins in those species which normally produce melanin for coloration as well as for defense against internal parasites. As applied to melanin, the original isotope procedure requires several milligrams of pigment with contemporary combustion apparatus and emission counters. The biologist can increase sensitivity 1000-fold by bringing the precursors to identical specific activities and assessing the relative incorporation by silver grain counts on radioautographs.

TABLE I

Predicted Distribution of Isotope in Melanin Following Administration of Specifically Labelled DOPA[a]

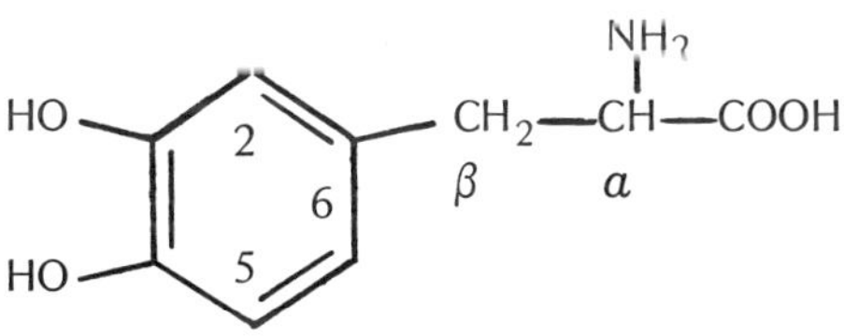

Label	Fate	Indication
DOPA		
^{14}C-Carboxylic	Retention	Portion not cyclized
^{14}C-α	Retention	Total pigment formed exclusive of catechol residues
^{14}C-Carboxylic, α, β	Loss	Catechol melanin
^{3}H-α	Loss (100%)	α - β unsaturation
^{3}H-$\beta(T_2)$	Loss (50%)	α - β unsaturation

TABLE I con't.

Label	Fate	Indication
^{3}H-2	Loss	2- linkage or cyclization at 2- position
^{3}H-5	Loss	5- linkage
^{3}H-6	Loss	Portion cyclized

a-Adapted from Hempel (1966).

BIOSYNTHESIS OF MELANIN

Melanogenesis can also be approached as an exercise in comparative enzymology. Several of the steps in pigment formation are known in sufficient detail to identify those enzyme systems subject to change during infection. Enzyme activity can be detected at levels significantly lower than the most sensitive procedure for polymer analysis. Enzyme histochemistry and cell-free systems can reveal those steps in the defense phenomenon that involve enzyme activation, inhibition, biosynthesis and secretion. Considerably more enterprise is evident on the part of enzymologists than cytologists, as evidenced by the latter group's failure to exploit recent developments facilitating the resolution of hemocyte groups (Rabinowitz, 1964; Peterson and Evans, 1967). Also in evidence is a reluctance to master those technically demanding avenues of histochemistry which require the visualization of relatively obscure oxidases, decarboxylases and polymerases.

Enzyme analysis. The biosynthetic scheme presented in Figure 2 is similar in most respects to that devised by Mason (1967). The need to distinguish between dopa formation via polyphenol oxidase ("tyrosinase") or via peroxidase has been empahsized by Patel *et al.* (1973) and Hearing (1973). A number of approaches to this destinction are described by these workers. Two additional details are particularly worthy of reinvestigation by invertebrate biologists. The stereo specificity of the initial oxidations has been established for fungi and vertebrates and clearly favors the L configuration. Arthropods, however, have special pathways for utilization of D-amino acids which may extend to melanogenesis (Corrigan, 1969). In addition, symbiotes supply certain invertebrates with partially aromatized precursors of tyrosine and phenylalanine that may contribute to melanin independent of tyrosine (Brunet, 1965). The major difficulty encountered in measurement of melanogenic enzymes, is the spontaneous formation of pigment in the absence of enzymes. Melanin is also synthesized by purified polyphenol oxidases, thereby taxing the ingenuity of the investigator studying the contribution of a single component of a multienzyme system. On the basis of monomer composition, King *et al.*, (1970) were unable to distinguish between the pigment formed by spontaneous oxidation in alkali and material produced enzymatically, but information on melanin formed *in vivo* is lacking.

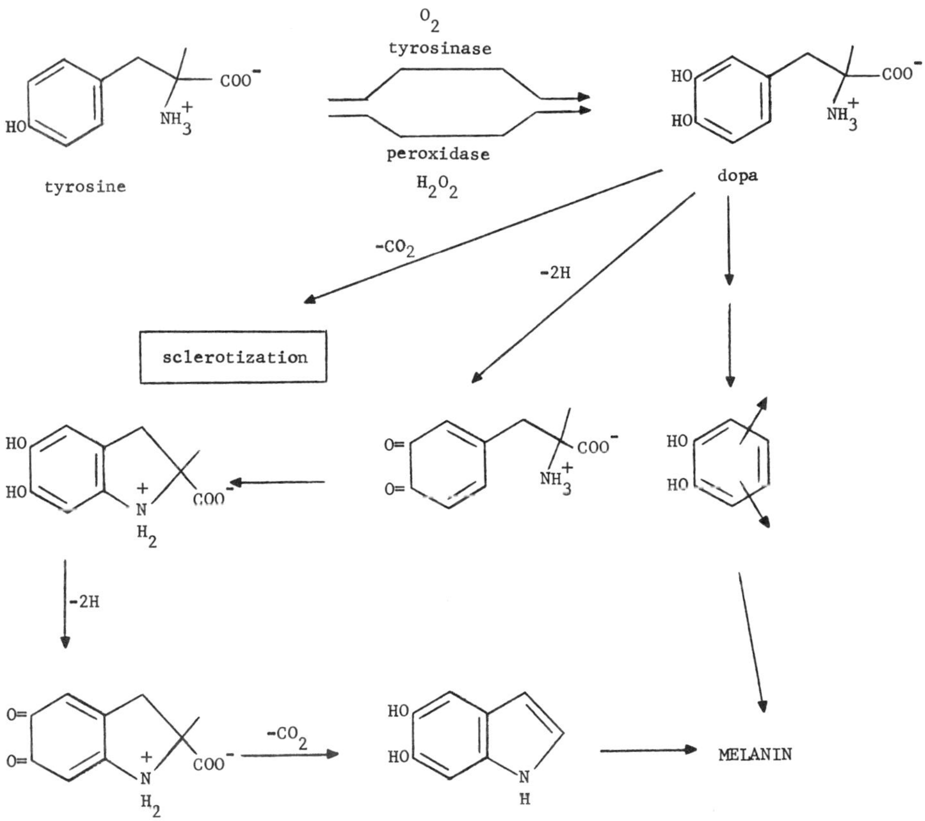

Figure 2. Biosynthesis of melanin. (After Mason, 1967).

Inhibitors of melanogenesis. Multienzyme enzyme systems which include spontaneous reactions in addition to those that are enzyme-dependent require probes that uncouple the constituent elements of the complex. Several steps leading to melanin can be followed *in vitro* from incubation mixtures composed of precursors, host enzymes, and fragments of parasites harvested from donor hosts. The more or less straightforward observations of substrate utilization and product formation can be augmented with more detailed insights into the nature of reaction intermediates fromed *in vitro* by electron spin resonance, temperature jump, solvent perturbation and rapid scanning spectrophotometric analysis (Mason and Peterson, 1965; Grady and Borg, 1968).

For situations more closely allied to natural parasitizations, namely stepwise analysis of melanogenesis during experimental infections, the most useful data has been obtained with inhibitors. Table II lists some common blocking agents and others perhaps worthy of future exploitation. In a sense these data are self-defeating in that they foster uncritical acceptance of inhibitor specificity. In practice, the mode of action of each agent should, of course, be established for the particular experimental situation.

TABLE II

Inhibitor Specificity for Melanogenesis

Reagent	Target	Remarks
Diethyldithiocarbamate	Tyrosinase	Catalase inhibition reversible
Anaerobiasis	Phenol, diphenol oxidases	Non toxic, invertebrate survival high
Catalase	Peroxidase	Consumes H_2O_2
Phospholipase	Bound tyrosinase	Disrupts hemocyte membranes
Triton X-100	Bound tyrosinase	Disrupts hemocyte membranes
Rabbit Anti-host hemocyte	Hemocytes	Distinguishes cellular from humoral response
Dihydroxyfumarate	Catalase	Stimulates catalase, not tyrosinase
Rabbit Anti-DOPA decarboxylase	Dopa decarboxylase	Blocks sclerotization
MDH (α-methyl-α-hydrazino-β-phenylpropionic acid)	Dopa decarboxylase	Blocks endocuticle formation
D-DOPA	DOPA decarboxylase	Not utilized by decarboxylase, good substrate for melanogenesis
KCN, phenylthiourea	Copper oxidases	Low specificity
2-CH_3 tyrosine	tyrosinase	Inhibits melanogenesis
3-CH_3 tyrosine	tyrosinase	Inhibits melanogenesis
2, 5, di-CH_3 tyrosine	tyrosinase	Inhibits melanogenesis

THE TOXICITY OF MELANIN

In addition to the acknowledged role of the melanized inclusion as a physical barrier, a number of other suggestions have been made concerning the adverse effect of melanin on parasite development (Salt, 1970). Suffocation, malnutrition, and disorientation can be observed in the light microscope, but physiological insights are lacking. The melanogenic apparatus is an ion-sink as well as a free-radical generator (Grady and Borg, 1968). Thus far the focus of this discussion has been the host reaction. Attention should also be drawn to the accessibility of the invader to biochemical analysis during encapsulation and black formation. The same tools available for the study of melanogenesis, namely, the labelled precursor in the intact animal, and the cell-free enzyme system can also reveal the pharmacology of the parasite. The surface of the parasite egg can be altered chemically following removal from donor hosts, and the egg "determinants" identified by artificial infection. Ultrasound and x-rays may also alter the course of the defense reaction with the added advantage that surgical manipulation becomes unnecessary.

REFERENCES

Acton, R., Evans, E. E., and Weinheimer, P. (1972). "Invertebrate Immune Defense Mechanisms." Mss. Information Corp., New York.

Benisek, W. F., Raftery, W. A., and Cole, R. D. (1967). *Biochemistry, 6*:3780-3788.

Brunet, P. (1965). "Aspects of Insect Biochemistry" (T. Goodwin, ed.), Biochemical Society Symposium, Vol. 25, pp. 49-50. Academic Press, New York.

Corrigan, J. J. (1969). *Science, 164*:142-149.

Fogal, W., and Fraenkel, G. (1969). *J. Insect. Physiol., 15*:1437-1447.

Grady, F. J., and Borg, D. C. (1968). *J. Am. Chem. Soc., 90*:2049-2952.

Hackman, R. H., and Goldberg, M. (1971a). *Anal. Biochem., 41*:279-285.

Hackman, R. H., and Goldberg, M. (1971b). *J. Insect. Physiol., 17*:335-347.

Hearing, V. J. (1973). *Archs. Biochem. Biophys., 158*:720-725.

Hempel, K., and Männl, H. F. K. (1966). *Biochim. Biophys. Acta., 124*:192-194.

Hempel, K. (1966). *Sympos. 6th. Intern Pigment Cell Conf., Sofia 1966*:162-175.

King, J. A. G., Percival, A., Robson, N. C., and Swan, G. A. (1970). *J. Chem. Soc., 1970*:1418-1422.

Kirby, G. W., and Ogunkoya, L. (1965). *Chem. Commun., 1965*:546-547.

Lipke, H., and Geoghegan, T. (1971). *Biochem. J., 125*:703-716.

Mason, H. S. (1967). In "Advances in Biology of Skin" (J. Montagra, ed.), Vol. VIII, pp. 273-301. Pergamon, London.

Mason, H. S., and Peterson, E. W. (1965). *Biochim. Biophys. Acta., 111*:134-146.

Mills, R. R., and Fox, F. R. (1972). *Insect Biochem., 2*:23-28.

Nicolaus, R. A. (1968). "Melanins" Herman, Paris.

Patel, R. P., Okun, M. R., Yee, W. A., Wilgram, G. F., and Edelstein, L. M. (1973). *J. Invest. Dermatol., 61*:55-59.

Peterson, E. A., and Evans, W. H. (1967). *Nature, 214*:824-829.

Piatelli, M., Fattorusso, E., Magno, S., and Nicolaus, R. A. (1962). *Tetrahedron, 18*:941-949.

Rabinowitz, Y. (1964). *Blood, 23*:811-820.

Riley, U., and Hutchinson, F. (1972). In "Pigmentation, Its Genesis and Biological Control." *Proc. 7th Conf. N.Y.* Appleton, N.Y.

Roark, D., and Yphantis, D. A. (1971). *Ann. N.Y. Acad. Sci., 234*:245-278.

Salt, G. (1970). "The Cellular Defense Mechanisms of Insects" Cambridge University Press.

Screaton, R. M., Cullen, P. F., Seeman, R. W., and Saunders, M. D. (1973). *J. Polymer Sci.* (Symposium No. 43) pp. 277-327.

Swan, G. A., and Waggott, A. (1970). *J. Chem. Soc., 1970*:1409-1418.

Veldkamp, H. (1970). In "Methods in Microbiology" (J. R. Norris and T. Ribbons, eds.), Vol. 3a, pp. 305-381. Academic Press, New York.

Witkop, B. (1968). *Science, 162*:318-326.

VECTOR DESTRUCTION OF PARASITES

Franklin G. Wallace
Department of Zoology
University of Minnesota
Minneapolis, Minnesota

In phagocytosis and encapsulation we have what may be the most primitive of defenses. But these functions were not always used for defense. Phagocytosis undoubtedly began as a way of eating. But very early in evolution it assumed other functions, excretion for example. In many invertebrates intracellular excretion, by which cells engulf waste products and carry them away, is widespread (Florey, 1966). The cells may simply sequester the waste material, or may carry it to a nephridium, or to the intestine or outer integument and migrate out. Richard Goldschmidt (1937) in his classic book on Ascaris recalls an observation from his travels. "In a Mediterranean port every evening after nightfall a silent procession of women went to the cliffs, each carrying a pail on her head, the unpleasant contents of which were poured into the surf. To make the comparison with the scavenger cells perfect, however, the poor women would have had to jump over the cliff's edge with their pails".

What more natural than that this excretory function should be adopted for the removal of foreign material so that suicide squads of cells are deployed to remove invaders. When the object is too big for phagocytosis the cells cooperate to enclose the foreign body, usually to sequester it but sometimes to eject the fragments out through the intestinal wall (Chen, 1934)

The processes of phagocytosis and encapsulation may appear primitive and simple but they are not. The foreign body is recognized as different from any part of the host. Only certain kinds of cells among the many in the invertebrate host are involved. The needed cells are propagated. They gather at the site of action. Some of them produce chemical changes, particularly the deposition of melanin. And the controls over the entire process are delicately adjusted to the normal controls of the life and development of the host. These processes are being subjected to critical analysis by ingenious experiments and our three speakers, Dr. Nappi, Dr. Lipke, and Dr. Riddiford are in the forefront of this work.

REFERENCES

Chen, H. T. (1934). *Z. Parasitenk.* *6*:603-637.

Florey, E. (1966). An Introduction to General and Comparative Physiology. Saunders, Philadelphia.

Goldschmidt, R. B. (1937). Ascaris, the biologist's story of life. Prentice-Hall, New York.

HOST HORMONES AND INSECT PARASITES

Lynn M. Riddiford
Department of Zoology
University of Washington
Seattle, Washington

Growth and development in insects is regulated by a number of hormones (for recent reviews see Doane, 1973; Gilbert and King, 1973; Wigglesworth, 1970; Willis, 1974; Wyatt, 1972). Then in the adult, maturation of the reproductive system in most females and some males is also controlled by hormones (see de Wilde and deLoof, 1973; Engelmann, 1970 for reviews). Many parasites of insects have exploited these hormonal controls in order to manipulate the host environment to make it more favorable to the growth and development of the parasite. In the examples studied thus far this manipulation can be classified into two broad categories:

1) Reliance directly on host hormones or indirectly through hormonally triggered changes in host physiology to synchronize the parasite's growth and/or development within the host.

2) Interference with the host's hormonal system or production of insect hormone mimics to ensure that the host lingers in the most beneficial stage for the parasite.

Other strategies may well be employed by the myriad of parasites of insects, but have not been investigated in detail from the endocrinological viewpoint. This review therefore will be concerned only with the two above categories.

On the other side of the picture are the means by which an insect host defends itself against invading parasites. Encapsulation and often melanization, the primary means of host defense (Salt, 1970), are covered in detail by two other authors in this volume (Nappi, 1974; Lipke, 1974). Very little is known about the hormonal control, if any, of these reactions.

THE INSECT NEUROENDOCRINE SYSTEM

Postembryonic development of an insect can be divided into 3 phases: growth, metamorphosis, and reproduction, all of which are under hormonal control. After the insect larva hatches from the egg, it feeds and grows. Periodically, it sheds (molts) its exoskeleton and grows eventually to a certain predetermined size at which time it undergoes metamorphosis to the adult stage. This metamorphosis may be directly from the larva to the adult as in the hemimetabolous insects such as cockroaches and locusts or it may be in two stages, first to a pupa, then to an adult as in the Lepidoptera and the Diptera. Depending on the insect, maturation of the reproductive system may occur either before or after adult emergence, although in most insects it occurs afterwards.

Figure 1 shows a generalized insect neuroendocrine system and its control of molting and metamorphosis. The median neurosecretery cells of the brain secrete prothoracicotropic hormone (PTTH) which travels down the axons to the corpora cardiaca where it is stored. At certain set times during the insect's life which are dictated by both extrinsic environmental signals and intrinsic signals relate to size, PTTH is released into the hemolymph and directs the prothoracic glands to secrete the hormone ecdysone. Ecdysone acts on the epidermal cells to initiate the molting process, i.e. to cause detachment of the epidermis from the overlying cuticle, cell division, and the synthesis of a new cuticle. The type of cuticle formed depends upon the presence or absence of another hormone, juvenile hormone (JH) which is produced by the corpora allata. When the JH titer is high, the new cuticle is larval; when there is no JH present at the time of ecdysone action, the cuticle is adult. In the holometabolous insects the JH titer is low but finite at the end of larval life so a pupal cuticle is secreted. Similarly, the simultaneous changes in the viscera as well as in external form are governed by the JH titer.

Once the adult stage has been attained, maturation of the ovaries in most insects is under hormonal control as portrayed in Figure 2. Juvenile hormone is necessary either to stimulate vitellogenesis (by initiating yolk protein synthesis by the fat body or by initiating yolk protein uptake into the oocytes) or to trigger a specific non-vitellogenic phase of oocyte development (Gwadz and Spielman, 1973; Nijhout and Riddiford, 1974). In many insects the neurosecretory cells of the brain also are necessary for complete ovarian maturation. They may either control the secretion of JH, stimulate metabolism (especially protein synthesis), or act directly on the ovary as a gonadotropin as in the mosquitoes (Lea, 1967; Hagedorn and Fallon, 1973).

USE OF HOST HORMONES BY INSECT PARASITES

For maximum reproductive success, parasites need to synchronize their life cycle with that of the host, so that the immature feeding stages or the adult reproductive stage may be in an optimum environment. One simple means of synchronization is to cue in on the host hormones which govern its growth and development.

One large class of insect parasites are other insects, particularly Diptera and Hymenoptera. Thus, it would seem that these parasitic insects could readily utilize the host's hormonal signals. But even if the parasite is in the host's hemolymph, its tissues

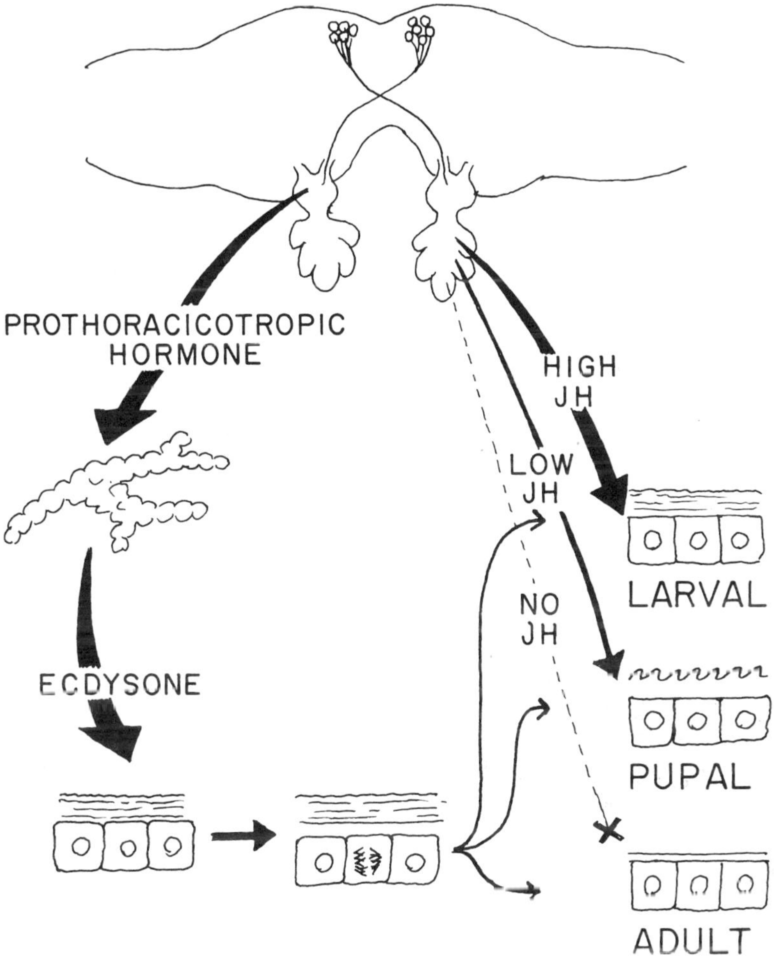

Figure 1. The neuroendocrine system of a generalized insect and the hormonal control of molting and metamorphosis.

can be exposed to the hormones of the host only if these hormones can penetrate into the parasite. Since ecdysone is a steroid, it cannot go through the cuticle; whereas JH, a homosesquiterpene compound, can. Thus, if the parasite must synchronize with

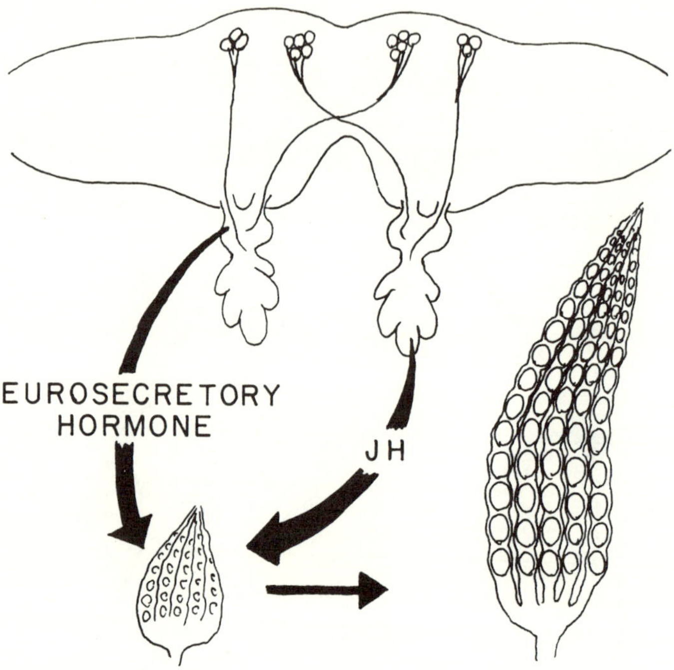

Figure 2. The neuroendocrine system of a generalized adult femal insect and the control of egg maturation.

the molting cycle of the host larva or pupa, it must be able to respond either to ingested ecdysone from the hemolymph or to ecdysone-mediated physiological or biochemical changes in the host.

Although there are many examples in which the insect parasite does seem to synchronize with its host (see Schoonhoven, 1962; Fisher, 1972 for reviews), there are very few instances in which this interrelationship has been pinpointed as hormonal. For instance, the diapausing tachinid fly parasite larva of the wild silkmoth *Hyalophora cecropia* begins rapid growth when the prolongedly chilled host pupa initiates adult development at room temperature. But C. M. Williams (personal communication) has found by various transplantation experiments that the parasite larva is not utilizing the hormones of the host. Rather the prolonged chilling activates the parasite's neuroendocrine system as well as that of the host; therefore, the parasite resumes larval development independently of the host.

By contrast, the tachinid fly parasite (*Eucarcelia rutilla*) of the pine looper moth *Bupalus pimarius* apparently is dependent on its host's ecdysone to trigger its development (Schoonhoven, 1962). Although chilling promotes the termination of diapause of both the host pupa and the first instar fly larva, it is not necessary for termination in either host or parasite since they both can develop in response to ecdysone from a parabiosed chilled pupa. Furthermore, the parasites will not initiate larval development even after 20 weeks of chilling if the abdomen of the host pupae is isolated immediately after removal from the cold. If, however, the abdomen is isolated 5 days after removal, the parasite larvae is activated and begins feeding, growth, and molting. By this time most of the abdomens have received sufficient ecdysone to complete adult development in the absence of the prothoracic glands. Schoonhoven (1962) concludes that the parasite is utilizing the ecdysone secreted by the host to initiate its own development and that it responds to lower titers than does the host. Normally the parasite begins feeding at the very early onset of adult development as signaled by an increase in respiration (before morphological signs are detectable). Therefore, presumably the low level of ecdysone in the hemolymph stimulates feeding; then the ecdysone in the ingested hemolymph either stimulates the parasite's tissues directly or activates the prothoracic glands of the parasite to secrete ecdysone. Such a positive feedback effect has been shown in several insects (Gilbert and King, 1973). Alternatively, changes in the host's hemolymph induced by ecdysone (Wyatt, 1972) may be the actual triggering factor(s) for the initiation of feeding and larval development of the parasite. This possibility has not been investigated.

Such hemolymph factors seem to govern the development of the hymenopteran parasite *Nemertis canescens* in its host *Ephestia kuhniella* (Corbet, 1968). The *Nemertis* larva remains in the first instar until the end of the last larval instar of the host. Once the host enters the wandering stage (characterized by the cessation of feeding, purging of the gut, and searching for a pupation site), the parasite larva begins to feed faster and develops rapidly to metamorphosis. Corbet correlates this dramatic increase in feeding rate with the decreasing osmolality of the blood due at least in part to the decline of free amino acids as they are used for protein synthesis. This decrease in osmolality and concomitant increased growth of the parasite does not occur on schedule in the debrained late last instar larva or in abdomens isolated from wandering larvae. However, after 8 days the parasite somehow induces these changes in the isolated abdomen and begins to feed. Although Corbet (1968) concludes that no endocrine control is involved, it seems likely that the changes in hemolymph osmolality are a direct result of hormone secretion.

An insight into the endocrine milieu in *Ephestia* during the critical part of the last larval instar is afforded by the recent studies of the endocrine events occurring in a comparable stage of the tobacco hornworm *Manduca sexta*. In the final instar larva the corpora allata cease secretion of JH when the animal has reached a certain critical weight (Nijhout and Williams, 1974a, b). Then about 24 hours later PTTH and ecdysone are released to initiate the entry into the wandering stage (Truman and Riddiford, 1974). Presumably many biochemical changes prerequisite for pupation occur at this time, since, for instance, after this first release of ecdysone, JH can no longer prevent pupation (Truman, *et al*, 1974). Two days later a second pulse of PTTH and ecdysone triggers the detachment of the epidermis from the cuticle and the subsequent transformation into a pupa (Truman and Riddiford, 1974).

Studies of ecdysone release in other Lepidoptera such as *Cerura vinula* (Buckmann, 1959) and *Bombyx mori* (Shaaya and Karlson, 1965) have obtained similar findings. Most likely the same holds true for the flour moth *Ephestia*. Consequently, the decrease in osmolality of the hemolymph probably is due to the first release of PTTH and ecdysone. Thus the parasite is responding to a change in the hormonal milieu of the host, albeit indirectly.

Also intriguing is the possibility that the maintenance of the parasite larva in the first larval instar is a consequence of the relatively high JH titer throughout larval life of the host. In the *Manduca* larva the JH titer fluctuates during the penultimate larval instar but remains at detectable levels (K. Judy, personal communication; Fain and Riddiford, unpublished), but in the final third of the last larval instar it declines drastically (Nijhout and Williams, 1974b). Similar changes in the JH titer in the hemolymph of *Ephestia* during the mid final instar could be instrumental in triggering the onset of feeding in the parasite. Similar hormonal interactions may be at work in the many other species in which the development of the parasite is arrested until very late in larval life of the host or at the beginning of its metamorphosis. Much more experimentation needs to be done in this area.

Since juvenile hormone mimics are being tested as possible insecticides, there is increasing concern that such compounds may also disrupt the development of the insect parasites which help to control the pest insects. At least in the case of the hymenopteran parasite *Muscidifurax raptor* of the stable fly *Stomoxys calcitrans*, these fears are unfounded. Wright and Spates (1972) have found that the two JH mimics which prevent metamorphosis of the stable fly have no effect on growth, metamorphosis, or reproduction of the parasite. Obviously, similar studies will have to be done with each host-parasite relationship.

Protozoa. When one considers other phyla of parasites which attack insects, the direct use of host hormones for synchronization of life cycles seems questionable. The best evidence comes from a study of the life cycles of the flagellate protozoan symbionts which inhabit the hindgut of the wood roach *Cryptocercus* (Cleveland, 1960). Normally gametogenesis in these protozoa is initiated each time the host nymph molts. When α-ecdysone was injected into an intermolt nymph or into an adult, gametogenesis was initiated, even though the injected nymphs did not subsequently molt (Cleveland *et al*., 1960). In some genera gametogenesis began as little as 3 hours after the injection of ecdysone. These data do not rigorously exclude the possibility that changes in the hindgut environment serve as the direct trigger of gametogenesis in the flagellates. However, the rapidity of response to injected ecdysone and the induction of the phenomenon in adult roaches strongly suggest that the symbionts are responding to ecdysone or its metabolities which have entered the hindgut.

Ilan *et al*, (1969) have reported a reversible growth retardant effect *in vitro* of the Williams-Law mixture of insect JH mimics on the trypanosome *Crithida fasciculata* which inhabits the mosquito gut. Since this mixture contains a number of chlorinated compounds, the retardant effect could be due to toxic agents and not a *bona fide* JH response. Ideally, the experiment should be repeated with one of the purified insect JH's. Even if it proves to be a true JH effect, the existence of such an effect *in vivo* and its relevance to the interrelationship between this commensal parasite and its host remain to be explored.

Several attempts have been made to examine the relationship of malarial parasites to the hormonal milieu of their mosquito hosts. Rozeboom (1961) demonstrated that decapitation of *Aedes aegypti* females immediately after a blood meal did not interfere with the development of *Plasmodium* to the oocyst stage. Gwadz (personal communication) obtained similar results after allatectomizing newly emerged females. Thus, development of the protozoan parasite to the oocyst stage is not dependent on either JH or the egg development hormone (EDNH) (Lea, 1967), both of which are necessary for egg maturation in the mosquito. Also, since oocysts injected into the haemocoel of unfed *Aedes* will mature normally into infective sporozoites (Weathersby, 1952), the further development of the oocyst does not depend on EDNH which is only released in response to a blood meal (Lea, 1967). Whether or not oocysts mature in allatectomized mosquitoes has not been investigated (Gwadz, personal communication).

Nematodes. Although nematodes undergo molting cycles, the endocrinological basis of these cycles is poorly understood (Davey and Hominick, 1972). With the increasing availability of the various ecdysones and of juvenile hormone and its mimics, there have been many studies concerned with the effects of these hormones on nematodes *in vitro*. In some instances the ecdysones seem to increase the rate of molting (Webster and Craig, 1969; Hitcho and Thorson, 1971), whereas in others they had no effect (Johnson and Viglierchis, 1970; Hansen and Buecher, 1971). In those nematodes were an inhibition of growth or adult development by JH was noted, the doses required were generally very high and often bore no relationship to the juvenile hormone activity (as tested on insects) of the compound (see for example, Shanta and Meerovitch, 1970; Dropkin *et al*., 1971; Hansen and Buecher, 1971). Thus most of these authors conclude that the effects are nonspecific. Hitcho and Thorson (1971) and Rogers (1973) have reported JH activity in lipid extracts of certain nematodes, and Rogers (1973) has tentatively identified one compound as farnesyl methylether which is a JH mimic for some insects. The only clue as to its possible role in nematodes is the observation of Davey (1971) that JH compounds may stimulate the release of a neurosecretion from the ventral ganglion which triggers the production and release of molting fluid.

Relatively little work has been done on the entomophilic nematodes and their endocrine relationships with their hosts. Thus far there seems to be no well-documented instance of the synchronization of a nematode parasite by the hormonal milieu of the host. Gordon (1970) reported that cautery of the median neurosecretory cells or removal of the corpus cardiacum of the adult female cockroach *Blatta orientalis* caused a significant decrease in the number of adult *Hammershcmidtiella diesingi* in the gut. He therefore concluded these nematodes required a neurosecretory substance from the host in order to complete their adult development. But Hominick and Davey (1972) in a similar study on another cockroach, *Periplaneta americana*, have shown that reduced food intake had the same effect as cardiaectomy on its two gut nematodes (the population of *H. diesingi* increases and that of *Leidynema appendiculata* decreases). Thus, they conclude that the nematodes are responding to the nutritive environment rather than to the hormones of their host and suggest that the same may be true in *Blatta*.

Recently, Gwadz and Spielman (1974) have shown that larvae of the filarial nematode *Brugia pahangi* mature normally after surgical ablation of the median neurosecretory cells or of the corpora allata or toal decapitation of the adult female *Aedes aegypti*. The parasite also developed normally in decapitated, unfed females and in males. Application of exogeneous ecdysone or juvenile hormone similarly had no effect. Thus this parasite apparently is completely independent of the endocrine system of its host.

By contrast, egg development of the nematode, *Heterotylenchus autumnalis*, is prevented as long as the host adult face fly *Musca autumnalis* remains in diapause (Stoffolano, 1967). When diapause of the host is broken (in this instance by an activation of the corpora allata, the ovaries of the nematode develop and eggs are shed into the hemolymph. The host's ovaries do not form mature eggs; instead the larval nematodes burrow into the ovaries and are subsequently "oviposited" by the host (Nappi, 1973c; Stoffolano, 1973). Stoffolano (1967) suggests that the maturation of the female nematode depends on JH or another substance from the corpora allata, but no experiments have been done to prove or disprove this hypothesis. Alternatively, maturation could be due to a change in host metabolism triggered either by neurosecretion from the brain or by JH.

PARASITE INTERFERENCE WITH ITS HOST'S ENDOCRINE BALANCE

The other way in which a parasite can insure its optimum development is to manipulate the host's endocrine milieu. Thus, it may prevent molting or metamorphosis with JH, accelerate molting with ecdysone or PTTH, or prevent maturation of the ovaries by interfering with the release of JH or of neurosecretory materials from the brain.

Molting and/or metamorphosis. One of the best known examples of this kind of interference is the production of a compound with juvenile hormone activity by the microsporidian parasite *Nosema*. Finlayson and Walters (1957) first noted that saturniid larvae infected with *Nosema* either formed pupae that retained some larval characters or formed moths that had patches of pupal cuticle. Since these effects were similar to those obtained by Williams (1956) with juvenile hormone, they suggested that *Nosema* either caused the corpora allata to secrete JH or produced the JH itself. Later Fisher and Sanborn (1964), using *Nosema*-infected *Tribolium* as a source of spores, showed that the *Nosema* itself was producing a substance with JH activity. This substance caused at least 4 supernumerary molts of *Tribolium* larvae, prevented metamorphosis in allatectomized penultimate instar larvae of both cockroaches and the silkworm *Bombyx mori*. The substance was blood-borne, diffusible and ether-soluble. Although to date the structure of this *Nosema* JH mimic has not been elucidated, it appears to be a more specific compound than farnesol (Fisher and Sanborn, 1964).

Recently, Milner (1972) has reported that he cannot confirm the results of Fisher and Sanborn (1964) on the increased growth of *Nosema*-infected *Tribolium* larvae. But an inspection of Milner's data fails to show any significant differences from that reported by the former workers.

Retnakaran and Retnakaran (1972) have reported a similar prolongation of larval life of the spruce budworm *Choristoneura fumiferana* infected with a pox virus. Since the larvae grow larger and many retain larval characters at pupation, they suggest that the virus is somehow causing a hormonal imbalance. However, other than the circumstantial evidence presented above, they have not done any of the proper endocrinological experiments to test the hypothesis.

Not as surprising perhaps are the effects of parasitic insects on their host's endocrine balance. Yet insects normally do not secrete or excrete their hormones; rather they rapidly inactivate their own hormones and these metabolites are excreted (Gilbert and King, 1973). Thus, insect parasites which manipulate their host environment in this way must have evolved special mechanisms to deliver these hormones to the host. Although these mechanisms have not been elucidated, it appears that the eggs of the parasitic wasp *Aphidius platinsis* prevent metamorphosis of their host *Aphis craccivora* (Johnson, 1959). The extent of juvenilization depends on the number of eggs oviposited and apparently depends on a factor released from the eggs within 1 to 2 days after oviposition (they hatch on the third day). Johnson concludes that the late embryos secrete JH into the hemolymph of the host, but no further experiments have been done to clarify this extremely tantalizing phenomenon.

The only clearcut instance in which a parasite triggers the molting of the host is the study by Schneider (1950) of the effects of a ichneumon wasp larva (*Diplazon pectoratorius*) on its syrphid fly host *Epistrophe*. Normally the presence of a wasp larva induces premature pupation. That it does this by secreting ecdysone or an ecdysone mimic was shown by ligation experiments in which the wasp larva was implanted into the posterior half of a ligated diapausing host larva. Only the posterior half pupated; the anterior half with the brain and ecdysone-producing ring gland remained in diapause. Here again the mechanisms whereby the parasite larva produces exogenous ecdysone are unknown.

Effects on reproduction. Development of parasites in adult insects depends mainly on an abundant nutrient supply. One way to insure this food supply, at least in the adult female, is to prevent egg maturation. Another way is to release mimics of the various neurosecretory hormones which govern metabolism (Highnam and Hill, 1969; Wigglesworth, 1970).

Wülker (1964) has reviewed parasite-induced changes in both the internal and external sex characters of insects, so I shall discuss only those studies in which a direct endocrinological effect is suspected. When the nematode *Sphaerularia bombi* infects the female bumblebee *Bombus*, the bee's ovaries fail to develop. Palm (1948), in a careful histological study of both the corpora allata and the ovaries of normal and infected *Bombus*, showed that the parasite caused atrophy of the corpora allata which then resulted in a suppression of the ovaries, and degeneration of any existing eggs. No histological changes were noted in the fat body or in the neurosecretory cells. Since the effect was not correlated with numbers of parasites and since aqueous extracts of the infecting parasites has similar effects, Palm concluded that the nematode secreted a toxic substance which inhibits the growth and function of the corpora allata. Pouvreau (1962) confirmed these findings and Brandenburg (1956) showed similar effects of the strepsipteran *Stylops* on the corpora allata and hence the ovaries of the sand bee *Andrena vaza*. But the nature of the inhibition has not been elucidated.

The nematode *Heterotylenchus autumnalis* which parasitizes the face fly *Musca autumnalis* somehow prevents the maturation of eggs by the adult fly. The nematode develops in the larva and pupa, then reproduces parthenogenetically in the adult female fly (Stoffolano, 1967). The larvae so produced enter the ovary and are oviposited by the fly after it has mated normally (Nappi, 1973c). Stoffolano (1967) suggests, but presents no evidence, that the nematodes secrete a toxic substance which somehow keeps the corpora allata inactive. Alternatively, the parasite could be somehow inactivating JH or utilizing the proteins secreted by the fat body so no egg development can take place.

The latter seems to be the case in the desert locust *Schistocerca gregaria* infected by the nematode *Mermis nigrescens*. In infected females vitellogenesis proceeds normally up through the beginning of yolk deposition (Gordon *et al*., 1973). But then very little yolk is deposited. Although many blood proteins were the same in noninfected and infected hosts, five of the eight yolk proteins were significantly decreased in the blood of infected hosts. Although the fat body is not degenerate, Gordon and coworkers (Gordon & Webster, 1971; Gordon *et al*., 1973) suggest that nematode stimulates catabolism of fat body and protein and then uses the amino acids produced for its rapid growth (which occurs mainly during the time of normal rapid yolk deposition). Gordon *et al*. (1973) also suggest that the nematodes stimulate this catabolism through interference with the host's hormonal balance, but show no evidence. It could either interfere with neurosecretory hormones regulating metabolism in the locust (Mordue *et al*., 1971) or it could interfere with JH release or action. None of these possibilities have been explored.

HORMONAL DEFENSES OF THE HOST INSECT AGAINST PARASITES

Wounding in insects triggers various biochemical changes throughout the insect as well as cytological events in the wounded epidermis which lead to wound healing (see Wigglesworth, 1970; Wyatt, 1972; Doane, 1973 for reviews). Although these events are the same as those triggered by ecdysone, they do not by themselves trigger molting. Furthermore, the same response can occur in the absence of the brain and prothoracic glands indicating that ecdysone release is not necessarily involved. Rather the wounded epidermis apparently releases an "injury factor" into the hemolymph which causes these reactions.

When an insect is parasitized, an injury response generally ensues. But the primary defense of the insect is encapsulation of the invading parasite (Salt, 1963, 1970). The blood cells surround the foreign body and form a cellular and often a connective tissue sheath around it. In many instances melanin is also deposited on the surface of the parasite. Most often this encapsulation occurs concurrently with the wound healing reaction.

In the flies, particularly *Drosophila*, Nappi (1973a, b) has suggested that this encapsulation response might be hormonally mediated. In *Drosophila* a specific type of blood cell called a lamellocyte is the encapsulating cell. Normally at the end of the last larval instar the circulating plasmatocytes transform into lamellocytes which then aid in puparium formation (Walker, 1959; Rizki, 1962). Rizki (1962) further showed that isolation of abdomens of the last larval instar of *D. melanogaster* caused a precocious transformation of these cells. Implantation of a ring gland (containing corpus allatum, corpus cardiacum, and prothoracic glands) into the abdomens then prevented this transformation.

Similarly, the eggs of the hymenopteran parasite, *Pseudeucoila bochei* induce in *Drosophila* the precocious formation of lamellocytes which then encapsulate the eggs. This induction occurs only during the last larval instar and not in the two preceding ones. Nappi (1973a, b) has suggested that this induction of hemocyte changes by the parasite is mediated through the host's endocrine system. He supports this hypothesis by experiments in which ligation of host *D. algonquin* larvae behind the brain-ring gland complex prevents the normal encapsulation of *Pseudeucoila* eggs in the posterior portion (Nappi, 1973d). These results are perplexing in view of Rizka's data cited above that a similar ligation of *D. melanogaster* provoked precocious transformation of the plasmatocytes into lamellocytes. Thus, one would have expected an enchancement of the encapsulation reaction rather than a reduction.

It certainly is possible that the injury response to the parasite's entry could trigger the appropriate changes in the host hemocytes which would lead to encapsulation. This triggering could arise either through a direct action of the "injury factor" on the blood cells or indirectly through the tropic effects of injury on the prothoracic glands. Obviously the hormonal milieu of the host must also be proper for the defense reaction to occur. For instance, it appears that the precocious transformation to lamellocytes in parasitized *Drosophila* larva can only proceed in the presence of low levels of JH as are found in the last larval instar in contrast to the presumed high levels in the first two instars. Much more work is necessary in this area to pinpoint the exact mechanisms by which hemocytes initiate the encapsulation reaction.

SUMMARY

This brief survey of the hormonal relationships between parasites and their insect hosts has indicated that two successful mean of parasitization are 1) Utilization of host hormones or their direct effects to trigger growth and development of the parasite when the host's environment is most beneficial for the parasite 2) Production of insect hormones or their mimics to manipulate the host's developmental stage for maximum utilization of nutrient resources by the parasite. Yet relatively few parasites seem to have capitalized on either of these alternatives. Although this review of the literature has not been exhaustive, it has shown that many potentially interesting relationships have not been explored from the endocrinological viewpoint. Also, even when the host-parasite interaction seems on the surface to have a hormonal basis, only experiments involving endocrine gland ablation and hormone replacement therapy, as well as careful observation of other possible physiological or biochemical events correlated with successful parasitism can elucidate the true nature of the relationship. Similarly, until the control of hemocyte transformation and of encapsulation is better understood, a hormonal basis for host defense mechanisms against parasites is questionable.

ACKNOWLEDGEMENTS

I with to thank Dr. James Truman and Dr. Robert Gwadz for helpful discussions during the preparation of this review and Dr. James Truman for a critical reading of the manuscript.

REFERENCES

Brandenburg, J. (1956). *Z. Morph. Oekol. Tier, 45*:343-364.

Buckmann, D. (1959). *J. Insect Physiol., 3*:159-189.

Cleveland, L. R. (1960). In "Host Influences on Parasite Physiology" (L. A. Stauber, ed.) pp. 5-10. Rutgers University Press, New Brunswick.

Cleveland, L. R., A. W. Burke, Jr., and P. Karlson. (1960). *J. Protozool., 7*:229-239.

Corbet, S. A. (1968). *J. Exp. Biol., 48*:291-304.

Davey, K. G. (1971). *Internatl. J. Parasitol., 1*:61-66.

Davey, K. G. and W. M. Hominick. (1973). *Exptl. Parasit., 33*:212-225.

deWilde, J. and A. deLoof. (1973). In "The Physiology of Insecta" (M. Rockstein, ed.), vol. I, 2nd edn., pp. 97-158. Academic Press, New York.

Doane, W. W. (1973). In "Developmental Systems: Insects" (S. J. Counce and C. H. Waddington, eds.), vol. 2, pp. 291-498. Academic Press, London.

Dropkin, V. H., W. R. Lower, and J. Acedo. (1971). *J. Nematol., 3*:349-355.

Engelmann, F. (1970). "The Physiology of Insect Reproduction." Pergamon Press, Oxford. 307 pp.

Finlayson, L. and V. A. Walters. (1957). *Nature, 180*:713-714.

Fisher, F. M. and R. C. Sanborn. (1964). *Biol. Bull., 126*:235-252.

Fisher, R. C. (1971). *Biol. Rev., 46*:243-278.

Gilbert, L. I. and D. S. King. (1973). In "The Physiology of Insecta" (M. Rockstein, ed.), vol. I, 2nd edn., pp. 249-370. Academic Press, New York.

Gordon, R. (1970). *Parasitology, 61*:101-110.

Gordon, R. and J. M. Webster. (1971). *Exptl. Parasitol., 29*:66-79

Gordon, R., J. M. Webster, and T. G. Hislop. (1973). *Comp. Biochem. Physiol., 46B*:575-593.

Gwadz, R. and A. Spielman. (1973). *J. Insect Physiol., 19*:1441-1448.

Gwadz, R. W. and A. Spielman. (1974). *J. Parasitol., 60*:134-137.

Hagedorn, H. H. and A. M. Fallon. (1973). *Nature, 244*:103-105

Hansen, E. L. and E. J. Buecher. (1971) *Experientia, 27*:859-860.

Highnam, K. C. and L. Hill. (1969). "The Comparative Endocrinology of the Invertebrates," 270 pp. American Elsevier Publ. Co., New York.

Hitcho, P. I. and R. E. Thorson. (1971). *J. Parasit., 57*:787-793.

Hominick, W. M. and K. G. Davey. (1972). *Can. J. Zool., 50*:1421-1432.

Ilan, J., J. Ilan, and S. Ricklis. (1969). *Nature, 224*:179-180.

Johnson, B. (1959). *Entom. exp. appl., 2*:82-99.

Johnson, R. N. and D. R. Viglierchia. (1970). *Exp. Parasit., 27*:301-309.

Lea, A. (1967). *J. Insect Physiol., 9*:793-809.

Lipke, H. (1974). This volume, pp.

Milner, R. J. (1972). *J. Invert. Pathol., 19*:248-255.

Mordue, W., K. C. Highnam, L. Hill, and A. J. Luntz. (1970). *Mem. Soc. Endocr., 18*:111-136.

Nappi, A. J. (1973a). *Parasitology, 66*:23-32.

Nappi, A. J. (1973b). *Exptl. Parasitol., 33*:285-302.

Nappi, A. J. (1973c). *J. Parasitol., 59*:963-969.

Nappi, A. J. (1973d). *Amer. Zool., 13*:1295.

Nappi, A. J. (1974). This volume, pp.

Nijhout, H. F. and C. M. Williams. (1974a). *J. Exp. Biol.*, in press.

Nijhout, H. F. and C. M. Williams. (1974b). *J. Exp. Biol.*, in press.

Nijhout, M. M. and L. M. Riddiford. (1974). *Biol. Bull.*, in press.

Palm, N. B. (1948). *Opuscula Entomologica Supp., 7*:1-101.

Pouvreau, A. (1962). *Ann Abeille, 5*:181-199.

Retnakaran, A. and F. T. Retnakaran. (1972). *J. Invert. Path., 20*:358-360.

Rizki, T. M. (1962). *Amer. Zool., 2*:247-256.

Rogers, W. P. (1973). *Parasitology, 67*:105-113.

Rozeboom, L. E. (1961). *J. Parasit., 47*:597-599.

Salt, G. (1963). *Parasitology, 53*:527-642.

Salt, G. (1970). "The Cellular Defense Mechanisms of Insects" Cambridge University Press, Cambridge. 118 pp.

Schneider, F. (1950). *Mitt. Schweiz. Entomol. Ges., 23*:155-194.

Schoonhoven, L. M. (1962). *Arch. Neerl. Zool. 15*:111-174.

Shaaya, E. and P. Karlson. (1965). *Develop. Biol., 11*:424-432.

Shanta, C. S. and E. Meerovitch. (1970). *Can. J. Zool., 48*:617-620.

Stoffolano, J. G., Jr. (1967). *J. Invert. Pathol., 9*:395-397.

Stoffolano, J. G., Jr. (1973). *Exptl. Parasitol., 33*:263-284.

Truman, J. W. and L. M. Riddiford. (1974). *J. Exp. Biol.*, in press.

Truman, J. W., L. M. Riddiford, and L. Safranek. (1974). *Develop. Biol.*, in press.

Walker, I. (1959). *Rev. Suisse Zool., 66*:569-632.

Weathersby, A. B. (1952). *J. Infect. Diseases, 91*:198-205.

Webster, J. M. and S. Craig. (1969). *J. Nematol., 1*:308.

Welch, H. E. (1965). *Ann. Rev. Entomol., 10*:275-302.

Wigglesworth, V. B. (1970). "Insect Hormones." W. F. Freeman Co., San Francisco. 159 pp.

Williams, C. M. (1956). *Nature, 178*:212-213.

Willis, J. (1974). *Ann. Rev. Entomol., 19*:97-116.

Wright, J. E. and G. E. Spates. (1972). *Science, 178*:1292-1293.

Wülker, W. (1964). *Exp. Parasit., 15*:561-597.

Wyatt, G. R. (1972). In "Biochemical Actions of Hormones" (G. Litwack, ed.), vol. II, pp. 386-490. Academic Press, New York.

COMPARISON OF IMMUNE MECHANISMS IN VERTEBRATES AND INVERTEBRATES

Frederik B. Bang

Department of Pathobiology
The Johns Hopkins University
School of Hygiene and Public Health
Baltimore, Maryland

In making a comparison between invertebrates and vertebrates in terms of immune response, the basic emphasis must be on determining whether a homologous or analogous reaction is occurring. Homologous clearly refers to reactions which are presumed to have evolved in the vertebrate directly from an invertebrate system of reactions. Analogous reactions are similar in action but have evolved separately and, therefore, may have used entirely different molecular constituents. For instance, the occurrence of a lysin in *Sipunculus* which destroys ciliates may be analogous to certain lytic activity in the vertebrate, but it is most unlikely that the phenomenon in the vertebrate has come from the *Sipunculus*. A similar comparison exists for the immune reaction to skin transplants in earthworms and similar reactions in vertebrates. There are no substances in the invertebrate with the general structure of gamma globulin which protect the invertebrate against infection. A variety of molecules is present in the invertebrate which can agglutinate red cells and bacteria, lyse foreign cells, etc. We have suggested that the molecules be given the specific name 'antisomes', so that they may be discussed, worked with and compared but not confused with antibodies.

There are four different systems in which comparisons between the vertebrates and the invertebrates may be made. From each of these sytems specific questions can be elicited. The first, phagocytosis, has been discussed fairly extensively in the TMP Workshop. Recognition of the foreign material by both vertebrates and invertebrates must occur and it is very likely that the phagocytic recognition in invertebrates evolved directly into the similar activity of macrophages in the vertebrate. The question remains whether macrophages can acquire the capacity to respond to a particular antigen and remember this, but not multiply. Lymphocytes in the vertebrate multiply following specific antigenic stimuli. This increase in specifically prepared cells is responsible for the accelerated secondary immune response. It would probably be very difficult to detect memory in cells if they did not increase greatly in number. In the vertebrate it is well established that the population of lymphocytes is separated into two: that of T and B cells. The T cells are responsible for cell-mediated immunity, are imprinted by the thymus, and produce killer cells which destroy tumor cells as well as foreign cells. The B cells are those which produce the familiar multiple variety of antibodies. T and B cells have the characteristic of remembering their previous contact with an antigen and are stimulated to multiply by contact with this particular antigen. These cells may also be stimulated to multiply by a series of plant lectins, such as phytohemagglutinin, conconavalin A, etc. The question for the invertebrate is then: Are there any cells which can be stimulated to multiply by administration of these mitogens? As fas as I know, relatively little work has been done on this.

The histological concomitant to this immune response is the formation of germinal centers. These are located in different areas of the lymphatic system of all vertebrates, and as far as we know, the association between antibody formation and germinal centers is complete. Are any germinal centers found in any invertebrate after antigenic stimulation? There is a variety of non-antibody mechanisms in vertebrates and a summary of these mechanisms has been included in the book "Non-Specific Factors Influencing Host Resistance: A Re-Examination." (W. Braun and J. Ungar, eds.), S. Karger, Basel, Switzerland, 1973. Among these is interferon, which is now so thoroughly studied in various infections of birds and mammals. We believe that we have some evidence that interferon occurs in the crab in response to a specific virus infection (Bang, 1974). This altered response of an infected animal raises the question whether an infected invertebrate can respond differently, particularly in relation to phagocytosis, than a non-infected animal. For instance: Is the phagocytic activity of the snail, which is full of metacercaria, different from that of a normal snail?

Finally, we come to the question of complement which was discussed by Dr. Anderson. This complex of substances, so important in immune mechanisms in vertebrates, has been discussed in the past by invertebrate zoologists but, as yet, no very exact definition of it in the invertebrate has been made. In the vertebrate, complement is activated by the combination of antigen and antibody which sets off a cascade of sequences of action. The initial step in complement activation may be bypassed so that the C_3 fraction can be directly activated by endotoxin and other substances, leading to the full activation of complement and the release of various inflammatory substances. There are some claims that invertebrate blood may substitute for some components in this, but the specificity of this part of the reaction needs to be demonstrated. However, at this point, we conclude much as we started, for Metchnikov defined inflammation in terms of macrophage activity and paid little attention to the serum and other factors that we now know to be involved. However, since many of these inflammatory substances are liberated by cells, it is entirely possible that we may again recognize the dominate role of the macrophage in inflammation, particularly as we concentrate on the invertebrate.

One of the most important methods of studying recognition and immunity is the recognition of various tissue grafts. The importance of this mechanism in all sorts of practical surgical efforts today has led to a renewed interest in the invertebrates. A summary of this work has been published by Hildemann and Reddi (1973). The question of whether a second set graft is rejected specifically and more rapidly than the first foreign graft is the basic problem that Cooper (1973) has undertaken. It is clear that foreign tissue grafts may either be completely rejected, as in the work of Théodor (1969, 1970) on sea fans, or sometimes accepted for long periods of time. The exact role of cells in causing such foreign recognition is under investigation by Chapron (1969), Duprat (1968), Valembois (1970), and others in France.

REFERENCES

Bang, F. B. (1974). *Infection & Immunity*, in press.

Chapron, C. (1969). "Contribution à l'étude histologique et infrastructural de la régénération céphalique chez le Lombricien *Eisenia foetida* (Sav.)." Thesis for the degree of Doctor of Natural Sciences, University of Bordeaux, France.

Cooper, E. L. (1973). In "Non-Specific Factors Influencing Host Resistance" (W. Braun and J. Ungar, eds.), pp. 11-23. S. Karger, Basel, Switzerland.

Duprat, P. (1968). "Etude des autogreffes et des homogreffes de paroi du corps chez le Lombricien *Eisenia foetida* (Sav.). Aspects histologiques et immunitaires de la prise et de l'acceptation ou du refus des greffons." Thesis for the degree of Doctor of Natural Sciences, University of Bordeaux, France.

Hildemann, W. H., and Reddy, A. L. (1973). *Fed. Proc., 32*:2188.

Théodor, J. L. (1969). *C. R. Acad. Sci. (D) Paris, 268*:2534.

Théodor, J. L. (1970). *Nature, 227*:690.

Valembois, P. (1970). "Etude d'une hétérogreffe de paroi du crops chez les Lombriciens. Aspects cytologiques, physiologiques et immunologiques de l'évolution du greffon (*Allolobophora caliginosa* Dugès) et de la réaction du porte-greffe (*Eisenia foetida* Sav.)." Thesis for the degree of Doctor of Natural Sciences, University of Bordeaux, France.

SUBJECT INDEX

A

B

C

D

E

F

A 5
B 6
C 7
D 8
E 9
F 0
G 1
H 2
I 3
J 4